Sustainable Civil Infrastructures

About this Series

Sustainable Infrastructure impacts our well-being and day-to-day lives. The infrastructures we are building today will shape our lives tomorrow. The complex and diverse nature of the impacts due to weather extremes on transportation and civil infrastructures can be seen in our roadways, bridges, and buildings. Extreme summer temperatures, droughts, flash floods, and rising numbers of freeze-thaw cycles pose challenges for civil infrastructure and can endanger public safety. We constantly hear how civil infrastructures need constant attention, preservation, and upgrading. Such improvements and developments would obviously benefit from our desired book series that provide sustainable engineering materials and designs. The economic impact is huge and much research has been conducted worldwide. The future holds many opportunities, not only for researchers in a given country, but also for the worldwide field engineers who apply and implement these technologies. We believe that no approach can succeed if it does not unite the efforts of various engineering disciplines from all over the world under one umbrella to offer a beacon of modern solutions to the global infrastructure. Experts from the various engineering disciplines around the globe will participate in this series, including: Geotechnical, Geological, Geoscience, Petroleum, Structural, Transportation, Bridge, Infrastructure, Energy, Architectural, Chemical and Materials, and other related Engineering disciplines.

More information about this series at http://www.springer.com/series/15140

Janusz Wasowski · Daniele Giordan
Piernicola Lollino
Editors

Engineering Geology and Geological Engineering for Sustainable Use of the Earth's Resources, Urbanization and Infrastructure Protection from Geohazards

Proceedings of the 1st GeoMEast International Congress and Exhibition, Egypt 2017 on Sustainable Civil Infrastructures

Springer

2017 GEOMEAST

SSIGE
SOIL-STRUCTURE INTERACTION GROUP IN EGYPT
المجموعة المصرية للهندسة الجيوتقنية والبنية التحتية

Editors
Janusz Wasowski
Institute for Geo-hydrological Protection (IRPI)
National Research Council (CNR)
Bari
Italy

Piernicola Lollino
Institute for Geo-hydrological Protection (IRPI)
National Research Council (CNR)
Bari
Italy

Daniele Giordan
Institute for Geo-hydrological Protection (IRPI)
National Research Council (CNR)
Bari
Italy

ISSN 2366-3405 ISSN 2366-3413 (electronic)
Sustainable Civil Infrastructures
ISBN 978-3-319-61647-6 ISBN 978-3-319-61648-3 (eBook)
DOI 10.1007/978-3-319-61648-3

Library of Congress Control Number: 2017946435

Printed on acid-free paper

This Springer imprint is published by Springer Nature
The registered company is Springer International Publishing AG
The registered company address is: Gewerbestrasse 11, 6330 Cham, Switzerland

Preface

Toward building sustainable and longer civil infrastructures, the engineering community around the globe continues undertaking research and development to improve existing design, modeling, and analytical capability. Such initiatives are also the core mission of the Soil-Structure Interaction Group in Egypt (SSIGE) to contribute to the ongoing research toward sustainable infrastructure. This conference series "GeoMEast International Congress and Exhibition" is one of these initiatives.

Ancient peoples built their structures to withstand the test of time. If we think in the same way, our current projects will be a heritage for future generations. In this context, an urgent need has quickly motivated the SSIGE and its friends around the globe to start a new congress series that can bring together researchers and practitioners to pursue "Sustainable Civil Infrastructures." The GeoMEast 2017 is a unique forum in the Middle East and Africa that transfers from the innovation in research into the practical wisdom to serve directly the practitioners of the industry.

More than eight hundred abstracts were received for the first edition of this conference series "GeoMEast 2017" in response to the Call for Papers. The abstracts were reviewed by the Organizing and Scientific Committees. All papers were reviewed following the same procedure and at the same technical standards of practice of the TRB, ASCE, ICE, ISSMGE, IGS, IAEG, DFI, ISAP, ISCP, ITA, ISHMII, PDCA, IUGS, ICC, and other professional organizations who have supported the technical program of the GeoMEast 2017. All papers received a minimum of two full reviews coordinated by various track chairs and supervised by the volumes editors through the Editorial Manager of the SUCI "Sustainable Civil Infrastructure" book series. As a result, 15 volumes have been formed of the final +320 accepted papers. The authors of the accepted papers have addressed all the comments of the reviewers to the satisfaction of the track chairs, the volumes editors, and the proceedings editor. It is hoped that readers of this proceedings of the GeoMEast 2017 will be stimulated and inspired by the wide range of papers written by a distinguished group of national and international authors.

Publication of this quality of technical papers would not have been possible without the dedication and professionalism of the anonymous papers reviewers. The names of these reviewers appear in the acknowledgment that follows. For any additional reviewers whose names were inadvertently missed, we offer our sincere apologies.

We are thankful to Dr. Hany Farouk Shehata, Dr. Nabil Khelifi, Dr. Khalid M. ElZahaby, Dr. Mohamed F. Shehata, and to all the distinguished volumes editors of the proceedings of the GeoMEast 2017. Appreciation is extended to the authors and track chairs for their significant contributions. Thanks are also extended to Springer for their coordination and enthusiastic support to this conference. The editors acknowledge the assistance of Ms. Janet Sterritt-Brunner, Mr. Arulmurugan Venkatasalam in the final production of the 15 edited volumes "Proceedings of GeoMEast 2017".

Contents

Site Investigation Using Engineering Geology Mapping and Geological Hazard Evaluation: Case Study of the New Hail Economic City, Hail Region, KSA

Turki E. Al-Sehly[1], Ahmed M. Youssef[1,2(✉)], Abdulla A. Al-Otaibi[1], and Hassan M. Al-Harbi[1]

[1] Geological Hazards Department, Applied Geology Sector, Saudi Geological Survey, P.O. Box 54141, Jeddah 21514, KSA
amyoussef70@yahoo.com
[2] Faculty of Science, Geology Department, Sohag University, Sohag, Egypt

Abstract. The New Hail Economic City in the Kingdom of Saudi Arabia is one of the pioneer projects and represents a corner stone for the new development activities for Hail region. It is located to the north of the Hail city, covering an area of ~156 km^2. Fundamental facilities of this project will include infrastructure, residential, commercial, industrial, and a new international airport. The present study aims to establish a new approach of site investigation using engineering geologic mapping and geological hazard evaluation. To achieve this approach, physical and engineering characteristics of surface and subsurface soil and/or rocks of the study area were evaluated. Several data sources were used including: geologic and topographic maps, satellite images, previous data and reports, and detailed field and laboratory investigations of soil and rocks. Our finding indicated that the study area is characterized by soil layer which varies in thickness from 0 to 12 m and rock outcrops. An engineering geology mapping of the new economic city was prepared using GIS environment. It classifies the area into six zones (4 for soils and 2 for rock outcrops) according to their physical and geotechnical properties. Three cross sections were prepared according to subsurface data showing the layers variations within the study area. Finally, the geological hazards causing negative impact on the area were identified. Various recommendations were proposed to help project decision makers to deal with the geological and geotechnical problems.

Keywords: Site investigation · Engineering geology mapping · Geological hazards · GIS · KSA

1 Introduction

Geotechnical properties of earth materials (soil and rocks) and evaluations of different geologic hazards play an essential role in designing engineering structures such as roads, bridges, and urban areas in order to avoid any impact on the project properties. Anonymous (1972) indicated that the conventional geological maps, from the point of view of civil engineer, have rocks and soils with different engineering properties as a

J. Wasowski et al. (eds.), *Engineering Geology and Geological Engineering for Sustainable Use of the Earth's Resources, Urbanization and Infrastructure Protection from Geohazards*, Sustainable Civil Infrastructures, DOI 10.1007/978-3-319-61648-3_1

single unit of the same age or origin. They also lack of quantitative information such as nature of different lineaments, weathering extent, ground water conditions, physical and geotechnical properties, and the potential geological hazards for engineering structures. Engineering structures are concerned with the physical and engineering characteristics of earth materials that will bring a dynamic equilibrium of the engineering structures with the geological conditions of the area (Waltham 1994).

Engineers and planners find difficulty to infer material properties using only geological maps. The evaluation of geotechnical properties of surface and subsurface earth materials and different geological hazards is important to provide more information that will help engineers for their design. Different specifications of engineering geological mapping of the major cities were prepared by the International Association of Engineering Geology (IAEG 1976, 1981a, b). Many authors developed different types of engineering geology/geotechnical maps including natural hazards and mechanical characteristics of the soils and rock masses (Dearman and Fookes 1974; UNESCO 1976; Xu and Li 1980; IAEG 1981a, b; Anonymous 2002; Rozos et al. 2004; Chacon et al. 2006; Jato et al. 2006; Que et al. 2008; Al-Harbi et al. 2008; Parry and Hart 2009; Okogbue and Aghamelu 2010; Mohammed et al. 2010; Commission Staff Working Document 2011; Chávez et al. 2012; Solberg et al. 2012; Graña et al. 2013). A major geotechnical problem in engineering geology mapping is related to the discrepancies in engineering description of soil and rocks. For that reason, the Geological Society of London (1977) and the International Association of Engineering Geology (IAEG) (1981a, b) proposed a unified and international classification system. Rock mass classification systems were also developed according to (Deere and Miller 1966; Wickham et al. 1972; Barton et al. 1974; Bieniawski 1973, 1989; ISRM 1977).

The aim of the engineering geology mapping is the selection of the geological and engineering characteristics of rocks and soils which are uniform, and group the layers consequently (Craa'chlev and Denness 1972; Graña et al. 2013). The quality of the engineering geology maps is important as these maps help considerably in the design and construction of engineering projects (UNESCO 1976). The degree of simplification of engineering geology map depends on the purpose and scale of the map, the accuracy of the information and the techniques of representation (Chávez et al. 2012). Engineering geology maps have qualitative and quantitative information of the physical and geotechnical properties of earth materials (soil and rocks).

In addition, geological hazards represent major impact on the extension of development in different cities all over the world. Most of the geological hazards associated with desert environments are principally related to natural and/or man-made conditions, such as erosion and deposition of sediments; corrosion, low bearing capacity, and unsuitable zone of sabkha soil; land subsidence due to loose soil; flash floods; and salt domes (Aly 1997; Aly et al. 2005; Youssef and Abdel Moneim 2006; Youssef 2008; El-Haddad and Youssef 2008; Youssef et al. 2011, 2016).

Many people have documented their experience about the Hail Region, among them, (Al-Rasheed 1991; Planet 2000). However, there is no previous work on engineering geology mapping and geologic hazards evaluation to help in site investigation in the study area. The aim of this work is to apply a new approach of site investigation using systematic engineering geology mapping and geological hazards evaluation for civil engineering purposes. The study area was chosen in the Hail region, because of

(1) pressure for establishing new Hail Economic city, (2) expansion toward prone hazard areas, and (3) presence of unknown areas in relation to their engineering characteristics. The main goal is to guide decision making on planning and management of the early stages of civil engineer design to reduce damage to constructions with consequent savings to society and the economy.

2 Study Area

The Hail region covers an area of $\sim 120{,}000\ \text{km}^2$ and represents $\sim 6\%$ of the Kingdom of Saudi Arabia. It is located in the middle north area of the KSA (Fig. 1a). The Hail region is mainly an agricultural area, with significant grain, date, and fruit production. Hail city is the capital of the region with a total population of 356,876 people. The sparsely inhabited An-Nafud Desert lies about 40 km north of the Hail city covering the entire northern third of Hail province. The new Hail Economic City is located to the north of Hail city by ~ 14 km and covers an area of $156\ \text{km}^2$ (Fig. 1b). The Main objectives of establishing the new Hail economic city is to attract investors, entrepreneurs and developers to establish logistic fields, information technology, building materials, infrastructure, agricultural industries, education, and smart infrastructure. The master plan of the New Hail Economic City includes residential/community facilities; commercial/high density residential areas; industrial areas; new international airport; dry port; arterial roads; and access point/gate.

Physiographically, Hail city lies between the peaks of Aja Mountain (about 1,400 m asl) and Salma Mountain (about 1,100 m asl). Wadi Aldyra is one of the most important geomorphological landmarks in the Hail area that runs through the Hail economic city from south-west to north-east.

The new Hail economic city has an elevation ranges from 860 m to 1065 m asl. It is characterized by different features including: (1) Part of the area is flat land covered by sediments (Fig. 2a). (2) The area has some wells for groundwater withdrawal (Fig. 2b). (3) In some parts, there are hills of granitic and sand stone rocks (Fig. 2c, d). (4) There are depressions filled with rainwater forming small lakes (Fig. 2e). (5) There is a railway line that runs through the area (Fig. 2f).

The area is characterized by an arid climate (high temperature and very low rainfall). The annual potential evaporation far exceeds the rainfall values. The rainfall intensity, duration, and frequency are irregular over time. The rainy season is from October to May with 19.7 rainy days per year. April is generally the wettest month followed by January, March and February. The coldest months are December, January, and February when temperature may fall to zero. The temperature gradually increases to reach its maximum value (44 C°) in June and July. The general wind direction is southwest. The humidity fluctuates has highest value in winter (52.8% in January) and lowest value in summer (14.4% in June and July).

Different geological studies were carried out for the Hail region including geological investigation of the area (Brown and Jackson 1960); a reconnaissance geological map of 1:500,000 scale was prepared by Brown et al. (1963); the Precambrian evolution of the northern part of the Shield (Delfour 1980); the Hail quadrangle sheet 27/41b

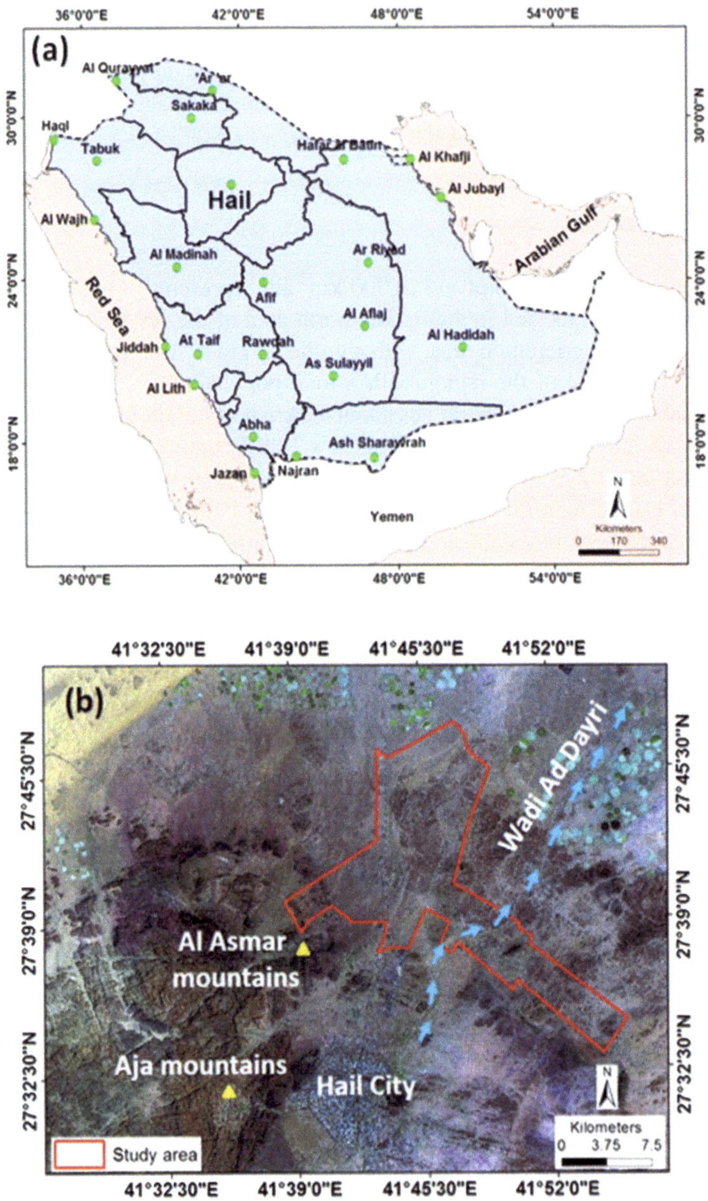

Fig. 1. (a) Location map of the Hail region in relation to the KSA. (b) Location of the New Hail Economic City in relation to the Hail City.

produced at a scale of 1:100,000 (Kellogg and Stoeser 1985). Five rock types were mapped in the Hail Economic City including the Saq sandstone (located in the central and eastern part of the study area); granite (located in the southwest corner of the study area, and as isolated hills in the eastern part of the study area and in the southern part of the study area; and extensive alluvial deposits which cover most of the study area.

Fig. 2. (a) Flat areas covered by quaternary sediments, (b) groundwater well, (c) granite hills, (d) Saq sandstone outcrops, (e) water bonds, (f) railway line.

3 Methodology

In order to manage the new approach for site investigation, engineering geology mapping and geological hazards evaluation were applied. Physical and geotechnical characteristics of the surface and subsurface earth materials (soil and rocks) and detailed investigations of the New Hail Economic city were required. Extensive field and laboratory programs were conducted. Additional data of the area in the form of geotechnical reports were obtained from different agencies. Physical and geotechnical tests were performed at the Engineering Geology laboratory at the Saudi Geological Survey in accordance to ASTM (1989). Drilling program was carried out where seventeen boreholes were drilled, ranging in depth between 10 and 15 m (Fig. 3a, b). In the upper soil layers, drilling was performed using 4 in. hollow stem augers through which standard penetration testing (SPT) was applied. Disturbed samples were obtained every 1.5 m using the Split spoon sampler during the Standard Penetration Tests (SPT) that was performed at various depths of the boreholes. Core drilling was performed using 3 m length NQ double core barrel with a diamond core bit. During the drilling operation, SPT testing was performed in the soil layers and the number of blows per foot was recorded. In the rock layers, the core recovery and the rock quality designation (RQD) (Deere 1964) were recorded for each drilling run. A drilling log was prepared for each borehole to present the soil and rock types. Twenty one surface soil samples were collected from the area (Fig. 3a, c) using hand auger and/or shovel at a depth ranges from 0.5 m to 2 m to avoid any disturbance in the top layer. Eighteen rock outcrop stations were investigated to determine the physical and engineering characteristics of the rock masses (Fig. 3a, c). The samples obtained from the boreholes, surface, and rock outcrops were examined, described and classified by a geotechnical engineer. The moist samples were placed in waterproof plastic bags before they were placed in the wooden boxes. The core samples were petrographically examined and identified. Field analysis of rock outcrop stations were performed considering (1) The degree of weathering which refers to the destructive processes of the rock mass caused by physical and chemical agents. Weathering effects are separated into mechanical disintegration and chemical decomposition. According to the Geological Society of London (1977), the descriptive terms for the state of weathering was used including fresh (F), slightly weathered (S), moderately weathered (M), highly weathered (H), and completely weathered (C). (2) The strength of intact rock is one of the most important index parameters used in describing and classifying the rock masses. The point load test was used for measuring the rock strength. A total of 100 readings for each station of intact rock samples were conducted. Following Deere and Miller (1966) the descriptive terms of rock strength included very low strength (VLS), low strength (LS), medium strength (MS), high strength (HS), and very high strength (VHS). (3) Fracture spacing between adjacent discontinuities was measured. They largely control the size of individual blocks of rock masses which govern the stability of rock structures. The closely spaced joints change the mode of failure completely more than the widely spaced joints which result in different interlocking conditions. The descriptive terms for joint spacing was characterized according to the Geological Society of London (1977) into extremely closely spaced, very closely spaced, closely spaced, moderately wide spaced, widely spaced, and extremely wide spaced. (4) The rock quality designation (RQD) is one of the most important parameters used in the classification of rock masses for engineering purposes. It is a modification to the core recovery (Deere and Miller 1966); one can use the empirical formula (Palmstrom 1982) for determining the RQD on the basis of the number of joints in a cubic meter in an outcrop. The RQD can be described as very poor (VP) (0–25%), poor (P) (25–50%), fair (F) (50–75%), good (G) (75–90%), and very good (VG) (90–100%).

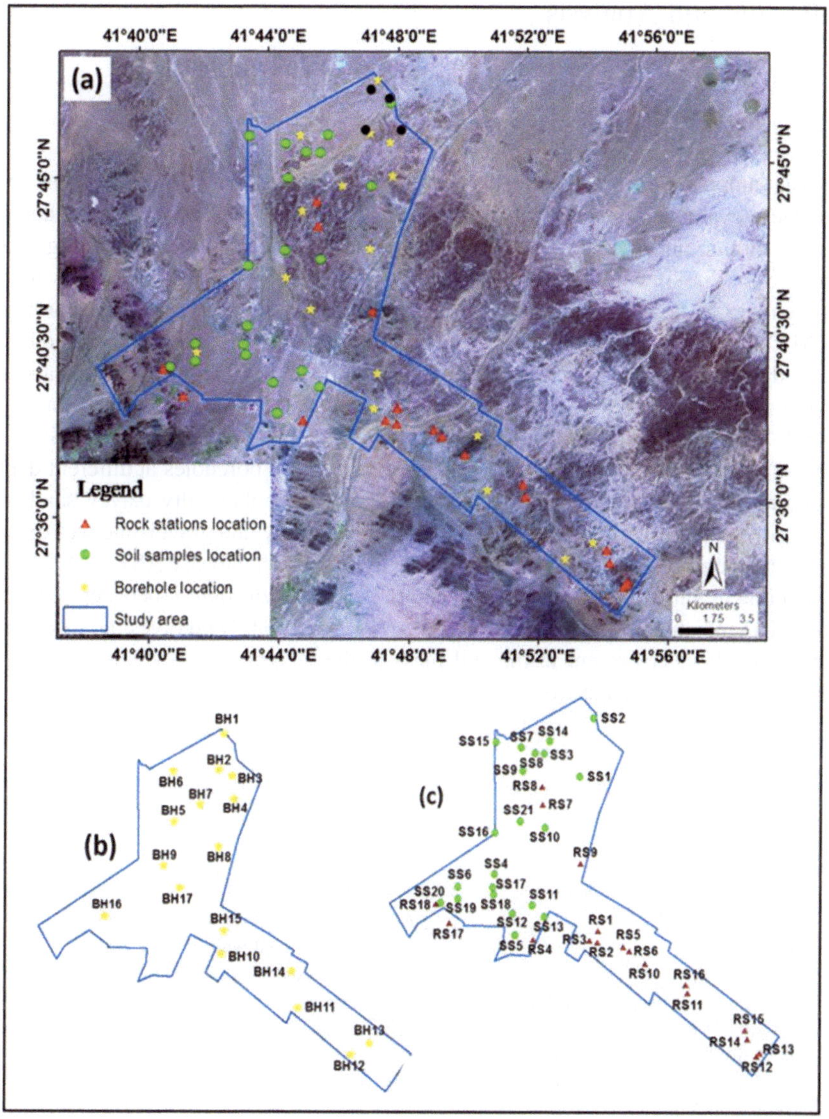

Fig. 3. (a) General map location; (b) Boreholes map, (c) Surface soil samples and rock cut stations map.

Finally, in order to determine the physical and engineering properties of the surface and subsurface materials, laboratory tests were performed on selected samples from the boreholes and surface augur samples. The following tests were performed according to American Society for Testing and Materials (ASTM) Standards including; grain size distribution, water content, liquid limit and plastic limit, and uniaxial compressive strength of rocks.

4 Results and Analysis

4.1 Physical and Engineering Properties of Soil and Rocks

Laboratory Properties of Soil Samples

The soil samples were systematically tested and their physical and engineering properties were evaluated. The obtained properties include grain size distribution, density, atterberg limits, and moisture content. Finally, soils were classified according to the Unified Soil Classification system (ASTM D2487). Twenty one surface soil samples, and twenty eight borehole soil samples were subjected to the gradation test according to ASTM D422-02, and classified based on the grain size analysis and plasticity indices as specified by the Unified Soil Classification System (ASTM D2487). The surface soil samples are characterized by clayey sand (SC), silty sand (SM), well–graded sand with silt (SW–SM), poorly–graded sand with silt (SP–SM), and silty clayey sand (SC–SM) (Table 1). Subsurface soil samples obtained from the nine boreholes at different depths are characterized by clayey sand, silty sand, sandy lean clay, silty clayey sand with gravel, silty sand with gravel, poorly graded sand with clay and gravel, and well-graded gravel with sand (Table 1). According to the hydrometer analysis (ASTM D1140-00) the silt content varies from 10% to 24% and the clay content ranges between 3.9% and 15.3%. The moisture content (according to ASTM D2216-98) involves the calculation of the ratio of water present in the soil mass to the weight of soil solids percent; the water content values range from 0 to 3.6%.

Table 1. Unified soil classification of the surface and boreholes samples.

Surface soil (sample no.)	Classification	USCS
SS 1, 2	Clayey sand	SC
SS 7, 8, 9, 10, 11, 16, 18, 24, 33	Silty sand	SM
SS 3, 4, 5, 17, 19	Well–graded sand with silt	SW–SM
SS 12	Poorly–graded sand with silt	SP–SM
SS 13, 14, 15	Silty clayey sand	SC–SM
Subsurface soil (Borehole no./Depth m)	Classification	USCS
BH1 (0.0–1.5 m, 3–4.5 m); BH2 (0–1.5 m, 3–3.7 m); BH4 (0–1.5 m); BH6 (0.5–1.5 m, 1.5–3 m, 3–4.5 m, 4.5–6 m, 6–8 m); HB16 (3–4.5 m)	Clayey sand	SC
BH1 (1.5–3.0 m, 4.5–5.4 m); BH5 (0–1.5 m)	Silty sand	SM
BH2 (1.5–3.0 m); BH9 (0–1.5, 1.5–3, 3–4.5 m)	Sandy lean clay	CL
BH3 (0–1.5 m, 1.5–3 m); BH8 (0–0.40 m); BH16 (0.0–1.5 m, 1.5–3 m)	Silty clayey sand with gravel	SC–SM
BH16 (4.5–6.0 m, 6–7.5 m)	Silty sand with gravel	SM
BH16 (7.5–9.0 m)	Poorly-graded sand with clay and gravel	SP–SC
BH16 (9.0–10.5 m, 10.5–12 m)	Well–graded gravel with sand	GW

Field Tests of Soil and Rocks
Standard penetration test

The Standard Penetration Test (SPT) is an important soil field test because of its simplicity, mobility, and low cost. According to the SPT N-values, soils are classified into loose soil (N < 10), medium dense soil (N from 10 to 35), dense soil (N from 30 to 50), very dense soil (N > 50). In the current study, according to the SPT values the subsurface materials were classified into three units; medium dense, dense, and very dense (Table 2). The results show heterogeneous variation of the SPT values depending on the variation in the subsurface soil characteristics in the area.

Table 2. Soil classification based on the SPT N values.

BH	Depth (m)	SPT (N)	Description	Density
BH1	0.0–1.5	20	Reddish brown, clayey sand, dry	Medium dense
	1.5–3.0	>50	White to reddish brown silty sand, dry	Very dense
	3.0–4.5	>50	White to reddish brown clayey sand, dry	Very dense
	4.5–5.4	>50	Reddish brown silty sand, wet	Very dense
BH2	0.0–1.5	>50	Reddish to brown clayey sand, dry	Very dense
	1.5–3.0	27	White to reddish brown sandy lean clay, dry	Medium dense
	3.0–3.7	15	Reddish to brown clayey sand, dry	Medium dense
BH3	0.0–1.5	>50	Reddish to brown silty, clayey sand, dry	Very dense
	1.5–2.80	24	Reddish to brown silty, clayey sand, dry	Medium dense
	2.80–3.5	24	Reddish to brown silty, clayey sand, dry	Medium dense
BH4	0.0–0.5	>50	Reddish brown clayey sand, dry	Very dense
BH5	0.0–1.0	22	Brown silty sand, dry	Medium dense
BH6	0.0–1.50	>50	Brown clayey sand	Very dense
	1.50–3.0	40	Brown clayey sand, dry	Dense
	3.0–4.5	25	Brown clayey sand	Medium dense
	4.5–6.0	31	Brown clayey sand	Dense
	6.0–7.50	>50	Brown clayey sand	Very dense
	7.50–8.00	41	Brown clayey sand	Dense
BH7	0.0–0.25	20	Top soil is silty sand with sandstone fragment	Medium dense
BH8	0.0–0.40	>50	Rock appears at 0.4 m, brown silty, clayey sand with gravel, dry	Very dense
BH9	0.0–1.5	>50	Brown sandy lean clay, dry	Very dense
	1.5–3.00	36	White to brown sandy lean clay, dry	Dense
	3.00–3.50	24	Reddish brown sandy lean clay, dry	Medium dense
BH16	0.0–1.5	22	Light brown silty, clayey sand dry	Medium dense
	1.5–3.0	32	Light brown silty, clayey sand, dry	Dense
	3.0–4.5	>50	Reddish clayey sand with gravel, dry	Very dense

(*continued*)

Table 2. (*continued*)

BH	Depth (m)	SPT (N)	Description	Density
	4.5–6.0	>50	Reddish silty sand with gravel, dry	Very dense
	6.0–7.5	>50	Reddish silty sand with gravel, dry	Very dense
	7.5–9.0	>50	Reddish poorly-graded sand with clay and gravel, dry	Very dense
	9.0–10.5	>50	Reddish well-graded gravel with sand, dry	Very dense
	10.5–12.0	>50	Reddish well-graded gravel with sand, dry	Very dense

Properties of Rock Cores

The core samples extracted from the boreholes are characterized by sandstone ranging in color from light brown, brown, dark brown, dark red, yellow and white. The rock recovery percent ranges from 52% to 100% and at most boreholes is above 70%. RQD values range from very poor at the top layer which is completely weathered, and the values increase with depth to the end of the boreholes (Table 3). In general, the degree of weathering ranges from completely weathered especially in the top layer and decreases with depth to be slightly weathered or fresh at each borehole. The strength and unit weight values are described in Table 4 in which the sample strength ranges from very low strength (4.32–24.7 MPa) to low strength (28.5–36.5 MPa). Density ranges from 1972.7 kg/cm^3 to 2299.4 kg/cm^3 (Table 4).

Table 3. Description and properties of rock core samples

BH	Depth (m)	RQD%	Type	REC%	W	Description
BH1	5.4–7.6	0	VP	100	H–C	Brown to white sandstone
	7.60–12.1	87–94	G–VG	94–100	S	White to yellowish sandstone
	12.1–15	91–100	VG	100	S–M	White to light brown sandstone
BH2	3.7–4.60	0	VP	70	C	Brown to gray sandstone with fragments
	4.6–12.15	31–51	P–F	78–100	M–C	Yellowish brown sandstone
	12.15–15.15	90	G	95	M	Yellowish to light brown sandstone
BH3	3.5–4.30	0	VP	68.7	H	Yellowish to brown sandstone with some gravel
	4.30–6.0	85	G	94	M	Yellowish to brown sandstone
	6.0–14.95	25–48	P	93.5–94.5	M	Yellowish/light brown/brown sandstone.
BH4	0.5–1.55	10.11	VP	59	M	Reddish brown sandstone with some gravel
	1.55–3.05	81	G	93	S	White to reddish sandstone
	3.05–15	37–60	P–F	97–100	M–H	Yellow to brown sandstone

(*continued*)

Table 3. (*continued*)

BH	Depth (m)	RQD%	Type	REC%	W	Description
BH5	1.0–1.60	0	VP	100	H	Dark brown sandstone fragment
	1.60–12.1	59–90	F–G	97–98	S–H	Yellow, Light brown sandstone
	12.10–15.10	27	P	100	H	Yellow to brown sandstone
BH6	8.00–8.90	16	VP	94	M–H	Brown to dark brown sandstone
	8.90–10.9	7	VP	60	M–H	Dark brown to yellowish sandstone
	10.9–15	51–74	F	85–90	M	Light to brown sandstone
BH7	0.25–4	44–64	P–F	88–100	M	Yellowish to light brown sandstone
	4.0–9	71–75	F	100	S	White to brown sandstone
	9.0–15	83–100	G–VG	100	S	Yellowish to brown sandstone
BH8	0.40–3.1	65–85	F–G	97.4–100	M	Yellowish to light brown sandstone
	3.10–6.10	71	F	100	S–M	Light brown to yellowish sandstone
	6.10–15.1	79–89	G	98–100	M	Brown to yellowish sandstone
BH9	3.50–5.40	15	VP	97	M–H	Dark brown to white sand stone
	5.40–9.10	0–6	VP	53–70	H	Brown/Light brown to reddish sandstone
	9.10–15.10	71–97	F–VG	100	S	White sandstone, dark/light brown Sandstone
BH10	0.0–6.1	69–88	F–G	93.5–100	S–M	Brown to light brown to yellowish sandstone
	6.10–9.10	66	F	100	M–H	Reddish to brown to yellow sandstone
	9.10–10.10	80	G	98	S–M	Yellowish to light brown sandstone
BH11	0.0–9.1	37–59	P–F	92–100	S–M	Light brown to dark reddish brown sandstone
	9.10–10.10	22	VP	100	H	Light brown sandstone include quartz grains
BH12	0.0–3.1	0–25	VP	100	M–H	Light brown - Brown sandstone
	3.10–10	0	VP	52–95	H	Brown – Dark red sandstone
BH13	0.0–1.55	15	VP	93	H	Light brown sandstone
	1.55–10.2	58–87	F–G	93–98	S–M	Light brown sandstone
BH14	0.0–1.55	6	VP	71	M–H	Brown sandstone
	1.55–3.10	35	P	84	M	Brown/white sandstone
	3.10–10.15	54–91	F–VG	100	S	Light brown to yellowish/white sandstone
BH15	0.0–3.1	19–26	VP–P	77–84	M–H	Brown, white sandstone
	3.10–10.5	18–85	VP–G	88–100	S–M	White to brown/Light brown sandstone

Weathering description; W = weathering, C = complete, H = high, M = moderate, S = slightly; RQD description; VP = very poor, P = poor, F = fair, G = good, VG = very good

Table 4. Strength values of the rock samples.

BH #	Depth (m)	Density (kg/m^3)	Compressive Strength (MPa)	Description
BH1	7.60–15.00	2104.7–2097.3	8.4–15.3	VLS
BH2	7.60–15.15	2083.5–2107.4	4.3–8.5	VLS
BH3	4.30–14.95	2157.3–2175.4	9.9–16.7	VLS
BH4	1.55–15.00	2104.1–2294.8	14.4–24.7	VLS
BH5	1.60–12.10	2069.9–2298.8	14.5–32.9	VLS–LS
BH7	3.00–15.00	2021.8–2182.9	16.4–24.1	VLS
BH8	0.40–15.10	2162.1–2284.9	19.6–30.9	VLS–LS
BH9	9.10–15.10	2167.6–2225.6	22.7–24.3	VLS
BH10	0.00–10.10	2079.6–2157.8	16.3–24.1	VLS
BH11	1.20–9.10	2059.9–2138.3	22.1–24.8	VLS
BH13	1.55–10.20	2147.8–2299.4	24.4–36.5	VLS–LS
BH14	1.55–9.10	2263.6–2268.9	15.5–31.7	VLS–LS
BH15	6.10–8.90	1972.7–2260.3	16.8–28.5	VLS–LS

Table 5. Results from the rock outcrop stations.

Name	Easting	Northing	Type	Grain size	Color	Weathering	UCS (MPa)	RQD
RS1	775778	3061188	Sandstone	Medium	Light Brown	Moderately/W3	29/S4	55.6/R3
RS2	775771	3060395	Granite	Coarse	Red	Moderately/W3	112/S2	85.3/R2
RS3	775247	3060497	Sandstone	Fine	Red	Moderately/W3	21/S4	58.9/R3
RS4	771062	3060504	Sandstone	Fine	White	Moderately/W3	15/S4	55.6/R3
RS5	777668	3060057	Sandstone	Fine	Yellow	Slightly/W2	59/S3	42.4/R4
RS6	778079	3059771	Sandstone	Medium	Yellow	Moderately/W3	19/S4	58.9/R3
RS7	771795	3070017	Sandstone	Medium	Yellowish Brown	Moderately/W3	15/S4	72.1/R3
RS8	771748	3071217	Sandstone	Medium	Yellowish Brown	Moderately/W3	11/S5	52.3/R3
RS9	774571	3065853	Granite	Coarse	Dark Brown	Moderately-Highly/W3-W4	47/S4	88.6/R2
RS10	779265	3058929	Sandstone	Fine	Dark Grey	Slightly/W2	47/S2	0/R5
RS11	782345	3056820	Sandstone	Medium	Yellow	Moderately/W3	15/S4	46/R4
RS12	787358	3052457	Sandstone	Medium	Yellow	Highly/W4	5/S5–S6	23/R5
RS13	787546	3052691	Granite	Coarse	Dark Brown	Moderately/W3	139/S2	69/R3
RS14	786672	3053673	Sandstone	Medium	Yellowish Brown	Moderately/W3	29/S4	0/R5
RS15	786485	3054313	Sandstone	Medium	Reddish Brown	Moderately/W3	68/S3	0/R5
RS16	782243	3057400	Sandstone	Medium	Yellow	Moderately/W3	54/S3	0/R5
RS17	764933	3061658	Granite	Coarse	Dark Brown	Highly/W4	14/S4	72/R3
RS18	763951	3062983	Granite	Coarse	Red to grey	Moderately/W3	44/S4	82/R2

Table 6. Grain size distribution and geotechnical properties of zones I, II, III, IV.

Zone	Grain size distribution				Physical and geotechnical properties			
	Gravel (%)	Sand (%)	Silt (%)	Clay (%)	Bulk density (g/cm^3)	Moisture content (%)	Plasticity index	SPT N value
I	9.2	66.1	12.7	12.0	1.67	1.99	–	–
II	10.0	69.5	10.0	10.5	1.92	2.55	12	–
III	8.5	86.5	5.0	0.0	1.94	0.28	–	–
IV	9.2	56.4	24.6	9.8	1.98	3.6	5	–

4.2 Rock Outcrop Properties

Eighteen rock outcrop stations were selected and investigated for different parameters including physical properties (rock type, color, and grain size), degree of weathering, uniaxial compressive strength (UCS), and rock quality designation (RQD) (Deere and Miller 1966). These collected data will help in understanding and classifying the rock masses. These tests are according to the geomechanics classification of rock masses (Bieniawski 1973, 1989). The results indicate that: (1) Sandstone is fine to medium grained. It has several colors (light brown, red, white, yellowish brown, yellow, and dark gray). Weathering degree ranges from slightly to moderately weathered (W2–W3). Uniaxial compressive strength (UCS) of sandstone samples ranges from 5 MPa (very low strength (S5)) to 68 MPa (moderately strength (S3)). Rock Quality Designation ranges from (0%) very poor rock (R5) to (59%) fair rock (R3). Joint spacing ranges from 6 to 20 cm (closely spacing (F4)). (2) Granite is coarse grained and dark brown, red and gray color. Weathering degree ranges from moderately to highly weathered (W3–W4). UCS of granite samples ranges from 14 MPa (very low strength (S5)) to 139 MPa (high strength (S2)). Rock Quality Designation ranges from poor rock (47%) (R4) to good rock (85%) (R2). Joint spacing ranges from 20 to 60 cm (Moderately wide spacing (F3)).

4.3 Geological Engineering Mapping

General
Field and laboratory investigations were carried out to evaluate the physical and engineering properties of soil and rock, to describe the geological and geomorphological features, to determine the hydrogeological conditions, and to understand the distribution of different units in the area. The interaction of the soil and rock characteristics and their conditions create various zoning areas of different physical and geotechnical properties; there are usually grouped them into individual geological engineering units that may or may not follow the traditional geological boundaries.

Comprehensive Engineering Geology Mapping for Surface Data
This approach is used primarily for engineering purposes and is particularly useful to geotechnical engineer. Therefore, it used for all structural-related projects; such as

bridges, retaining walls, buildings, etc. Precise classification requires grain size distribution and atterberg limits values for each sample. It is useful in placing soils and rocks into various groups and/or subgroups according to their characteristics. In the current study, the developed engineering geological map was divided into six main zones (four zones for soil and two zones for rocks) based on the geologic conditions and the geotechnical properties of the materials in the area (Fig. 4). Zone (I): This zone is characterized by silty sand (SM). It covers most of the study area (Fig. 4). The geotechnical characteristics of the zone are also shown in the pie diagram. The periphery of the pie diagram shows the soil fraction percentage including bulk density, moisture content, plasticity index, and SPT N value. Zone (II): This zone is characterized by clayey sand (SC). This zone covers some isolated areas in the north and south west of the project area (Fig. 4). The physical and geotechnical characteristics of this zone are shown in the pie diagram. Zone (III): This zone is characterized by well graded sand with silt (SW-SM). This zone covers some isolated areas in the north and center part of the study area. The physical and geotechnical characteristics of this zone are shown in the pie diagram. Zone (IV): This zone is characterized by silty clayey sand (SC-SM). This zone covers a small area in the middle of the study area. The physical and geotechnical characteristics of this zone are shown in the pie diagram. Zone (V): This zone is characterized by Saq Sandstone which contains small isolated hills in the northern and eastern parts of the study area (Fig. 4). This zone is characterized by (1) slightly - moderately weathered (W2–W3); (2) Rock strength of this zone is characterized by very low strength (5 MPa) - moderate strength (68 MPa) (S5–S3); (3) RQD values range between zero (Very poor) to 59 (Fair) (R5–R3); and (4) closely spaced joints range from 6–20 cm (F4). Zone (VI): This zone is characterized by granite rocks.

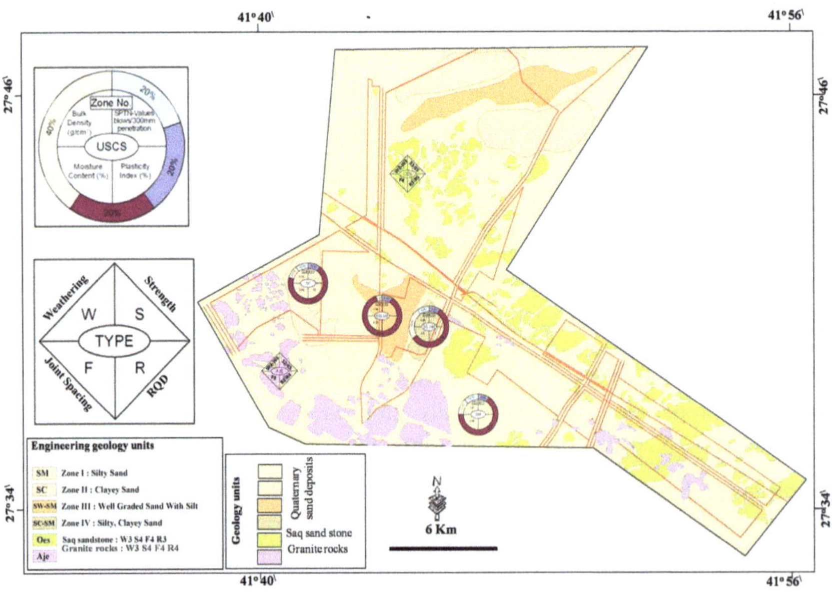

Fig. 4. Comprehensive surface geological map of the study area.

It contains small isolated hills in the south western side of the study area (Fig. 4). This zone is characterized by moderately - highly weathered (W3–W4); Very low strength (14 MPa) - high strength (139 MPa) (S5–S2); RQD values range between 47 (Poor) - to 85 (Good) (R4–R2); and closely spaced joints range from 20–60 cm (F3).

Preparation of Stratigraphic Sections
Based on the interpretation and analysis of the borehole data, three stratigraphic sections were prepared (Fig. 5). Profile (1) between boreholes BH9, 15, 11 and 12 (Fig. 5a, b); Profile (2) between boreholes BH01, 02, 08, 15, and 10 (Fig. 5a, c); and profile (3) between BH01, 06, 05, 09, and 16 (Fig. 5a, d). Two layers were identified based on the soil characteristics (Fig. 5b, c, d). The upper surface layer consists of sand deposits and the second layer is characterized by sandstone rocks.

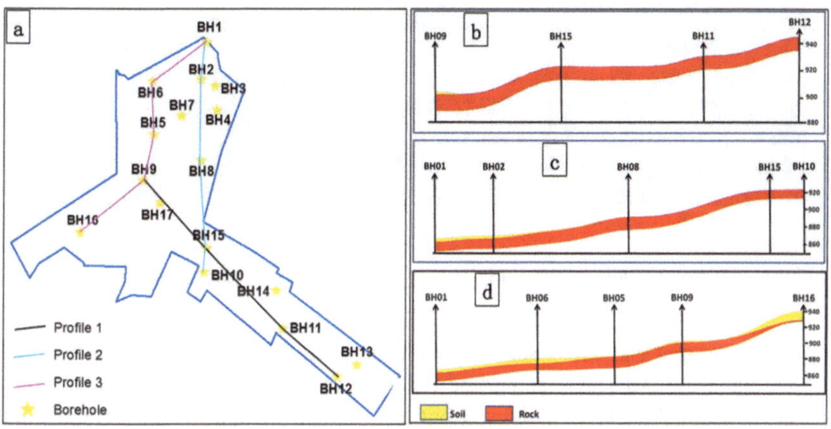

Fig. 5. (a) Location of the stratigraphic sections in the study area; (b) Profile (1) along boreholes 9, 15, 11, and 12; (c) Profile (2) along boreholes 1, 2, 8, 15, and 10; (d) Profile (3) along boreholes 1, 6, 5, 9, and 16.

4.4 Geological Hazards Evaluation

In recent years, the Saudi government encourages the private sectors and companies to invest in The New Hail Economical City. However, in most projects, potential geological problems that might affect areas for dedicated urbanization are not fully recognized. The New Hail Economical city is affected by flash floods where some wadis move through the area. Flood water leads to dangerous situations such as the inundation of urban and agricultural lands, especially in arid regions. It is, therefore, necessary to manage and control these problems. In case of floods, such controls are of prime importance to protect the area (urban areas and industrial sites). Flood hazards can be divided into primary (direct) and secondary (indirect) categories. The former is due to direct contact with water and the latter occurs as disruption of services, health impacts such as famine and disease. Flood water can transport garbage, debris, and

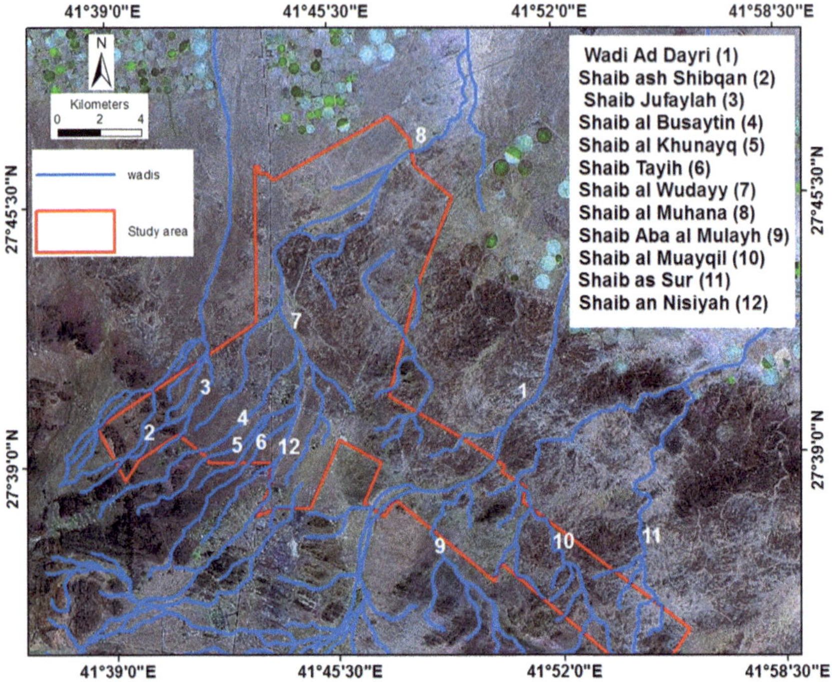

Fig. 6. Different wadis in the study area extracted from topographic map and DEM.

toxic pollutants that can cause secondary effects of health hazards. The current study indicates that wadi Ad Dayri and some other small wadis have major impact on the study area. Wadi Ad Dayri and its tributries were mapped from topographic maps (1:50000) and digital elevation model (DEM 90 m) (Fig. 6).

5 Conclusion and Recommendation

The study area is characterized by flat areas with some isolated hills including three geological units; Saq sandstone, granite, and alluvial deposits. The geological map, prepared for the New Hail Economic city, was used as a base map for producing engineering geology map at a scale of 1:100,000. Six zones were recognized based on physical and geotechnical engineering properties. Four zones were identified for surface soils and two zones for the outcropping rocks. Soil zones include; zone (I) silty sand (SM), zone (II) clayey sand (SC), zone (III) well-graded sand with silt (SW-SM), and zone (IV) silty clayey sand (SC-SM). Rock zones include Zone (OES) Saq Sandstone and zone (AJE) granite.

Different boreholes were drilled in the study area and three cross sections were prepared. The results indicate that there are two layers. The surface layer is characterized by soil; its thickness ranges from 0.5 m to 12 m. The second layer is characterized by sandstone rocks; its thickness ranges from zero to more than 12 m.

According to the geological hazards that affect the study area, it was found that flash floods consider to be major hazards due to the presence of many wadis.

Finally, recommendation and consideration were introduced as follow: engineering geology mapping and geological hazard evaluation represent crucial work before begging of any infrastructure project as they provide information that could help engineers and decision makers in understanding the study area. Wadis may cause flood hazard if they are left without any solution. Channels need to be constructed to accommodate each wadi peak discharge for 100 years return period.

References

Aly, M.H.: Integration of remote sensing and GIS for geo-environmental assessment and monitoring of the New Minia City, Egypt. M.Sc. thesis, Geology Department, Faculty of Science, Zagazig University, Zagazig, Egypt (1997)

Aly, M.H., et al.: Suitability assessment for New Minia City, Egypt: a GIS approach to engineering geology. Environ. Eng. Geosci. **XI**(3), 259–269 (2005)

Al-Harbi, H.M., et al.: Engineering geological map of Ad Dammam, Dhahran and Al-Khobar cities, Kingdom of Saudi Arabia. Saudi Geological Survey technical report SGS-TR-2008-4, 39 p. (2009)

Al-Rasheed, M.: Politics in an Arabian Oasis: the Ibn Rashid tribal dynasty. I.B. Tauris & Co. Ltd., London, New York (1991). ISBN 1-85043-320-8. Based on a Ph.D. thesis presented to Cambridge University, 1988

Anonymous: The preparation of maps and plans in terms of engineering geology. Report of a working party of the engineering group of the geological society. Q. J. Eng. Geol. **5**, 293–381 (1972)

Anonymous: BS EN ISO 14688-1-Geotechnical investigation and testing. Identification and classification of soil—part 1: identification and description. British Standards Institution, London (2002)

ASTM: Soil and rock building stone, peats, annual book of ASTM standards, part 19. American Society of Testing and Material (1989)

Barton, N.R. et al.: Engineering classification of rock masses for the design of tunnel support. Rock Mech. 6, 189–236. Springer, Heidelberg (1974)

Bieniawski, Z.T.: Engineering classification of jointed rock masses. Trans. S. Afr. Inst. Civ. Eng. **15**, 335–344 (1973)

Bieniawski, Z.T.: Engineering Rock Mass Classifications, p. 251. Wiley, New York (1989)

Brown, G.F. et al.: Geologic map of the wadi ar-Rimah quadrangle, Kingdom of Saudi Arabia. U.S. Geological Survey Miscellaneous Geologic Investigation Map 1–205A, scale 1: 50,000 (1963)

Brown, G.F., Jackson, R.O.: The Arabian shield. In: 21st Proceeding International Geology Congress, Copenhagen, Sec. 9, pp. 69–77 (1960)

Chacon, J., et al.: Engineering geology maps: landslides and geographical information systems. Bull. Eng. Geol. Environ. **65**, 341–411 (2006)

Chávez, J.A., et al.: Engineering geology mapping in the southern part of the Metropolitan area of San Salvador. Revista Geológica de América Central **46**, 161–178 (2012)

Commission Staff Working Document: Implementation plan of the EU strategy for supporting disaster risk reduction in developing countries 2011–2014, SEC (2011), vol. 215. European Commission, Brussels, Belgium (2011)

Craa'chlev, C.R., Denness, B.: Engineering geology in urban planning with an example from the new city of Milton Keynes. In: International Geological Congress, 24th Session, Montreal, Section 13 Proceedings, pp. 13–22 (1972)
Dearman, W.R., Fookes, P.G.: Engineering geological mapping for civil engineering practice in the United Kingdom. Q. J. Eng. Geol. **7**, 223–256 (1974)
Deere, D.U.: Technical description of rock cores for engineering purposes. Rock Mech. Eng. Geol. **1**(1), 17–22 (1964)
Deere, D.V., Miller, R.D.: Engineering classification and index properties of intact rock. Technical reference AFNL-TR-65-116. University of Illinois (1966)
Delfour, J.: Geologic, tectonic and metallogenic evolution of the Northern part of the Precambrian Arabian Shield (Kingdom of Saudi Arabia). Bulletin BRGM, **II**(1–2), 1–19 (1980a)
El-Haddad, A.A., Youssef, A.M.: Overview of the potential geo-environmental hazards in the low desert zone, Egypt. Case study: new projects West of El-Kawamel village, Sohag, Egypt. Assiut Univ. J. Geol. **37**(2) (2008)
Geological Society of London: The description of rock mass for engineering purposes. Q. J. Eng. Geol. **10**, 355–388 (1977)
Graña, A.M. et al.: Engineering geology maps for planning and management of natural parks. Las Batuecas-Sierra de Francia and Quilamas (Central Spanish System, Salamanca, Spain). Geosciences **3**, 46–62 (2013). doi:10.3390/geosciences3010046
International Association Enginering Geology and Environment (IAEG): Engineering Geological Maps: A Guide to Their Preparation, p. 79. UNESCO Press, Paris (1976)
International Association Enginering Geology and Environment (IAEG): Recommended symbols for engineering geological mapping. Bull. Intern. Assoc. Eng. Geol. **24**, 227–234 (1981a)
International Association Enginering Geology and Environment (IAEG): Rock and soil description for engineering geological mapping. Bull. Intern. Assoc. Engn. Geol. **24**, 235–274 (1981b)
ISRM: Quantitative description of discontinuity in rock masses. Int J. Rock Mech. Min. Sci. Geomech. Abstr. **15**, 319–368 (1977)
Jato, J.N. et al.: Engineering geological mapping for the urban area of Salamanca (Spain). IAEG2006 Paper number 315 (2006)
Kellogg, K.S., Stoeser, D.B.: Reconnaissance geology of the Ha'il Quadrangle, sheet 27/41 B, Kingdom of Saudi Arabia. U.S. geological survey, Open-file report 85–618, vol. ii, 35 p. (1985)
Mohammed, A.M.S., et al.: Geotechnical risks map of Saïda City. Algeria. Electron. J. Geotech. Eng. **15**, 403–413 (2010)
Okogbue, C.O., Aghamelu, O.P.: Comparison of the geotechnical properties of crushed shales from Southeastern Nigeria. Bull. Eng. Geol. Environ. **69**, 587–597 (2010)
Palmstrom, A.: The volumetric joint count - a useful and simple measure of the degree of rock mass jointing. In: IAEG Congress, New Delhi, pp. V.221–V.228 (1982)
Parry, S., Hart, J.R.: Engineering geology and the reduction of geotechnical risk: challenges facing the profession in Hong Kong. Q. J. Eng. Geol. Hydrogeol. **42**, 499–510 (2009)
Planet, L.: The Middle East, 3rd edn. (2000). ISBN 0-86442-701-8
Que, J.S., et al.: Geotechnical properties of the soft soil in Guangzhou College City. Bull. Eng. Geol. Environ. **67**, 479–483 (2008)
Rozos, D., et al.: Engineering-geological map of the wider Thessaloniki area. Greece. Bull. Eng. Geol. Environ. **63**, 103–108 (2004)
Solberg, I.L., et al.: Combined geophysical and geotechnical approach to ground investigations and hazard zonation of a quick clay area, mid Norway. Bull. Eng. Geol. Environ. **71**, 119–133 (2012)

UNESCO: Engineering Geological Maps: A Guide to Their Preparation. UNESCO press, Paris, 79 p. (1976)

Waltham, A.C.: Foundations of Engineering Geology; Blackie Academic and Professional. Chapman and Hall, London (1994)

Wickham, G.E. et al.: Support determination based on geologic predictions. In: Lane, K.S., Garfield, L.A. (eds.) Proceedings of North American Rapid Excavation Tunneling Conference, Chicago, New York, pp. 43–64 (1972). Soc. Min. Engrs, Am. Inst. Min. Metall. Petrolm. Engrs

Xu, B., Li, Y.: Engineering geological mapping for an open pit. J. Bull. Eng. Geol. Environ. **21**(1), 131–138 (1980). doi:10.1007/BF02591553

Youssef, A.M.: Mapping the Pliocene clay deposits using remote sensing and its impact on the urbanization developments in Egypt: case study, East Sohag area. Geotech. Geol. Eng. **26**, 579–591 (2008). doi:10.1007/s10706-008-9191-6

Youssef, A.M., Moneim, A.A.A.: Evaluation of the geo-environmental hazard in relation to the future development using geographic information systems, East Sohag area, Egypt. In: The Third International Conference of Environment and Development in the Arab World, Assiut, Egypt, vol. 1, pp. 673–692, March 2006

Youssef, A.M., et al.: Coupling of remote sensing data aided with field investigations for geological hazards assessment in Jazan area, Kingdom of Saudi Arabia. Environ. Earth Sci. **65**(1), 119–130 (2011). doi:10.1007/s12665-011-1071-3

Youssef, A.M., et al.: Natural and human-induced sinkhole hazards in Saudi Arabia: distribution, investigation, causes and impacts. Hydrogeol. J. **24**(3), 625–644 (2016). doi:10.1007/s10040-015-1336-0

Pre-design Site Investigation for an Oil and Gas Facility in the Marginal Lands of the Nigerian Niger Delta Sub-region

So-ngo Clifford Teme[1,2(✉)]

[1] Institute of Geosciences and Space Technology (IGST), Rivers State University of Science and Technology, P.M.B. 5080, Port Harcourt, Nigeria
teksgeotech@yahoo.com, cliffordteme@yahoo.com

[2] Department of Geology, Rivers State University of Science and Technology, P.M.B. 5080, Port Harcourt, Nigeria
teme.clifford@ust.edu.ng

Abstract. A pre-design site investigation for Oil and Gas Storage Tanks to be located at a project site between Latitudes N04° 47′32.2″ to N04° 47′49.5″ and Longitudes E006° 55′ 35.9″ to E006° 56′ 35.9″ within the Marginal Lands of the Niger Delta sub-region has been carried out for an Oil & Gas Company in Nigeria. A total of fourteen (14) 152.4 mm diameter, 40 m deep borings, using the Percussion Drill Rig were carried out along the shoreline of the site. Eight (8) of the borings were situated along the shoreline at high tide while the remaining six (6) borings were situated 20.00 m offshore. The lithology along the coastline consisted of an uppermost dark-greyish Organic Clay layer (OH) [7–10 m thick]; underlain by greyish Silty-Sands (SM) [3–5 m thick]. These are further underlain by Clayey Silts (ML) [2.00–4.50 m thick], Clayey Sands (SC) [3–10 m thick] and Well-graded Sands and gravels (SW) at the bottom of the borings [>40 m thick]. The borings off-shore had similar lithologies but with preponderance of Sands and Gravels at the lower depths of the borings. While the upper dark organic clays (OH) have friction angles (ϕ) ranging from 0°–6° and cohesion intercepts (C_u) from 14.50 kPa to 28.0 kPa, the Silty-Sands (SM) and well-graded Sands and gravels (SW) at depths of between 30 m–40 m have friction angles (ϕ) ranging from 10°–36° and cohesion intercepts (C_u) of 0.00 kPa. The average Standard Penetration Test (SPT) N-values of the Sands (SP) and Gravel layers (SW) at depths varied from 15 to >50 with corresponding Bearing Capacity values (based on Factor-of Safety [FS = 3.0]) of 100.00–478.50 kPa, indicating high to very high values. A proposed Quay wall to protect the project site is to be constructed with reinforced concrete tubular steel piles. These are to be embedded at depths of between 30.00 to 40.00 m with computed average Total Pile Loads [$(Q_b + Q_s)] = q_{ult}$ varying from 323.25 MN to 1044.50 MN, [where: Q_b = total Pile base load and Q_s = Pile Skin friction]. Work on this project has since commenced to about 60% completion at the time of preparing this paper.

Keywords: Oil and gas storage tanks · Quay wall · Marginal lands · SPT N-values · Bearing capacity and pile load tests

J. Wasowski et al. (eds.), *Engineering Geology and Geological Engineering for Sustainable Use of the Earth's Resources, Urbanization and Infrastructure Protection from Geohazards*, Sustainable Civil Infrastructures, DOI 10.1007/978-3-319-61648-3_2

1 Introduction

The Niger Delta sub-region hosts the major crude oil exploration and exploitation activities in Nigeria. There is therefore the need to establish oil storage facilities to accommodate the refined petroleum products. However, the Marginal Lands consisting of

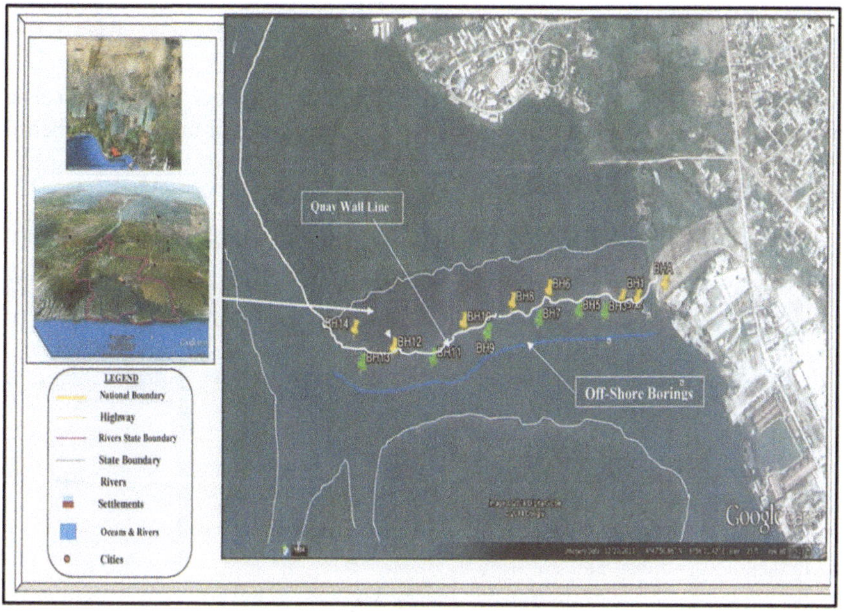

Fig. 1. Satellite imagery showing location of Quay Wall Line at project site.

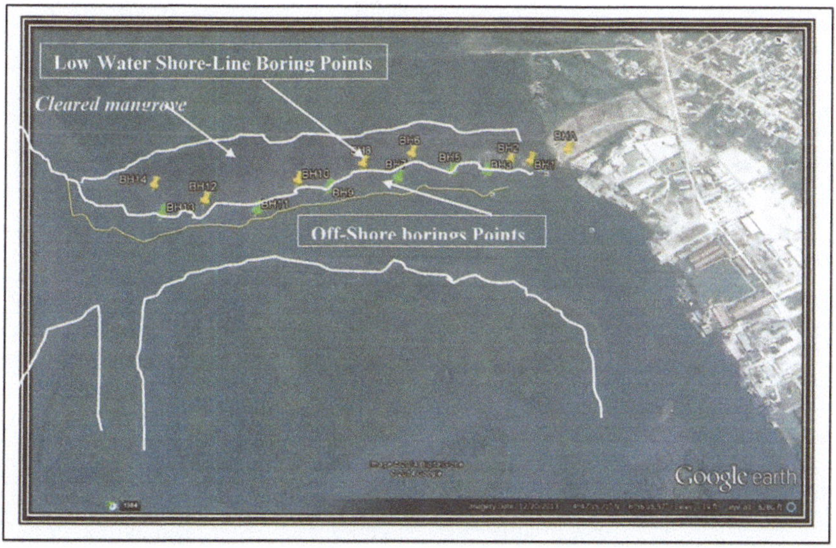

Fig. 2. Locations of boring points along Quay Wall site.

the mangrove and the freshwater forests, (Teme 1992a, b) do not have firm foundations to support the Oil and Gas storage facilities, thus necessitating the need to carry out proper geotechnical investigations for placement of facilities such as Quay walls to retain them.

This paper describes the pre-design site investigations for such an oil and Gas facility employing a Quay Wall structure for shore protection.

1.1 Project Location and Geologic Settings

A Quay Wall is an earth retaining structure which is used to dock floating vessels and transfer goods. Quay Walls are of various types and are used for mooring and berthing

(a)

(b)

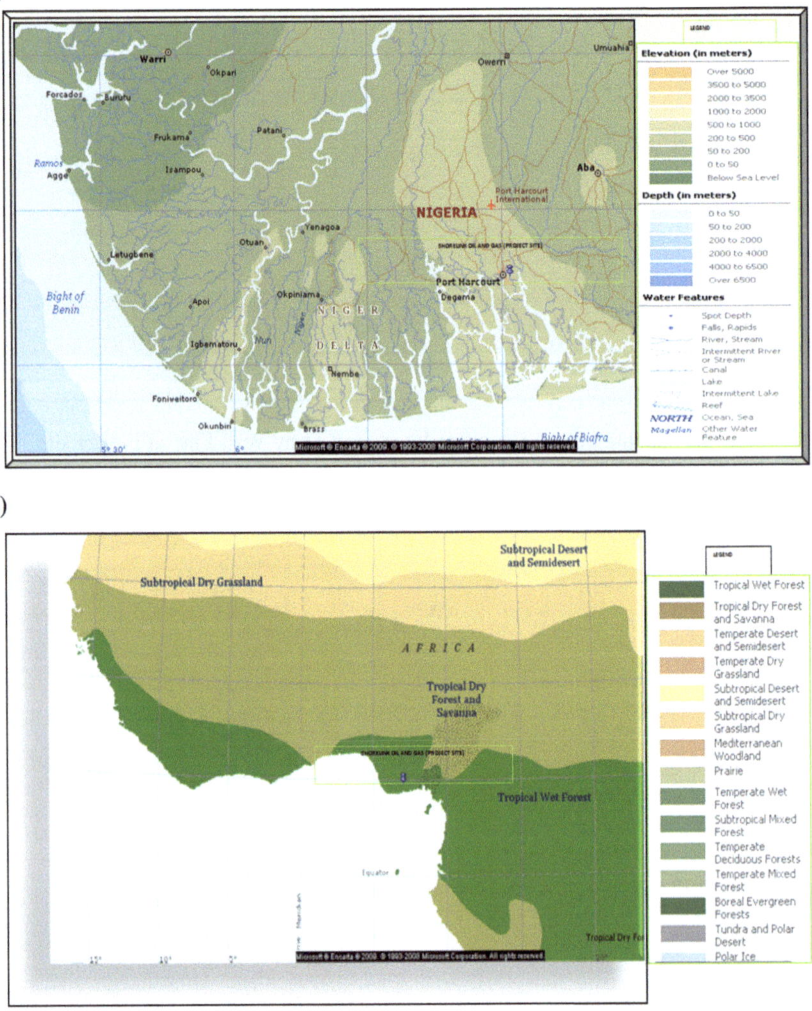

Fig. 3. (a) Topography at the project site. (b) Vegetation at the project site.

floating vessels such as barges, container vessels, ships, boats etc. *(NSCC International SoilMechanics of Abu Dhabi* - 2014*)*.

The Quay Wall project is located geographically between latitudes 04°47′43.2″ and 04°47′ 49.5″ North of the Equator and Longitudes 006°55′35.9″ and 006°56′35.9″ East of the Greenwich Meridian (Fig. 1).

Along the proposed outline of the Quay Wall along the shoreline were a total of fourteen (14) 152.4 mm diameter borings drilled by Percussion Drill Rig, each 40 m deep. Eight (8) of the borings were situated along the shoreline at high tide while the remaining six (6) borings were situated 20.00 m offshore of the quay wall (Fig. 2).

The project area is generally very low, often inundated by diurnal tidal ebb-and-flow waters and vegetated by mangrove trees (Figs. 3a and b).

The local geology of the study area consists of Clays, Sands, gravels and alluvium typical of the uppermost Benin Formation of the Niger Delta (Short and Stauble 1967; Malomo 2006) as shown in Fig. 4.

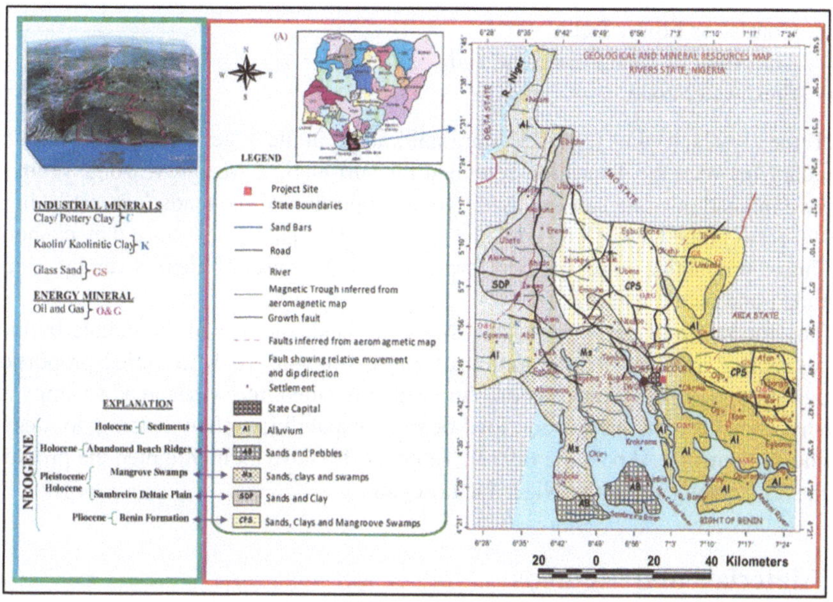

Fig. 4. Local geology of the project site.

2 Literature Review

Marginal land according to Rutledge (1970) is any land which is unsuitable for development in its original condition. However, others such as Forgione et al. (2008) see "Marginal land' as idle, degraded, waste and abandoned lands. There is also the general belief "marginal lands" have low bio-diversity due to their emptiness (Christopher 2012).

According to the Watch Indonesia (2014), "Marginal lands" in Indonesia are considered to be highly acidic. Indonesia "Marginal land" therefore includes swamp-lands, wetlands, peat forests as well as dry land on acidic soil such as mountain area, whereas "critical lands" is land that has been ecologically degraded as a result of intensive agricultural practices and which is no longer suitable for farming.

The UN Food and Agriculture Organization (FAO) see "Marginal land" as land that is exhausted and abandoned (Rossi and Lambrou 2008) while Gallagher Review (2008) believes that "ïdle" and "Marginal lands" are synonymous with grasslands.

Shujiang et al. (2013) have described "Marginal lands" as potential lands for development of improved food security, bioenergy and land use for infrastructural facilities such as Oil and gas which forms the subject matter for this paper.

2.1 Marginal Lands in the Niger Delta

Several morphological units have been identified in the Niger Delta due to the complex hydro-meteorological processes. Fubara et al. (1988) recognized three of such units as:

(i) Coastal or Lower Delta
(ii) Transition or Mangrove zone and
(iii) Freshwater zone.

Teme (1992a, b) thus referred "Marginal Lands" in the Niger Delta region to be the boundary between the Transition zone and the surrounding Mangrove zone. From the above explanations, one can say that a good proportion of the lands, especially in the coastal areas in the Niger Delta could be termed "Marginal Lands" since they cannot be put into "productive" use in their present state because of their swampy, peaty, degraded and acidic nature.

The formation of the Niger Delta "Marginal lands" is part of the deltaic build-up processes that involved the deposition of clays, silts and sands in varied proportions brought in during the "transgression" floods that resulted into widespread swampy and marshy grounds that have poor load-bearing capabilities often leading to uneven ground settlements. Akpokodje (1979), Teme (2002) have all highlighted the problems of uneven settlements in the Niger Delta region.

3 Materials and Methods

A total of fourteen (14) 40-m deep borings were carried out by manually operated Standard Penetration Test (SPT) drilling rigs during the project investigations. While the eight (8) on-shore borings were carried out on land at high water tide (Fig. 5a), the remaining six (6) off-shore borings were carried out about 20.00 m from the shoreline using floating pontoons specially constructed for that purpose (Fig. 5b). Floating pontoons were constructed with wooden planks encasing a series of empty thick-walled plastic drums as shown in Fig. 6. Where the ground was firm as at point #1 (Fig. 5a), a mechanized dando-percussion rig similar to that described by Tomlinson (1980) was utilized in the drilling process. Also, for the six (6) off-shore borings, a dinghy boat was

Fig. 5. (a) Standard Penetration Rig used for boring on land. (b) Standard Penetration Drilling Rig used for boring at over-water (off-shore) sites.

Fig. 6. Floating pontoon used in off-shore boring exercise.

used to ferry personnel and working materials to and from the drilling platform. Anchoring of the floating drilling platform was done by means of marine ropes tied to anchored poles or to sub-river bed metal anchors.

4 Results of Field and Laboratory Investigations

From the results of the Percussion drilling exercises, the following subsurface lithologies were observed. For the **on-shore borings**, the subsurface successions were as follows:

(i) An uppermost dark-greyish Organic Clay layer (OH) [7–10 m thick];
(ii) A greyish Silty-Sands (SM) layer, [3–5 m thick].
(iii) A Clayey Silts (ML) layer, [2.00–4.50 m thick],
(iv) Clayey Sands (SC) layer [3–10 m thick] and
(v) Well-graded Sands and gravels (SW) at the bottom of the borings [$\geq$40 m depth].

For the **off-shore borings**, the subsurface lithology down to a depth of 40.00 m is as follows:

(i) Dark greyish Organic Clay (OH) layer below the water column [5–6 m thick],
(ii) Poorly-graded Sands (SP) layer [4–23 m thick]
(iii) Well-graded Sands (SW) layer [0–17 m thick]
(iv) Dark greyish Clayey Silt (ML) layer [0–4 m thick]
(v) Silty Sands and Gravels (SM-SW) layer [$\geq$3–4 m thick].

The Lithological sequence of the soil profiles both On-shore and Off-shore are represented in Fence diagrams in Figs. 7a and b, respectively. In both on-land and off-shore borings, the Standard Penetration Test N-values increased with depth, indicative of the increase of the bearing capabilities of the subsurface materials with depth along the Quay Wall alignment (Figs. 8a and 8b, respectively).

4.1 Laboratory Investigations and Test Results

The following laboratory tests were carried out on the field obtained soil samples:

(i) Grain Size Distribution Analyses,
(ii) Atterberg Consistency Limits tests [namely, Liquid Limit (LL) Tests; Plastic Limit (PL) Tests, and their corresponding Plasticity Indices (PI),
(iii) Shear Strength Tests comprising Triaxial Compression Tests to determine the friction angles (ϕ) and Cohesion intercepts (c) of the subsurface soils.
(iv) Bearing Capacity Tests comprising Oedometer tests to determine the Coefficients of Consolidation (C_v), Pre-consolidation pressure (p'_c or σ'_p) and coefficient of Volume change ($M_v = 1/D$), *where* D = constrained or oedometric modulus

Tables 1 and 2 show the summary of the geotechnical characteristics of the subsoils along the alignment of the Quay Wall obtained both from field soundings and results of Laboratory tests carried out on soil samples obtained from borings at the project site.

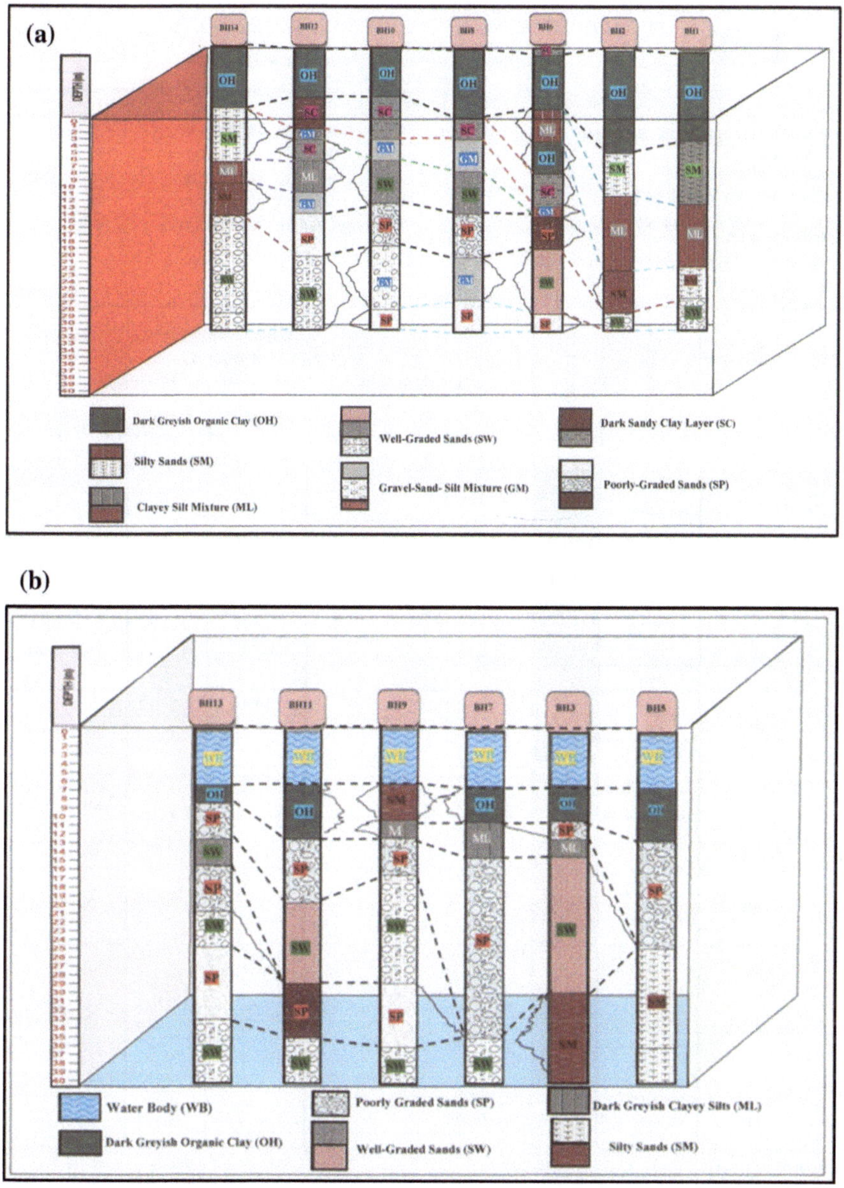

Fig. 7. (a) Lithological Fence Diagram of on-shore boring sites. (b) Lithological Fence Diagram of off-shore boring sites.

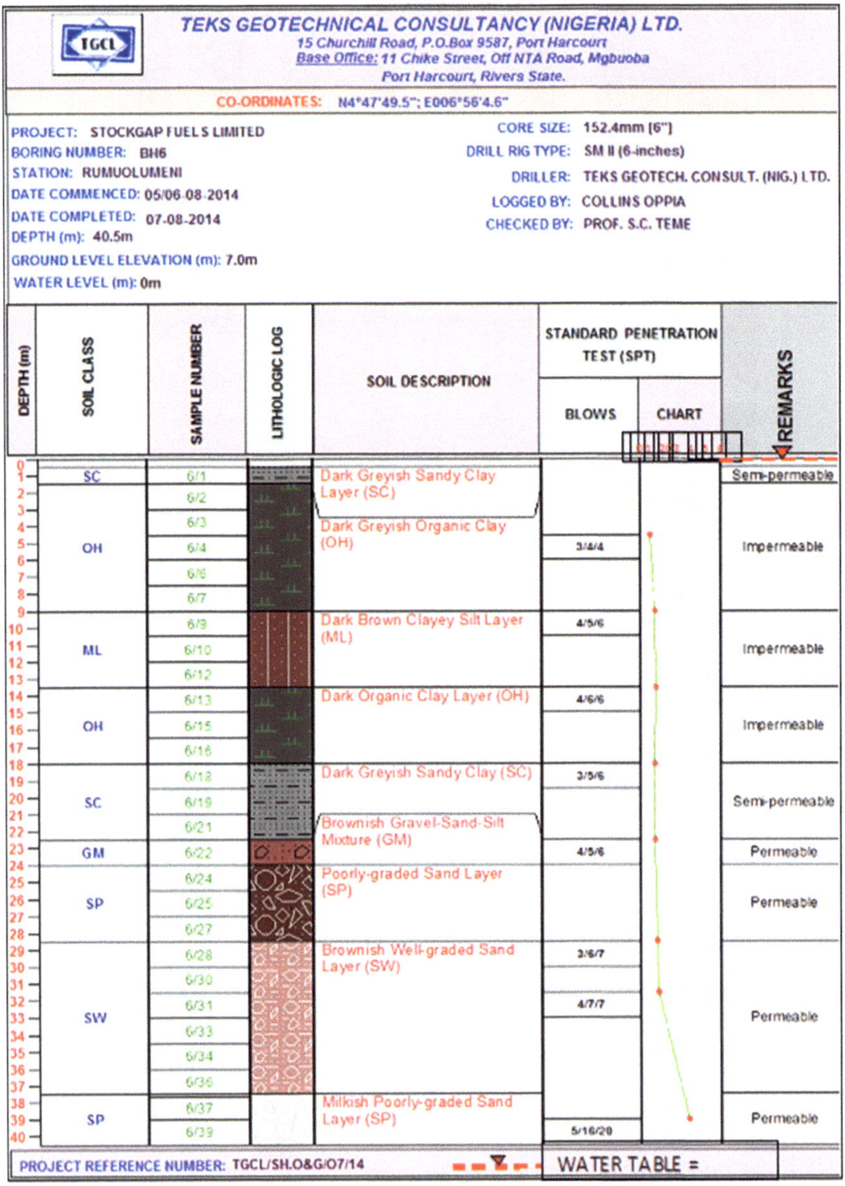

Fig. 8a. Typical Soil Log of on-shore drill sites indicating variations of SPT.

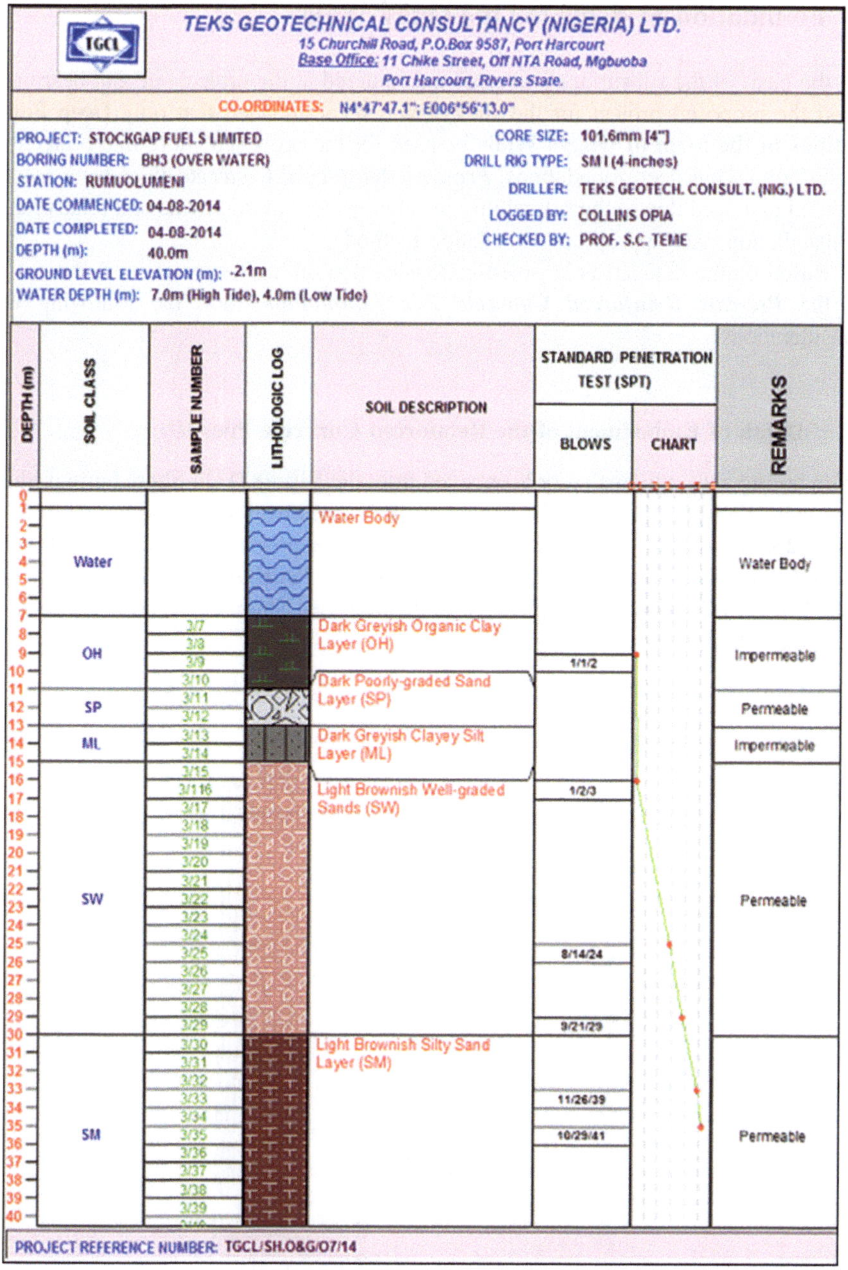

Fig. 8b. Typical Soil Log of off-shore drill sites indicating variations of SPT N-values with depth.

5 Foundation Systems for the Quay Walls

On the basis of the subsurface soil types encountered at the project site and bearing in mind the proposed project for the sites, it was recommended that only **Deep Foundations in the form of Quays Walls** be used for the proposed shore line protection.

In employing deep foundations, ***Pre-cast Reinforced Concrete Piles*** were recommended and used due to their durability in saline water environments and relative ease of installation with appropriate installation methods.

Based on the detailed field investigations carried out at the project site, the design of the ***Pre-cast Reinforced Concrete Pile Foundation*** took the following into considerations.

5.1 Depth of Embedment of the Reinforced Concrete Piles (D_f)

From boring and sounding records obtained from the *Seven* (7) On-Shore borings along the Quay Wall line (Fig. 7a), the depths of pile embedment were determined to be at about **25.0 m** below ground level. This depth corresponds to the competent ***Well-graded sands and Gravels (SW); Silty Sands (SM); Poorly-graded Sands (SP); and Silty Gravels (GM)*** **layers.** A schematic of the proposed Shore Protection structures for the **Quay Wall** is shown in Fig. 9.

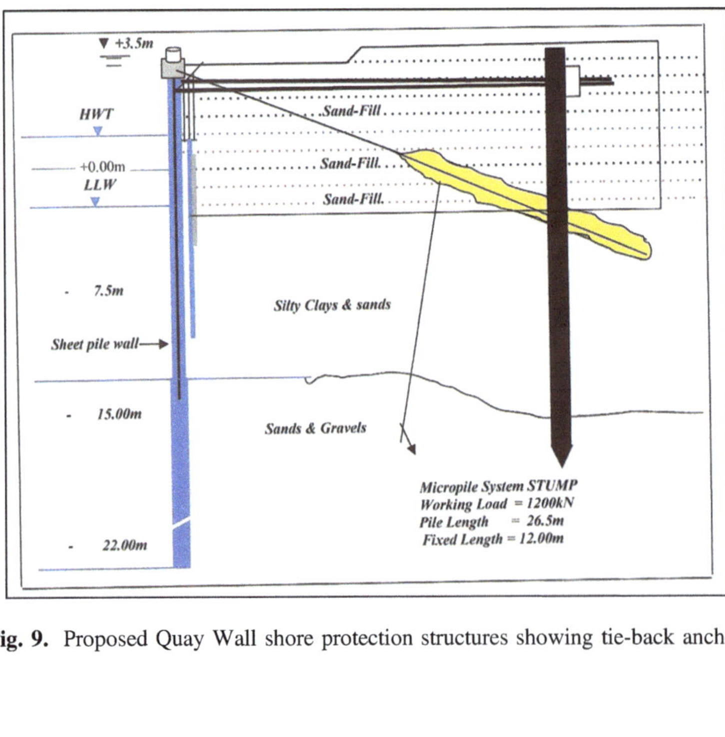

Fig. 9. Proposed Quay Wall shore protection structures showing tie-back anchor.

5.2 Soil Bearing Pressures

Based on soundings from field investigations and laboratory tests results on soil samples, the safe bearing loads obtained, based on the methods of Meyerhof (1963) and Bowles (1996), range from 62.90 to 210.50 MN for 25.00 m long piles and between 90.60 to 338.65 MN for 30.00 m long piles. These values are comparable to values obtained in similar soil conditions elsewhere within the Niger Delta sub-region and adjoining flanks (Teme 1989; 1991a, b; 1992a, b; 1993a, b; 1994a, b; 2002; 2004; 2007; 2008; 2012; 2014a, b).

5.3 Ultimate Carrying Capacities of the Reinforced Concrete Piles at the Project Site

Based on the ***Meyerhof (1963) Method***, the Bearing Capacity of a reinforced concrete pile in a soil possessing both cohesion and friction (Fig. 10) can be taken as:

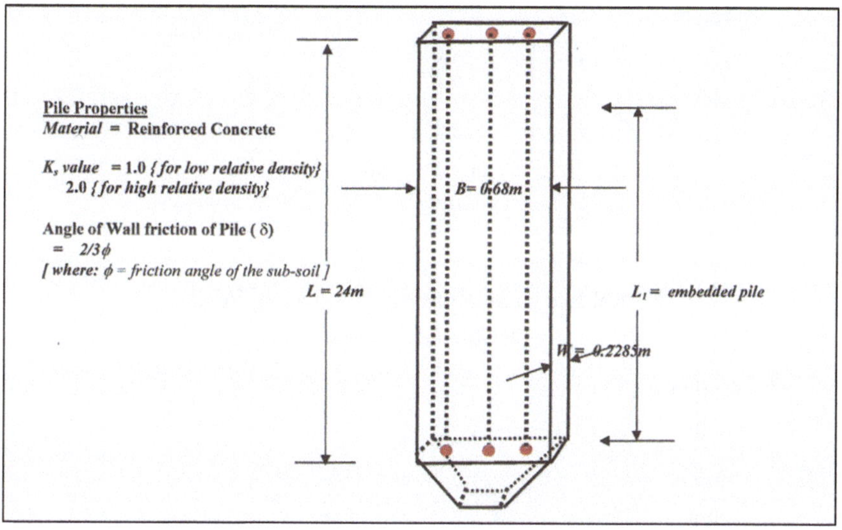

Fig. 10. Schematic diagram of a reinforced concrete pile used at the Quay Wall shore protection site.

$$\mathbf{P_u} = Q_s + Q_b = \mathbf{A_s}\{\mathbf{c_a} + \mathbf{K_s}\gamma\,\mathbf{D}/2\,\mathbf{tan}\,\delta\} + \mathbf{A_b}\{\mathbf{cN_c} + \mathbf{K_s}\gamma\,\mathbf{D}\,\mathbf{N_q} + \gamma\,\mathbf{B}/2\,\mathbf{N_\gamma}\} \quad (1)$$

where:

$$\text{Pile Base Load } (Q_b) \text{is given as: } \mathbf{A_b}\{\mathbf{cN_c} + \mathbf{K_s}\gamma\,\mathbf{D}\,\mathbf{N_q} + \gamma\,\mathbf{B}/2\,\mathbf{N_\gamma}\} \quad (1a)$$

$$\text{Pile Shaft Load } (Q_s) \text{is given as: } \mathbf{A_s}\{\mathbf{c_u} + \mathbf{K_s}\gamma\,\mathbf{D}/2\,\mathbf{tan}\,\delta\} \quad (1b)$$

K_s the coefficient of earth pressure on the shaft within the failure zone (varying from 0.50 for loose soils to 1.00 for dense soils)

N_c, N_q *and* **N_γ** Terzaghi bearing capacity factors that are dependent on (ϕ) and the embedment Ratio (**L/B**)

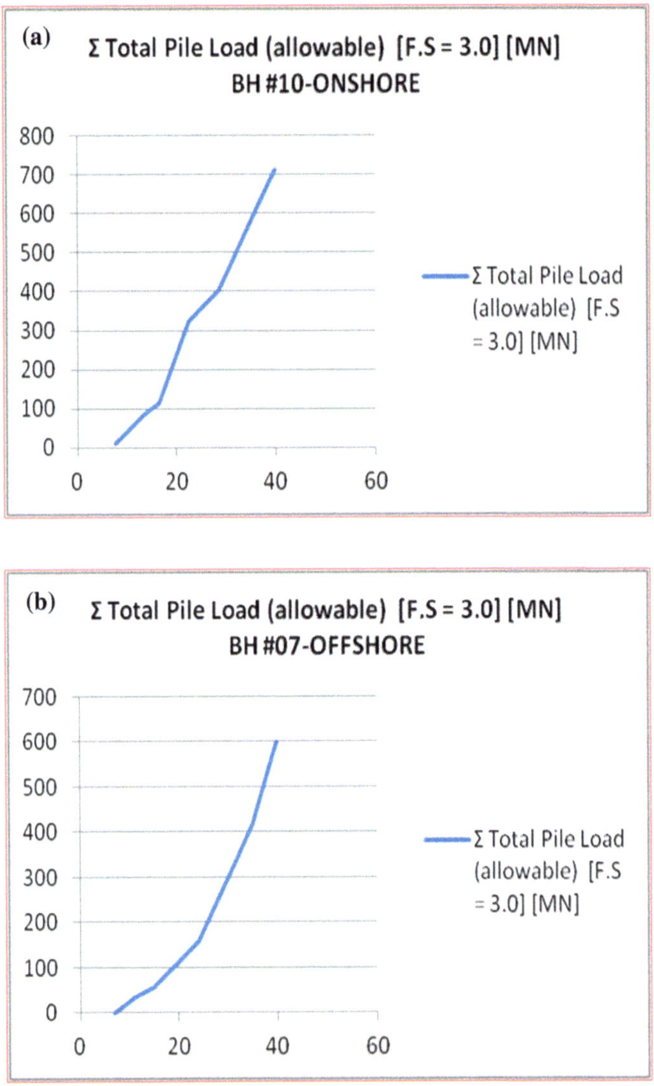

Fig. 11. (a) Plot of Total Pile Load allowable versus Depth of Pile embedment for on-shore Quay Wall alignment at project site. (b) Plot of Total Pile Load allowable versus Depth of Pile embedment for off-shore Quay Wall alignment at project site.

where:

W Width of Pile base;
B Breadth of Pile
L Length of reinforced concrete pile = 25.00 m
$\mathbf{A_s}$ Area of shaft of pile = {(**2B + 2 W**) × **L**} = 45.425 m^2
$\mathbf{A_b}$ x-sectional Area of base of pile = {(**B** × **W**)} = 0.1554 m^2
$\mathbf{c_a}$ the adhesion per unit area of the pile
δ the angle of friction of the soil on the shaft of the pile [~2/3 of soil friction angle (ϕ)];

Note that for clays: $\boldsymbol{\delta} = \boldsymbol{0^o}$,
γ = unit weight of soil at foundation level.

For piles of normal proportions, B/W ~ 30 so the term with the ***W*** becomes negligible and may be ignored!

$$\therefore \mathbf{Q_u} = \mathbf{P}_u = \boldsymbol{Q_s} + \boldsymbol{Q_b} = \mathbf{A_s}\{\mathbf{c_a} + \mathbf{K_s}\gamma\,\mathbf{L}/2\,\mathbf{tan}\,\delta\} + \mathbf{A_b}\{\mathbf{cN_c} + \mathbf{K_s}\gamma\mathbf{L}\,\mathbf{N_q}\} \quad (2)$$

where:

$$\boldsymbol{Q_s} = \mathbf{P_d}\{\mathbf{A_b N_q} + \mathbf{K_s tan}\,\delta\mathbf{A_s}\} \quad (2a)$$

$$\boldsymbol{Q_b} = \mathbf{A_b}\{\mathbf{cN_c} + \mathbf{K_s}\gamma\,\mathbf{L}\,\mathbf{N_q}\} \quad (2b)$$

Plots of Pile Loads versus Pile Embedment for the Quay Wall site, for both on-land and off-shore positions are given in Figs. 11a and b for a Factor of Safety (FS) = 3.0).

The computations of **Pile Loads for the recommended Reinforced Concrete Piles** along the proposed ***Quay Wall line***, based on the Eqs. (2a and 2b) - (derived from **Meyerhof (1963)**) **-** and the soil properties as obtained from the ***On-Shore Borings*** along the ***Quay Wall line*** during the field investigations, are contained in Tables 1 and 2 below.

Table 1. Summary of geotechnical properties for the Subsoils at BH#O1 along the proposed on-shore alignment of the Quay wall Line, at the project site.

	Soil Type	Depth (m)	USC Class	Soil Consistency			Sat. Unit Wt	U–U triaxial Tests		Oedometer Consolidation Tests				Sound-ings	Grain size Distribution Pattern (% Passing Sieve Sizes)			
				.wn (%)	LL (%)	PI (%)	kN/m³	ϕ_u (°)	C_u kPa	M_v (m^2/MN) x 10^{-4}		C_v (m^2/yr.)		SPT (N)	> 4.75 mm	4.75 .Mm	75μ	2μ
										50.00 kPa	400.00 kPa	50.00 kPa	400.00 kPa					
BH #1	*Dark-Greyish Organic Clay layer (OH)*	0.50 - 11.0	OH OH OH	36.4 - 37.8	64.8 - 65.9	27.4 - 27.8	26.4 - 27.8	4 - 6	6.0 - 28.0	0.08 - 0.12	0.12 - 0.18	0.45 - 0.78	- - 0.42	- - -	82 - 98	62 - 78	21 - 32	4 - 4
	Light Greyish Silty Sand Layer(SM)	12.0 - 20.0	SM SM SM	6.2 - 8.4	NP NP NP	NP NP NP	18.6 - 22.8	10.0 - 12.0	20.0 25.0	- - -	- - -	- - -	- - -	2/3/3 4/6/8 3/5/7	34 28 26	12 14 14	10 10 8	2 1 4
	Reddish-Brown Clayey Silt layer(ML)	21.0 - 29.0	ML ML ML	5.0 - 6.8	NP NP NP	NP NP NP	27.0 - 27.5	8.0 - 8.0	0.0 0.0 0.0	- - -	- - -	- - -	- - -	4/4/8 3/5/7 4/5/6	20 - 25	14 10 11	8 9 6	2 5 2
	Milkish Silty Sand Layer(SM)	30.0 - 33.0	SM SM SM	5.0 - 6.4	NP NP NP	NP NP NP	10.6 14.8 18.8	14.0 - 16.0	0.0 0.0 0.0	- - -	- - -	- - -	- - -	- 3/5/7 -	24 22 24	14 10 12	8 9 6	2 5 2
	Milkish Well-graded Sands Layer (SW)	34.0 - 40.0	SW SW SW	5.8 - 6.2	NP NP NP	NP NP NP	19.6 - 23.8	32.0 - 36.0	0.0 0.0 0.0	- - -	- - -	- - -	- - -	- 8/22/2 5/15/2	20 20 18	14 10 12	8 9 6	2 5 2
	Water Table(WT) = 0.0 meters-																	

Table 2. Summary of geotechnical properties of the Subsoils at BH#O4/05 along the proposed off-shore alignment of the Quay wall Line, at the project site.

BH#	Soil Type	Depth Range (m)	USC Class	Soil Consistency			Sat. Unit Wt	U–U triaxial Tests		Oedometer Consolidation Tests				Soundings	Grain size Distribution Pattern (% Passing Sieve Sizes)			
				.wn (%)	LL (%)	PI (%)	kN/m³	φu (°)	Cu kPa	Mv (m²/MN) x 10⁻⁴		Cv (m²/yr.)		SPT (N)	> 4.75 mm	4.75 mm	75μ	2μ
										50.00 kPa	400.00 kPa	50.00 kPa	400.0 kPa					
	WATER BODY	7.00	-	-	-	-	-	-	-	-	-	-	-	-	-	-	-	-
	Dark Greyish Organic Clay (OH)	7.0 - 12.0	OH OH	36.4 - 37.8	64.8 - 65.8	27.2 - 28.2	26.4 - 27.8	0 - 6	14.5 - 28.0	0.08 - 0.12	0.12 - 0.18	0.45 - 0.78	0.42 - 0.34	- - -	- - -	12 - 14	8 - 10	1 - 4
BH #04/05	*Dark Poorly-graded Sands layer(SP)*	13.0 - 24.0	SP - SP	4.0 - 5.2	NP NP NP	NP NP NP	25.0 - 27.8	26.0 - 30.0	0.0 0.0 0.0	- - -	- - -	- - -	- - -	1/2/3 - 3/6/18	12 - 28	4 - 14	4 - 9	2 - 5
	Milkish Silty Sand Layer(SM)	26.0 - 36.0	ML ML	34.0 - 36.0	62.5 - 64.0	20.8 - 16.8	18.6 - 19.4	4 - 6	0.0 0.0 00	- -	- -	- - -	- - -	5/8/14 - 9/18/31	14 - 28	2 - 12	2 - 4	- - 3
	Milkish Silty Sand Layer(SM)	37.0 - 40.0	SW SW	4.0 - 5.2	NP NP NP	NP NP NP	24.6 24.8 23.8	24.0 28.0 30.0	0.0 0.0 0.0	- - -	- - -	- - -	- - -	8/32/39 - -	20 28 26	14 8 6	8 6 4	2 5 2

Water Table = 7.00m above River bed

Base Resistance of Piles

$$\text{Net unit base resistance } q_{ult} = q_f = p = p_d(N_q - 1) \\ = \mathbf{A_b\{cN_c + K_s\gamma\, L\, N_q + \gamma B/2\, N_\gamma\}} \quad (3)$$

where:

$$P_d = \textit{effective overburden pressure at pile base level} \quad (4)$$

The proposed Pile depth *(L_1)* for fore-shore is given in the sections that follow.

Shaft Resistance {Unit Skin Friction} of Piles

The total ***Ultimate Skin Friction*** on pile shaft is given as:

$$f = Ks.p_d.\tan\delta = \mathbf{A_s\{\, c_a + K_s\gamma\, L/2 \tan\delta\,\}} \quad (5)$$

However, the failure load of the pile: $\boldsymbol{Q_u}$ = load at failure applied to the pile

$$= \mathbf{Q}'_u + \text{Wt of the pile}\,(\boldsymbol{W_p}) \quad (6)$$

i.e.:

$$\mathbf{Q}'_u = \mathbf{Q_u - W_p = Q_b + Q_s} \\ = \mathbf{A_b P_d (N_q - 1) + Q_s} \\ = \mathbf{A_b P_d N_q - A_b P_d + Q_s}$$

But weight of concrete in pile $\leq$ weight of soil displaced,

$$\therefore \mathbf{Q_s} = \mathbf{P_d}\{\mathbf{A_b N_q} + \mathbf{K_s} \tan \delta\, \mathbf{A_s}\} \quad (7)$$

where:

A_b cross-sectional area of base of pile = 0.1554 m^2

A_s embedded surface area of pile = 45.425 m^2 = $\{(B + W)L_1\}$

P_d average effective overburden pressure over embedded depth of pile

$$\mathbf{Q_s} = \mathbf{P_d}\{(0.1554\,m^2)(\gamma_s)\} + \{1.1 \times \tan(2/3\,(\phi))\,(B + W.L_1)\} \quad (8)$$

In Eq. (8), P_d can be determined from the soil profile and properties, γ_s is known; ϕ is known for the particular soil; L_1 = depth of pile embedment is known. A sensitivity analysis on Eq. (8) above can be carried out for different values of Depth of Pile Embedment (**L_1**) thus giving a plot such as shown on Figs. 11(a) and (b) {for this particular Quay Wall Shore Protection in the Niger Delta.} 45.425 m^2 = $\{(B + W)L_1\}$.

The various Pile Shaft Loads *(Q_s)* estimated for the **Quay Wall Shore Protection** studied in the course of this investigation are given in the Pile design Table 3 below.

Table 3. Components of Pile Loads at various Depth horizons and Soil Layers at Boring sites along the Quay Wall line (BH#14-On-Shore). (Based on Meyerhof 1953) © TGCL 2014.

F.S = 1.0				
Depth (m)	Soil type penetrated	Pile shaft load (Qs) [MN]	Pile end load (Qb) [MN]	Total pile load (Qs + Qb) [MN]
13.5	1-OH	0.0063	0.0189	0.0253
22.5	2-SM	0.169	0.0225	0.1918
31.5	3-ML	0.0414	0.0134	0.0548
36	4-SM	0.00855	0.0043	0.0128
38	5-SW	0.597	0.310	0.9066
40	**6-SW**	**0.6234**	**0.307**	**0.9305**

6 Discussion of Presented Data

The data obtained from the field investigations and Laboratory tests indicate both similarities and differences in terms of lithological thicknesses and geotechnical properties of the soil materials retrieved.

6.1 Relations of Lithological Variations on-Shore Versus off-Shore

It is observed from the Fence Diagrams of both on-shore and off-shore borings that the off-shore borings seem to exhibit better sorting of the coarser grain soils than their equivalents in the on-line borings, probably due to the influence of the hydrodynamics of the overlying water body. Generally the coarser soil materials seem to be deposited at the deeper portions of the borings, with the finer-grained soils lying on top.

The lithology on-shore along the coastline consisted of an uppermost dark-greyish Organic Clay layer (OH) [7–10 m thick]; underlain by greyish Silty-Sands (SM) [3–5 m thick]. These are further underlain by Clayey Silts (ML) [2.00–4.50 m thick], Clayey Sands (SC) [3–10 m thick] and Well-graded Sands and gravels (SW) at the bottom of the borings [>40 m thick]. The borings off-shore had similar lithologies represented as follows: Dark greyish Organic Clay (OH) layer below the water column [5–6 m thick]; Poorly-graded Sands (SP) layer [4–23 m thick]; Well-graded Sands (SW) layer [0–17 m thick]; Dark greyish Clayey Silt (ML) layer [0–4 m thick] and Silty Sands and Gravels (SM-SW) layer. [$\geq$ 3–4 m thick].

6.2 Strength Distribution with Depth

In both sets of borings (Figs. 7a and b), bearing pressures along the vertical soil profiles tend to increase with depth as the coarser soil fractions, with higher soil densities, tend to increase with depth. The upper dark organic clays (OH) have friction angles (ϕ) ranging from 0°–6° and cohesion intercepts (C_u) from 14.50 kPa to 28.0 kPa. The Silty-Sands (SM) and well-graded Sands and gravels (SW) at depths of between 30 m–40 m have friction angles (ϕ) ranging from 10°–36° and cohesion intercepts (C_u) of 0.00 kPa. The average Standard Penetration Test (SPT) N-values of the Sands (SP) and Gravel layers (SW) at depths varied from 15 to >50 with corresponding Bearing Capacity values (based on Factor-of Safety [FS = 3.0]) of 100.00–478.50 kPa, indicating high to very high values.

6.3 Anchors for the Quay Wall Shore Protection System

Owing to the thickness of the overlying soft organic clays [range in thickness from 7.50 m–15.00 m], there is the need to erect steel tie-backs from the Quay Wall to a "Dead man" away from it. The "Dead man" anchors consisted of driven rods to refusal before the steel tie-backs installations (Figure 9).

6.4 Bearing Capacities Beneath Tank Positions; [Weight of Oil Tanks When Full]

The Oil tanks for which the Quay Wall is being built should be borne on foundations with bearing capacities that will be capable of withstanding the dead-weights of these Oil tanks and subjected to tolerable minimal settlements ($\sim$30 cm) as recommended by Bowles (1996) and Tomlinson (1980, 2001).

6.5 Allowable Bearing Capacities Versus Allowable Computed Settlements

Using Terzaghi's (1943) classical equation for settlement computation:

$$S_c = C_c/1 + e_o[H_o \mathrm{Log}_{10}\{\sigma_{vo} + \Delta_{sv}\}/\sigma_{vo}] \tag{9}$$

where:

S_c final settlement (in cm) of layer of thickness H (m).

H thickness of compressible layer beneath base of foundation

σ_{vo} vertical stress in kN/m^2 induced at the center of layer by the net foundation pressure $q_n = ½ (h_1) (\gamma_1)$. Tank loads on the soil ~ T *kPa* ~ 245 MPa,

The final computed settlements at four (4) boring points at the project site are as shown on Table 4, while computed rates of settlements for $\mathbf{T_{50}}$ (time to 50% consolidation settlements) and for $\mathbf{T_{90}}$ (time to 90% consolidation settlements) are as shown on Table 5.

Table 4. Final computed settlements at the Project site [Initial settlement (ρ_I) = 0.3088 m].

Borings #	1	2	3	4
H_o	2.50	2.50	2.50	2.50
σ_{vo} (kPa)	32.00	32.00	32.00	32.00
Cc ~ 0.009 (w_L – 10)	0.0954	0.0576	0.0648	0.0954
$\Delta\sigma_v$ (MPa) [assumed]	245 MPa	245 MPa	245 MPa	245 MPa
$Sc = Cc/1 + e_o\ H_o.\ Log_{10}\{\sigma_{vo} + \Delta_{\sigma v}\}/\sigma_{vo}$ [m]	0.02262	0.0126558	0.015363	0.022617
Total settlement for site $\rho_{total} = \rho i + S_c$ (m)	0.33145	0.32248	0.32345	0.33145
Average total settlement for site $\rho_{total} = \rho i + S_c$ (cm)	0.3272075 m			

Table 5. Summary of the computed Rates of Settlements for the on-shore boring points at site.

S/No.	Items		Borings						Average
			1.	2.	3.	4.	5.	6.	
1.	Thickness of layer (m)		1.10	1.10	1.50	4.50	0.10	1.05	1.55
2.	.d = $H_o/2$		0.55	0.55	0.75	2.25	0.05	0.525	0.78
3.	Coefficient of consolidation (Cv)		0.60	0.78	0.78	0.60	0.60	0.78	0.69
4.	Estimated time rate of consolidation (t (yrs.) = $[T.d^2)/C_v]$	T_{50}	0.10	0.078	0.144	1.687	0.00083	0.0707	0.3475
		T_{90}	0.428	0.329	0.613	7.172	0.0035	0.300	1.474

7 Conclusions

Boring of 40-meter deep holes along the proposed Quay Wall line revealed the following subsurface lithologies on-land at the project site: (i) an upper dark-greyish organic clay layer (OH) [7–10 m in thickness], underlain by (ii) greyish silty-sands (SM) layer [3–5 m thick]. These are further underlain by (iii) clayey silts (ML) layer

[2.00–4.50 m thick], which are further underlain by 3–10 m thick Clayey sands (SC) at depths of 40.00 m at the bottom of the holes.

At a distance of about 20.00 m off-shore, the lithologies consist of (i) Dark greyish organic clays (OH) below the 7.00 m column of saline water at high tide, followed by (ii) poorly-graded Sands (SP) layer. These are underlain by (iii) Well-graded Sands (SW) layer, followed by (iv) Dark-greyish Clayey-silts (ML) layer and finally by (v) Silty-Sands and gravels (SM-SW) layer.

Laboratory tests carried out on the soil samples indicated that the Dark-greyish organic clays had Liquid Limits and Plasticity Indices (PI) of between 64.8–65.9% and 27.2–28.2% respectively.

The saturated unit weights (γ_{sat}) of the greyish clays varied from 16.4–17.8 kN/m^3 while those of the silty-sands varied from 16.8–19.4 kN/m^3 and those of the poorly-graded sands (SP) from 18.5–19.8 kN/m$^{3.}$

Unconfined-undrained (U-U) triaxial test results showed that the undrained friction angle values (ϕ_u) and undrained cohesion intercepts (c_u) for the Clays ranged from 4° to 6° and 14.5–28.0 kPa, respectively. For the Silty-sands (SM) the (ϕ_u) and (c_u) values were 10°–12° and 0.00 kPa, respectively. The well-graded sands and gravels had (ϕ_u) and (c_u) values of 24°–35° and 0.00 kPa, respectively.

The Coefficient of Consolidation (C_v) for the soft clays ranged from 0.44–0.78 m^2/yr under an imposed stress of 50 kPa and from 0.34–0.42 m^2/yr under an imposed stress of 400 kPa.

Standard Penetration Test N-values obtained in cohesionless soil horizons along the Quay Wall line showed that the Light greyish **Silty-Sands (SM)** at depths of between 12.00–20.00 m had N-values of 6–14, while the **Well-graded sands (SW)** at depths of between 34–40 m had N-values of 17–24. These values when converted to **allowable net soil pressures**, employing the relationship .$\mathbf{q_a}$ **= 0.22 N (0.1073) MPa** (according to Peck et al. 1974), give **0.142–0.330 MPa** for the Silty-Sands and **0.4013–0.567 MPa** for the Well-graded Sands.

Deep foundations for the Quay Walls to depths of 25.00 m with appropriate tie-backs were employed to protect the project site from flooding and coastal erosion.

Acknowledgements. The authors are sincerely very grateful to the Managing Director of our Clients - Stopgap Fuels Nigeria Limited – for permission to use the data for the preparation of this paper.

References

Akpokodje, E.G.: The importance of engineering geological mapping of the Niger Delta. Can. Geotech. J. **5**(6), 239–242 (1979)

Bowles, J.E.: Foundation Analysis and Design, 5th edn., 1175 pages. McGraw-Hill, New York (1996)

Christopher, B.F.: Biomass energy: the scale of the potential resource. Trends Ecol. Evol. **23**(2), 28 (2012)

Forgione, J., Hill, J., Tillman, D., Polasky, S., Hawthorne, P.: Land clearing and the biofuel carbon debt. Am. Assoc. Adv. Sci. (AAAS) **319**(5867), 1238–1240 (2008)

Fubara, D.M.J., Teme, S.C., Mgbeke, T., Gobo, A.E.T., Abam, T.K.S.: Master plan design of flood and erosion control measures in the Niger Delta. IFERT Technical Report No. 1 (1988)
Gallagher Review: Indirect Effects of Biofuel Production, Renewal fuel Agency Report, pp. 24–31 (2008)
Malomo, S.: Geology Maps of the States of Nigeria, Revised Edition Series. Nigerian Geological Survey Agency (Rivers State Sheet) (2006)
Meyerhof, G.G.: Some recent research on bearing capacity of foundations. Can. Geotech. J. **1**, 16 – 26 (1963)
Peck, R.B., Hanson, W.E., Thornburn, T.H.: Foundation Engineering, 2nd edn., 514 pp. Wiley, New York (1974)
Rossi, A., Lambjou, Y.: Gender and equity issues in liquid biofuels production. In: Minimizing the Risks to Maximize the Opportunities, FAO, April, pp. 64–68 (2008)
Rutledge, P.C.: Utilization of marginal lands for urban development. Soil Mech. **96**, 2–22 (1970). No. S.M.I. ASCE
Short, K.C., Stauble, R.J.: Outline of geology of the Niger Delta. Bull. Am. Assoc. Pet. Geol. **54** (5), 761–779 (1967)
Shujiang, K., Post, W.M., Nicholas, J.A., Wang, D., West, T.O., Bandaru, V., Izaurralde, R.: Marginal lands: concept, assessment and management. J. Agric. Sci. **5**(5), 129–139 (2013)
Teme, S.C.: Subsurface investigations for Okrika ATC Ferry Terminal in the marginal lands of the Niger Delta sub-region, Nigeria. Technical report submitted to the Nigerian Federal Ministry of Transportation, Inland Waterways Department, Lagos, 27 pages (1989)
Teme, S.C.: Subsurface investigations for shore protection structures at Oloma-Ayaminima in the Coastal zone of the Niger Delta Sub-region, Nigeria. Technical report submitted to Kydon Nigeria Limited, Port Harcourt, 48 pages (1991a)
Teme, S.C.: Subsurface investigations for shore protection structures and reclamation project at Asarama, in the Mangrove zone of the Niger Delta Sub-region, Nigeria. Technical report submitted to Dredging International Company, BV. Port Harcourt, 126 pages (1991b)
Teme, S.C.: Susceptibility of steel piles to corrosion in the marginal lands of the Niger Delta. In: Proceedings of the International Conference on the Implications of Ground Chemistry and Microbiology for Construction, University of Bristol/UK/1992, pp. 369–382 (1992a)
Teme, S.C.: subsurface investigations for failures of bank protection measures at Yenagoa, in the freshwater zone of the Niger Delta Sub-region, Nigeria. A Technical report submitted to Schwartz-Krystoffel Engineering (Nigeria) Limited, Port Harcourt, 86 pages (1992b)
Teme, S.C.: Subsurface investigations to determine the consolidation characteristics of land reclamation sites around Buguma City in Asari-Toru Local Government Area of Rivers State in the Mangrove zone of the Niger Delta Sub-region, Nigeria. A Technical report submitted to Etteh Aro and Partners, Port Harcourt, 98 pages (1993a)
Teme, S.C.: Subsurface investigations and analyses for Piles Anchor Points along the shoreline of Buguma City in Asari-Toru Local Government Area of Rivers State in the Mangrove zone of the Niger Delta Sub-region, Nigeria. Technical report submitted to Pile Foundation Limited, Port Harcourt, 42 pages (1993b)
Teme, S.C.: Geotechnical investigations and analyses for Landing Jetty Foundation at Angulama in Asari-Toru Local Government Area of Rivers State in the Mangrove zone of the Niger Delta Sub-region, Nigeria. Technical report submitted to Cypress Konsult, Mini Ezeku Layout, off Aba Road, Port Harcourt, 58 pages (1994a)
Teme, S.C.: Geotechnical investigations and analyses for the Bassambiri Reclamation and Sandfilling Project, Bassambiri in Brass Local Government Area of Bayelsa State in the Mangrove zone of the Niger Delta Sub-region, Nigeria. Technical report submitted to Cypress Konsult, Mini Ezeku Layout, off Aba Road, Port Harcourt, 87 pages (1994b)

Teme, S.C.: Geotechnical Considerations on Foundation Design in the Niger Delta: A Lead Paper at Plenary Technical Session at the 38th Annual International Conference of the Nigerian Mining and Geological Society, March, pp. 18–25 (2002)
Teme, S.C.: Geotechnical investigations and analyses for the shore protection, dredging and reclamation of the port harcourt ports complex project, Port Harcourt in the Mangrove zone of the Niger Delta Sub-region, Nigeria. Technical report submitted to AIM Consultants Limited, Victoria Island, Lagos, 120 pages (2004)
Teme, S.C.: Geotechnical investigations and analyses for the Shore protection Works and Jetties for the Bille, Bonny and Nembe Water Fronts, Port Harcourt in the Mangrove zone of the Niger Delta Sub-region, Nigeria. Technical report submitted to AIM Consultants Limited, Victoria Island, Lagos, 127 pages (2007)
Teme, S.C.: Geotechnical investigations and analyses for the proposed AMG Terminal and Landing Jetty at Marina Street, Calabar, Cross River State in the Mangrove zone of the Niger Delta Sub-region, Nigeria. Technical report submitted to AMG Petro-Energy Limited, Maitama, Abuja, Nigeria, 165 pages (2008)
Teme, S.C.: Geotechnical investigation for a 3-Storey Hotel Building at Olaka Street, off Abacha Road, Port Harcourt, Technical report submitted to Brass Engineering Construction Co. Ltd., Port Harcourt, Rivers State Nigeria, 62 pages (2012)
Teme, S.C.: Geotechnical investigations and analyses for the proposed Arogbo Town Landing Jetty, Ese – Odo Local Government Area of Ondo State in the Freshwater zone of the Niger Delta Sub-region, Nigeria. Technical report submitted to Skylink Engineering Consult, Ibadan, Nigeria, 165 pages (2014a)
Teme, S.C.: Geotechnical investigations and analyses for the proposed Shore Link Oil and Gas Facility, Rumuolumeni, Obio-Akpor Local Government Area of Rivers State in the Mangrove zone of the Niger Delta Sub-region, Nigeria. Technical report submitted to Shore Link Oil and Gas Company, No. 15 Aba-Port Harcourt Expressway, Port Harcourt, Nigeria, 165 pages (2014b)
Terzaghi, K.: Theoretical Soil Mechanics. Wiley (1943)
Tomlinson, M.J.: Foundation Design and Construction, 4th edn, 793 pages. Pitman Advanced Publishing Programme, London (1980)
Tomlinson, M.J.: Foundation Design and Construction, 7th edn. Prentice Hall, Upper Saddle River (2001)
Watch Indonesia: Marginal Lands in Indonesia, Agro-fuels and the Management of the Marginal Lands: a briefing by the Gala Foundation, pp. 3–4 (2014)

Geotechnical Properties of Expansive Clay Shale in El-Mahrowsa, Qena, Egypt

M. Abd Ellatief[1], M. Mahmoud[2(✉)], and H. Abdo[3]

[1] Mining Engineering Department, Al Azhar University, Cairo, Egypt
[2] Civil Engineering Department, Faculty of Engineering, Sohag University, Sohag, Egypt
[3] Mining Engineering Department, Faculty of Engineering, Al-Azhar University, Cairo, Egypt

Abstract. This study aims to investigate the geotechnical properties of expansive clay shale in El-Mahrousa region, Qena, Egypt. For this purpose, two sites were chosen and three undisturbed samples were taken within the active zone at depths 1.0 m, 2.0 m and 3.0 m below the ground surface for each site. Different laboratory tests were carried out including, natural water content, natural dry density, grain size distribution, Atterberg limits, unconfined compressive strength, free swelling test, swell potential test and swelling pressure tests. These tests were supported by chemical and mineralogical investigations such as X-ray diffraction (XRD) and X-ray fluorescence spectrometer (XRF). The main results of this study that there are various forms of empirical equations have been proposed which relate swelling indices to certain physical properties of soils, such as chemical components of clay, consistency limits (liquid limit, plastic limit and plasticity index and linear shrinkage which are well help geotechnical engineers to predict the behavior of expansive soil from it's physical properties.

Keywords: Expansive soil · Qena · Swelling pressure · Swell potential · Atterberg limits · Free swelling

1 Introduction

Expansive strata are soil and rock which contains clay minerals that have the potential for swelling and shrinkage under changing moisture conditions. Clay minerals originate from the weathering of shale, slate, sandstone and limestone. Another source is the diversification of volcanic ash that was deposited under marine conditions during

J. Wasowski et al. (eds.), *Engineering Geology and Geological Engineering for Sustainable Use of the Earth's Resources, Urbanization and Infrastructure Protection from Geohazards*, Sustainable Civil Infrastructures, DOI 10.1007/978-3-319-61648-3_3

geologic times and settled alone or mixed with shale or limestone [1]. Subtropical climate influences the development of this type of soil and accelerates the desiccation and weathering processes of the source rock. If they contain montmorillonite (smectite group) or a certain type of illite, they will have significant swelling potential when they contact with water. This soil causes a lot of problems if it is found under light structures. There is no statistical data confirmed about the economic losses due to structures founded on expansive soils in Egypt. Qena governorate which is situated at 594 km south Cairo has several areas where this kind of problematic soil exists in the form of variable-thickness layers in the new development areas, such as new Qena city, Qeft city and El-Mahrowsa region. These problematic soils are considered the major obstacle for development in these areas. Study of geotechnical properties of this soil plays an important role in its engineering classification, evaluation of its mechanical behavior, prediction and avoiding of its geotechnical problems. There are some previous studies were conducted in Qena governorate and in upper Egypt, [2–8]. El-Mahrowsa region which is considered as an important region in Qena governorate has not any geotechnical studies on its expansive soil in spite of there are a lot of construction damages due to existing of this soil. For these reasons, we think of this research to identify the characteristics of the soil in El-Mahrowsa region to take all precaution when executing any projects in this region. Also, the aim of this study is to investigate the relationship between swelling indices to certain physical properties of soils, such as chemical components of clay, consistency limits (liquid limit, plastic limit and plasticity index), shrinkage limit and linear shrinkage which are well understood by engineers.

2 Location and Geological Setting of Study Area

The studied area, El-Mahrowsa region is located at the west side of Qena city. It lies between latitudes 26° 15′ 10″ to 21° 8′ 18″ N and longitudes 32° 22′ 13″ to 32° 21′ 34″ E, and 22 km from Qena city. It consists of a series of marls and shales which vary in thickness from one place to another. It is subdivided into two members, the lower member is usually composed of argillaceous gray or reddish limestone with gypsum veinlets, the upper member consists of gray to green shale, yellowish in some parts, and fissle with gypsum veinlets. The soil of El-Mahrowsa region consists of cretaceous-plaeocene (Dakhla formation). The distribution of the sediments is mapped in some details in a geological map as shown in Fig. 1.

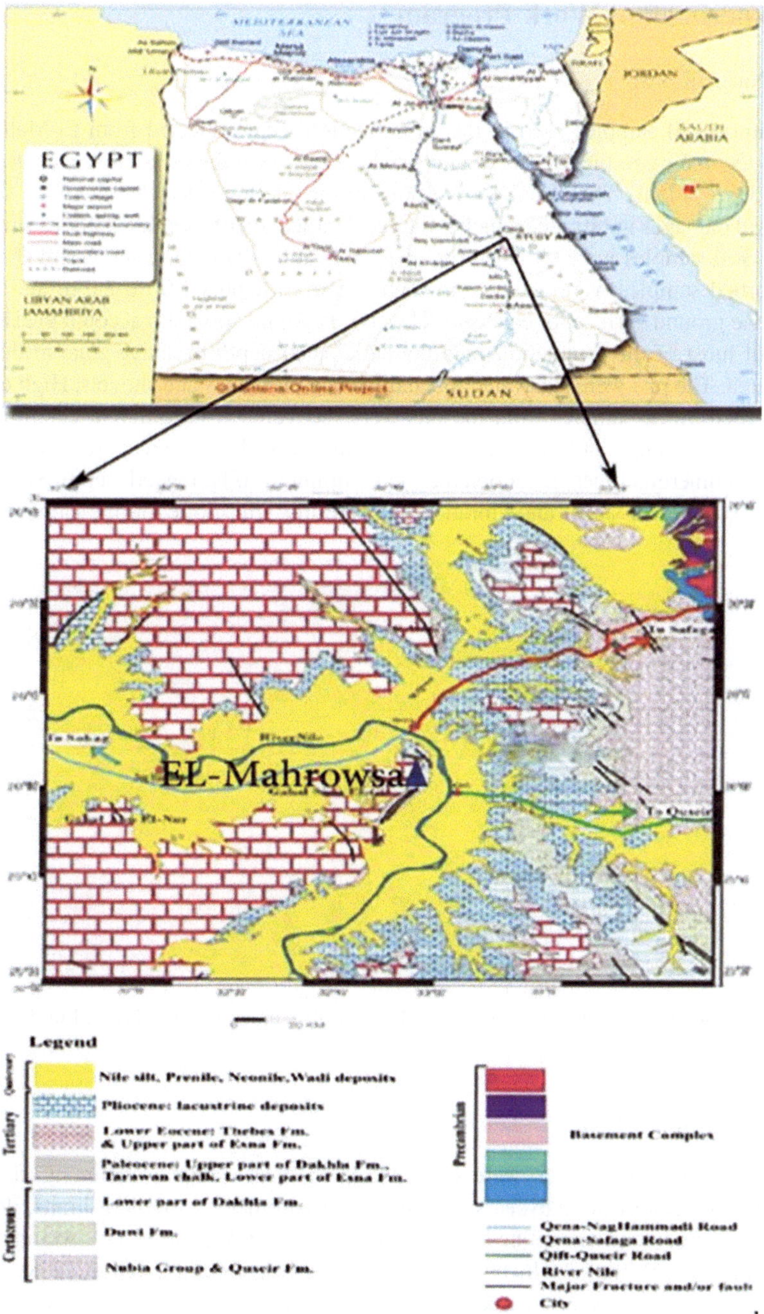

Fig. 1. Location and geological map of El-Mahrowsa area

3 Experimental Work Program

3.1 Natural Soil

The natural soil samples used in this investigation were collected from El-Mahrowsa region, Qena, Egypt. Field visits were carried out to collect the soil samples and to make visual observations. The visual observations of the site indicated that there are several structural damages in this area which is a typical indication of swelling soils, as shown in Fig. 2. Soil samples were obtained by open excavation from two sites A and B. Three undisturbed samples were taken from each site from depths 1.0 m, 2.0 m and 3.0 m below the ground surface for each site. The chosen depths took into account the shallow depth of most of the foundations and the anticipated depth of active zone in the case study area. The pits were excavated manually using pick-axes and shovels. High quality undisturbed block samples from the pits were recovered from different carefully ear-marked depth using hand tools such as knives and trowels. The recovered blocks were wrapped numerous times in cellophanes and aluminum foils; logged; set on Styrofoam boards to minimize shock and vibration and carefully transported to the laboratory.

Fig. 2. A site photograph of a fractured road founded on an expansive clay bed in the studied area

3.2 Laboratory Tests

Different experimental tests were carried out on these samples including chemical analysis, mineralogical analysis, sieve analysis, hydrometer, Atterberg limits, density, natural water content, unconfined compression tests, shrinkage tests and swelling index tests. The experimental results obtained from these tests are presented in Tables 1, 2, 3 and 4. These results were used in the predicted equations for swelling potential from soil index properties and validated with suggested empirical equations proposed by various researches.

3.3 Chemical Analysis

Chemical analysis was carried out by X-ray fluorescence spectrometer (XRF) technique using the instrument JEOL, JSX 3222, Japan. The results of these tests are shown in Table 1. From these results, it can be observed that the total sum of the various chemical components of the studied samples are up to only 74.68% to 78.97% for site A and up to 81.27% to 84.66% for site B, and not 100%. This is relatively large difference which could be partly due to the presence of accessory amount of heavy metal compounds, and partly by the loss on ignition of organic matter compounds. Also, the studied samples have high content of ferric oxide (Fe_2O_3) and aluminum oxide (Al_2O_3), which are the main elements of clay minerals, where the increase of swelling strain of these samples is related to the increase of ferric oxide (Fe_2O_3) and aluminum oxide (Al_2O_3). The effect of these minerals on the geotechnical behavior of studied soils will be discussed later.

Table 1. Results of chemical analysis

Site	Sample no.	Depth (m)	Chemical composition								
			MgO	Al_2O_3	SiO_2	K_2O	CaO	TiO_2	Fe_2O_3	MnO	SO_3
A	1	1.0	1.07	15.61	20.37	1.34	24.08	1.45	12.03	-	-
	2	2.0	1.18	16.33	16.95	1.25	23.32	1.32	14.33	-	-
	3	3.0	1.20	18.16	16.13	1.47	22.73	1.74	17.11	0.43	-
B	4	1.0	0.98	16.76	22.52	1.55	24.08	1.73	14.93	0.31	1.80
	5	2.0	1.70	17.28	20.02	1.61	23.02	1.90	15.38	0.36	-
	6	3.0	1.77	18.24	19.95	1.22	22.5	1.66	17.57	-	-

3.4 Mineralogical Analysis

The mineralogical analysis of six representative samples was carried out using PW1710 BASED diffractometer with a generator operating at voltage 40 kV and current 30 mA. The mineralogy of studied samples provides the basis for understanding their geotechnical behavior. It also helps to identify types of clay minerals such as high activity smectite minerals, calcite, quartz and other minerals. It was found that the clay minerals present in El-Mahrowsa soil were montmorillonite as a major constituent, as well as kaolinite and discrete illite as minor constituents in contrast sample No.1 montmorillonite as minor constituent. The clays minerals as montmorollonite in the soil is a good indication of the swell potential. These problems of excessive expansive characteristics lead to much damage to the structures built in and on these soils. The test results are shown in Table 2.

Table 2. Results of mineralogical analysis

Site	Sample no.	Major const.	Minor const.	Trace const.
A	1	Calcite, Kaolinite	Quartz, Montmorillonite	Dolomite
	2	Calcite, Montmorillonite	Quartz, Kaolinite.	-
	3	Calcite, Montmorillonite, Kaolinite	Quartz	-
B	4	Calcite, Montmorillonite	Quartz	Illite
	5	Calcite, Montmorillonite	Quartz, Kaolinite	-
	6	Calcite, Montmorillonite	Quartz, Kaolinite	-

3.5 Geotechnical Soil Tests

The geotechnical soil tests were performed in accordance with Egyptian code of practice for soil mechanics and foundation design, part 2 [9]. The results of these tests are shown in Table 3.

Table 3. Results of geotechnical tests

Site	Sample no.	Grain size			W_c	γ_b (g/cm^3)	γ_d (g/cm^3)	q_u (kg/cm^2)
		Sand %	Silt %	Clay %				
A	1	2.25	73.45	24.30	5.00	2.25	2.14	6.81
	2	2.20	72.20	25.60	5.05	2.25	2.14	7.12
	3	3.00	69.80	27.20	5.22	2.26	2.15	8.05
B	4	1.80	72.10	26.10	6.12	2.20	2.06	6.97
	5	1.70	71.90	26.40	6.56	2.23	2.10	7.10
	6	2.50	70.10	27.40	6.61	2.23	2.09	7.87
Site	Sample no.	Atterberg limits			Activity	RC	LS (%)	SL (%)
		LL	PL	PI				
A	1	51.02	20.2	29.82	1.08	1.64	9.99	8.5
	2	51.2	22.1	29.1	1.14	1.59	10.02	8.53
	3	53.2	23.8	29.4	1.19	1.55	10.09	8.56
B	4	51.1	21.8	29.3	1.09	1.59	9.98	8.43
	5	51.7	22.8	28.9	1.10	1.56	10.00	8.46
	6	53.1	23.3	29.8	1.12	1.55	10.07	8.45

Where:-

- W_c = Natural water content (%)
- γ_b = Bulk density, γ_d = Natural dry density, q_u = Unconfined compressive strength
- LL = Liquid limit, P.L = Plastic limit, P.I = Plasticity index
- Activity is defined as the ratio of plastic index to percent of clay fraction finer than 0.002 mm
- RC = Relative consistency = $((L.L - W_c)/P.I)$
- LS = Linear shrinkage
- SL = Shrinkage Limit

3.6 Swelling Index Tests

Swelling index tests point out the potential expansiveness of soils. Three common laboratory tests are used to determine the swelling index and are called free swelling test, swelling potential test and swelling pressure test.

3.7 Free Swelling Test

Free swelling test is a user-friendly approach quickly carried out to indicate the potential expansiveness of soils. The test is performed according to Holtz and Gibbs [10] by slowly pouring 10 cm^3 of dry soil passing from the 0.42 mm sieve into a graduated cylinder filled with distilled water. The tested specimens swell and reach an equilibrium state after 24 h.

The free swell is defined as the ratio of the increase in volume of the soil from a loose dry powder form to the equilibrium sediment when it is poured into water, expressed as the percentage of the original volume. The percent of free swell is expressed as:-

$$F_s = (V - V_0)/V_0 \times 100 \tag{1}$$

Where:-
F_s = Free swelling
V = Final volume after swell
V_0 = Volume of dry soil, 10 cm^3

Results of the free swelling tests are given in Table 4. The results illustrated that the free swelling ratios of the studied soil are ranging from 69.3 to 73%. This means that these soils have ability to swell and there are expected problems due to these soils according to the Egyptian Code [11].

Table 4. Results of swelling index tests

Site	Sample no.	FS (%)	S (%)	SP (kPa)
A	1	69.3	9.83	445
	2	71.2	10.16	450
	3	73	10.37	483
B	4	69.6	9.89	400
	5	70.1	10.09	437
	6	72.5	10.32	450

3.8 Swelling Potential Test

The swelling potential S (%) of a soil is determined through the one-dimensional restrained swell test by utilizing the oedometer apparatus. The undisturbed soil specimen is cut at its in situ moisture content, put in an oedometer, a small surcharge of about 6.9 kPa is applied, water is then added to the specimen, and the expansion of the volume of the specimen is measured until an equilibrium is reached. The swelling potential is calculated as follows:-

$$S\,(\%) = \frac{\Delta H}{H} \times 100\% \tag{2}$$

Where:-
S (%) = Swelling potential
ΔH = Height of swell due to the saturation
H = Original height of the specimen

Figure 3 shows the relationship between swelling potential versus time for all tested samples. When the soil reaches its maximum heave, the final heave was measured and the swelling potential was calculated according to Eq. 2. It was found that the value of swelling potential ranges from 9.83 to 10.37% and the time required for equilibrium is about 1000 min for all tested samples.

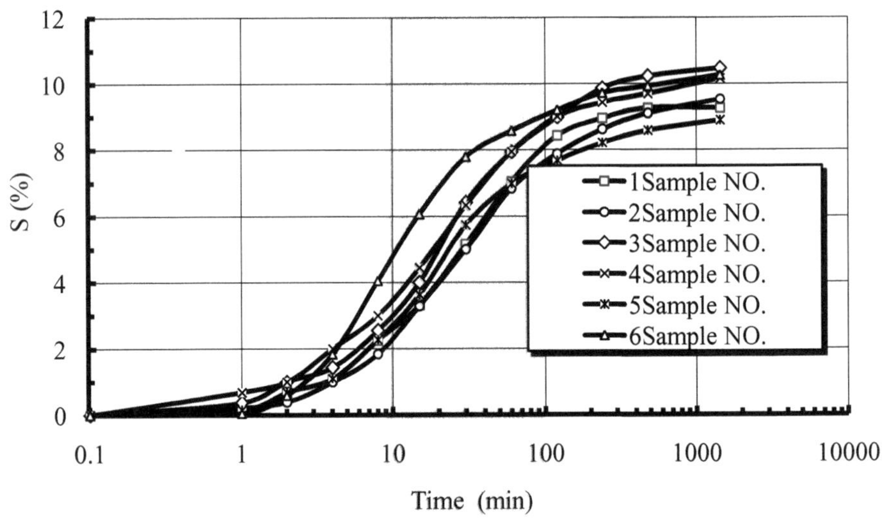

Fig. 3. Swelling potential vs. time at stress level 6.9 kPa for tested samples

3.9 Swelling Pressure Tests

Swelling pressure (SP) of a soil is defined as the external pressure that needs to be placed over a swelling soil sample to prevent volume increase. The most reliable mean of measuring swelling pressure is laboratory determination using one-dimensional consolidometer. This method is called direct measurement and the test is conducted on

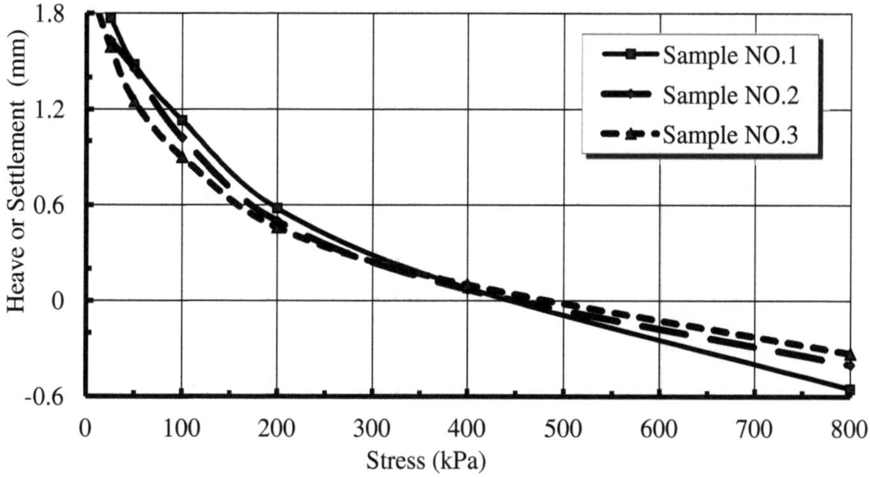

Fig. 4. Vertical displacement vs. different stress levels for studied soils in site A

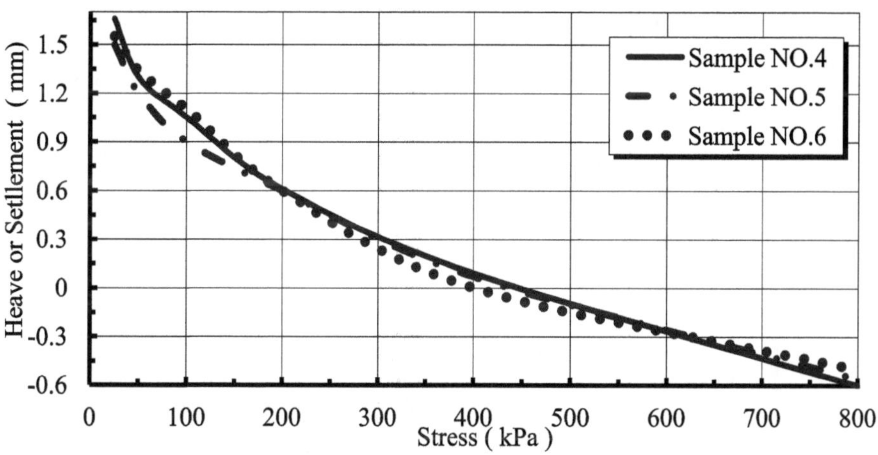

Fig. 5. Vertical displacement vs. different stress levels for studied soils in site B

the soil of the studied area. Figures 4 and 5 show the change of the vertical displacement (heave or settlement) of the studied samples with different stress levels recorded at the end of tested period of 24 h for each stage of loading. It could be noticed from these figures that the heave of the studied samples decrease with increasing the stress levels. The swelling pressure values of the studied soil samples are shown in Table 4.

4 Results and Discussion

4.1 Effect of Chemistry on the Geotechnical Properties of Studied Clay Samples

The chemical composition of the clay depends mainly on the chemistry of the main minerals, cementing materials and adsorbed cations and anions on the surfaces of clay minerals. Mitchell [12] suggested that the swelling and other geotechnical properties of the soil are controlled by the chemistry of soil water and soil components. For the same soil mineralogy, more swelling would occur in a sample having exchangeable Na+ cation than in a sample with Ca2+ or Mg2+ cations. In the present study, Al_2O_3 is more related to clay contents where it shows positive correlations with liquid limit (LL), plastic limit (PL), linear shrinkage (LS) and free swelling (FS) as shown in Fig. 6(a, b, c, d).

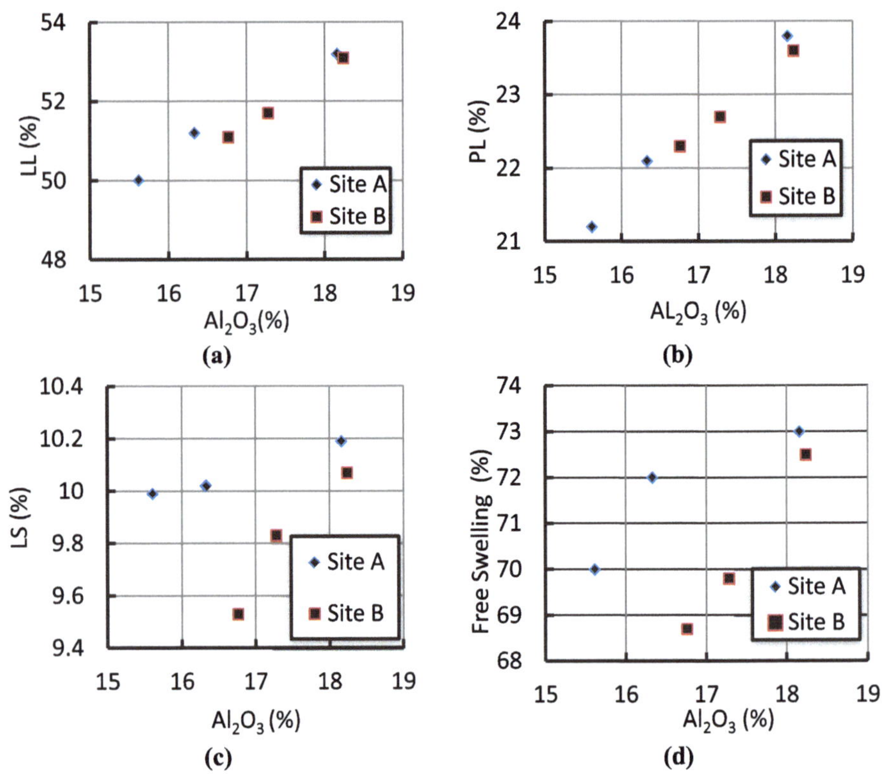

Fig. 6. Effect of alumina oxide (Al_2O_3) on the geotechnical properties of studied soils

On the other hand, SiO_2 is mainly derived from sand and silt fractions and partially from the clay fraction. Therefore, the SiO_2 content is negatively correlated with the parameters related to the clay content such as liquid and plastic limits, linear shrinkage and free swelling as shown in Fig. 7(a, b, c, d). In addition, Fe_2O_3 contents show positive correlations with liquid limit, plastic limit, linear shrinkage and free swelling as shown in Fig. 8(a, b, c, d).

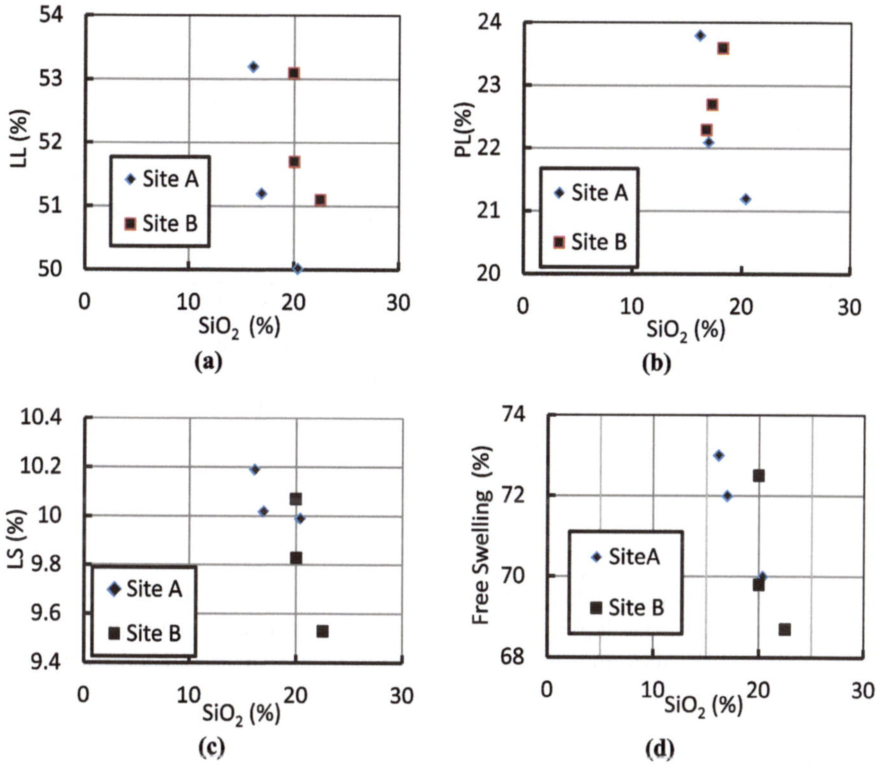

Fig. 7. Effect of silica oxide (SiO_2) on the geotechnical properties of studied soils

4.2 Soil Classification

Soil classification is an important aspect of laboratory tests, which tells the characteristics of the soil under interest. There are different methods of classification based on the identification tests performed on the soil. These methods are discussed in details as the following:-

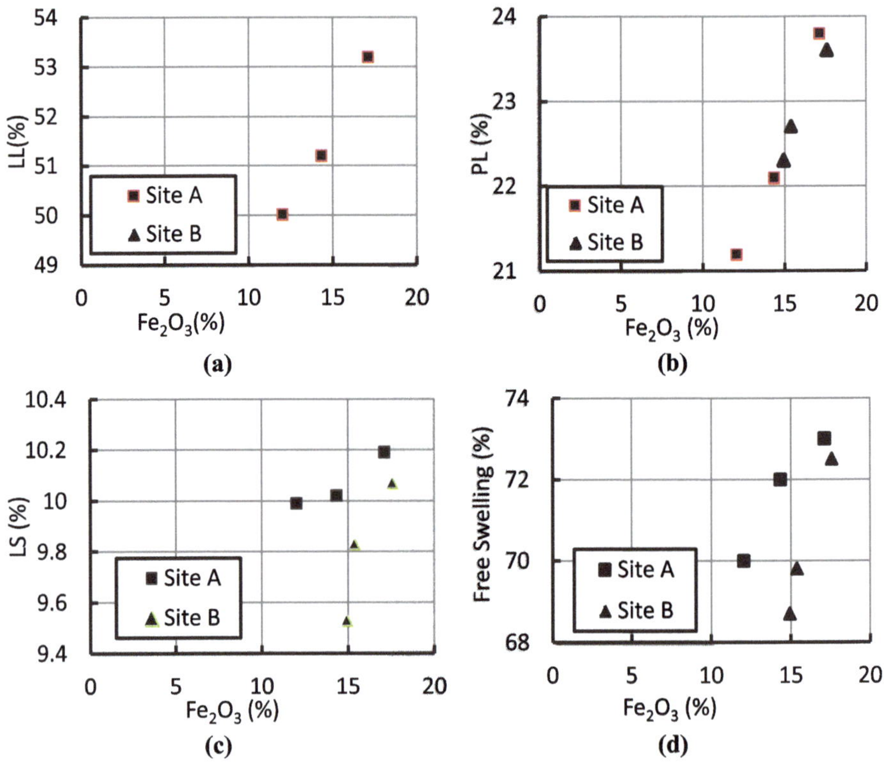

Fig. 8. Effect of ferric oxide (Fe_2O_3) on the geotechnical properties of studied soils

4.2.1 Classification According to Unified Soil Classification System (USCS)

The basis for USCS is the liquid limit and plasticity index (PI)of a soil as proposed by Getaneh Biru [13]. According to this classification scheme most of studied soils are considered high plasticity (CH), as shown in Fig. 9.

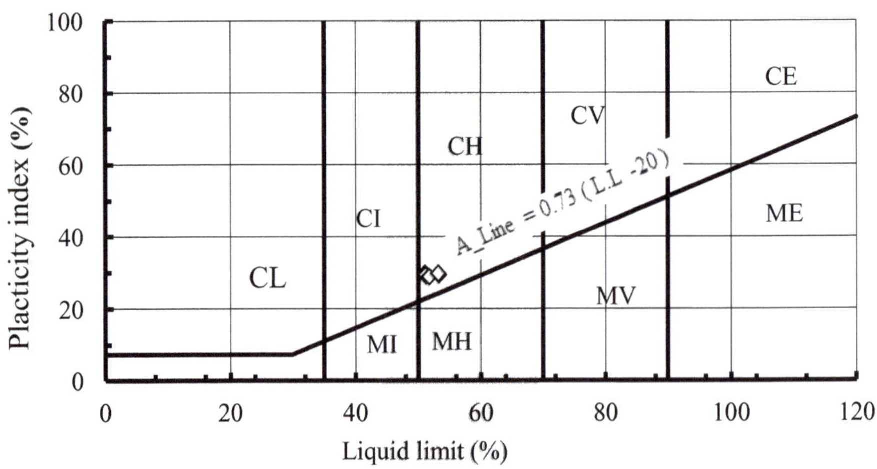

Fig. 9. Plasticity chart for classification of fine grained soils.

4.2.2 Classification Based on Activity

According to Skempton [14] clays are classified with respect to their activity, as shown in Table 5. The activity in Table 5 is taken as the dimensionless ratio of plasticity index to clay contents, both taken in percent. Thus;

$$\text{Activity } (A_c) = \text{plasticity index (PI) in } \%/\text{clay fraction finer than } 2\ \mu\text{m in } \% \quad (3)$$

Table 5. Classification of expansive soil according to activity [14]

Degree of activity	Activity
Inactive	Clay less than 0.75
Normal	Clay 0.75–1.25
Active	Clay greater than 1.25

According to this classification, the studied soils are considered as normal active clay where the values of activity range between 1.00 to 1.12.

Another way of identifying the expansive soil is to use the activity method quoted by Carter and Bentley [15]. The proposed classification chart is shown in Fig. 10. The activity term in Fig. 10 is defined a bit different from Eq. (3) as follows:

$$Ac = \frac{PI}{C - 5} \quad (4)$$

where PI is plasticity index and C is colloids (or clay) content.

Based on this classification the degree of soil expansion in the studied area is medium to high.

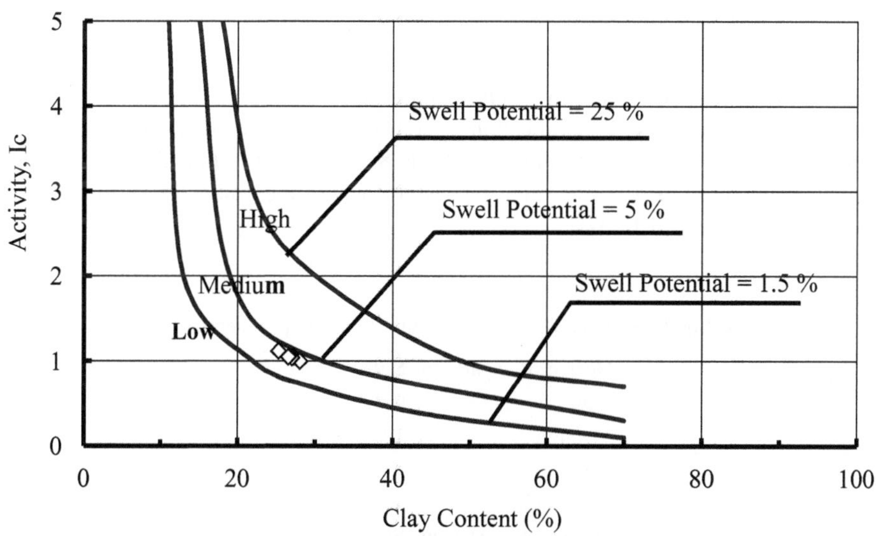

Fig. 10. Classification chart based on clay content and activity

4.2.3 Classification Based on Plasticity Index

Plasticity index is a parameter which can be used as a preliminary indicator of the swelling characteristics of a soil. The following values as shown in Table 6 were proposed by Chen [16], to relate soil expansivity and plasticity index. Relating the plasticity index of the studied soils with the above given range reveals that the soil falls in the range of medium to high expensivity.

Table 6. Classification of expansive soil according to plasticity index [16]

Soil expansivity	Plasticity Index (PI)
Low	≤ 15
Medium	10–35
High	20–50
Very high	>50

4.2.4 Classification Based on Shrinkage Limit and Linear Shrinkage

It was suggested by Altmeyer [17] as a guide to the determination of potential expansiveness of various values of shrinkage limit and linear shrinkage as shown in Table 7. According to this proposal, the studied soils have critical degree of expansion

Table 7. Values of shrinkage limit and linear shrinkage for the determination the degree of expansion [17]

Shrinkage limit (%)	Linear shrinkage (%)	Degree of expansion
Less than 10	Greater than 8	Critical
10–12	5–8	Marginal
Greater than 12	0–5	Non-critical

4.2.5 Classification Based on Swelling Potential

Seed et al. [18] proposed classification system for expansive soils according to the values of swelling potential as shown in Table 8. According to this classification, the studied soils are considered as high expensivity where the values of swell potential range between 9.83 to 10.37%.

Table 8. Soil classification according to swelling potential [18]

Degree of expansion	S (%)
Low	0–1.5
Medium	1.5–5
High	5–25
Very high	>25
Degree of expansion	Swell potential

4.2.6 Classification Based on Swelling Pressure

Geraid [19] proposed classification system for expansive soils according to the values of swelling pressure as shown in Table 9. According to this classification, the studied soils are considered as high expensivity where the values of swelling pressure range between 411 to 476 kPa.

Table 9. Classification of expansive soil according to the swelling pressure [19]

Degree of expansion	SP (kPa)
Low	<196
Medium	196–392
High	392–687
Very high	>687

From the previous different methods used in the classification of expansive soil, it is clear that all methods of soil classification agree that the soil of studied area has high degree of expansion.

5 Correlation of Index Properties

The evaluation of swell behavior of a soil using undisturbed samples and specialized swell tests is a difficult. Therefore, there is a need for simple routine tests that can be performed to achieve the same purposes. The empirical models appearing in the literature are primarily related for prediction of swelling potential and swelling pressure from index properties of expansive soils. Sometimes, the proposed empirical models can not be applied appropriately to all soils due to different soil conditions and testing procedures.

5.1 Liquid Limit and Plastic Limit

These two relationships could be especially useful in the approximate estimation of the plastic limit from measured values of liquid limit for the more silty varieties of clay soils in which determination of the plastic limit could otherwise be difficult. A correlation of values of liquid limit and plastic limit is presented diagrammatically in Fig. 11. From the diagram, it can be seen that the plastic limit increases with the increase of liquid limit and the plastic limit of studied soil could be approximately estimated by the strong (R^2 = 0.9845) relationship:-

$$PL = 0.02LL^{1.795} \tag{5}$$

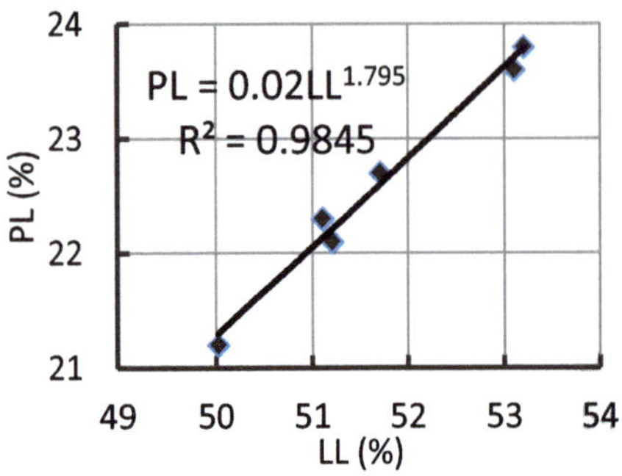

Fig. 11. Liquid limit vs. plastic limit

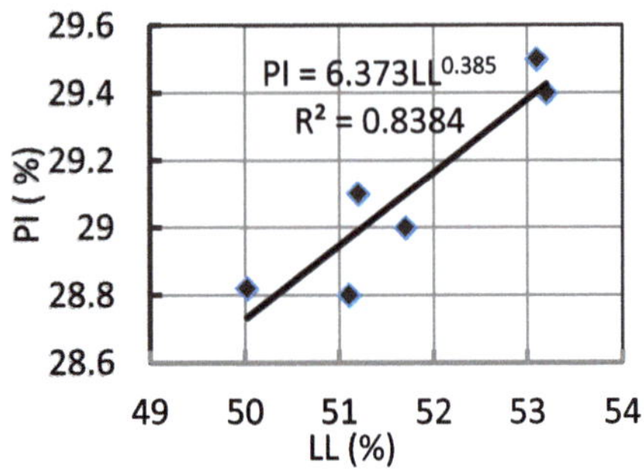

Fig. 12. Liquid limit vs. plasticity index

5.2 Liquid Limit and Plasticity Index

A correlation of values of liquid limit and plasticity index is presented diagrammatically in Fig. 12. From the diagram, it can be seen that the plasticity index increases with the increase of liquid limit and the plasticity index of studied soils could be approximately estimated by the strong ($R^2 = 0.8384$) relationship:-

$$PI = 6.37LL^{0.385} \tag{6}$$

5.3 Liquid Limit and Linear Shrinkage

A correlation of values of liquid limit and linear shrinkage is presented diagrammatically in Fig. 13. From the diagram, it can be seen that the linear shrinkage increases with the increase of liquid limit and the linear shrinkage of studied soils could be approximately estimated by the strong ($R^2 = 0.7925$) relationship:-

$$LS = 5.18LL^{0.1673} \tag{7}$$

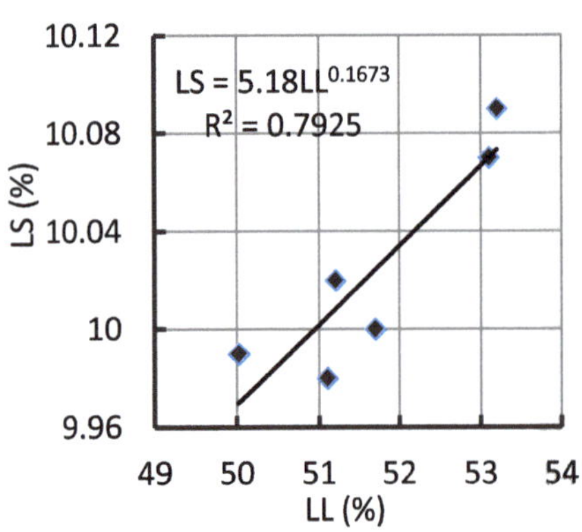

Fig. 13. Liquid limit vs. linear shrinkage

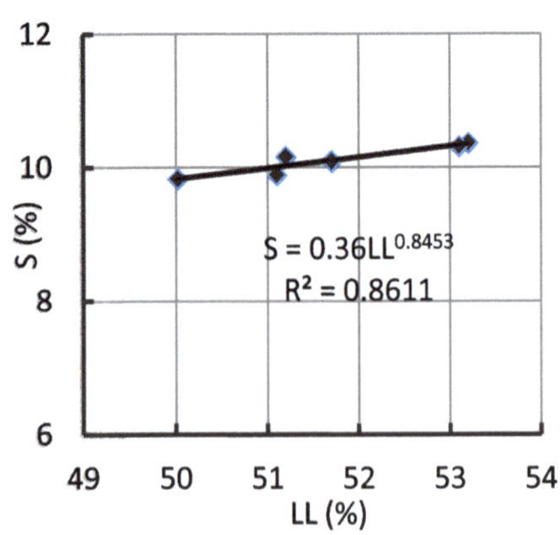

Fig. 14. Liquid limit vs. swell potential

5.4 Liquid Limit and Swell Potential

A correlation of values of liquid limit and swell potential is presented diagrammatically in Fig. 14. From the diagram, it can be seen that the swell potential increases with the increase of liquid limit and the swell potential of studied soils could be approximately estimated by (R^2 = 0.861) relationship:-

$$S = 0.36LL^{0.8453} \tag{8}$$

5.5 Liquid Limit and Swelling Pressure

A correlation of values of liquid limit and swelling pressure is presented diagrammatically in Fig. 15. From the diagram, it can be seen that the swelling pressure increases with the increase of liquid limit and the swelling pressure of studied soils could be approximately estimated by relationship, (R^2 = 0.27):-

$$\mathrm{SP\ (kPa)} = 2.38\mathrm{LL}^{1.325} \tag{9}$$

5.6 Swelling Pressure and Plasticity Index

The relationship between swelling pressure (SP) and plasticity index (PI) is shown in Fig. 16. From this figure it can be seen that there is a tendency of increment of the swelling pressure as the plasticity index increases and manifested in a linear relationship. A suggested equation was developed as, (R^2 = 0.8812):-

$$\mathrm{SP\ (kPa)} = 0.0335\mathrm{PI}^{2.8336} \tag{10}$$

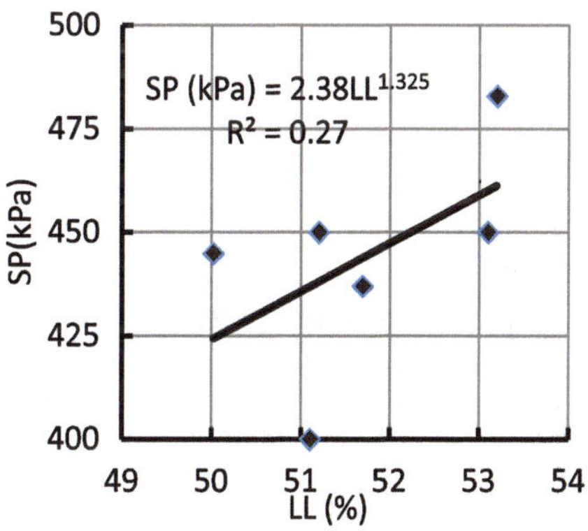

Fig. 15. Liquid limit vs. swelling pressure

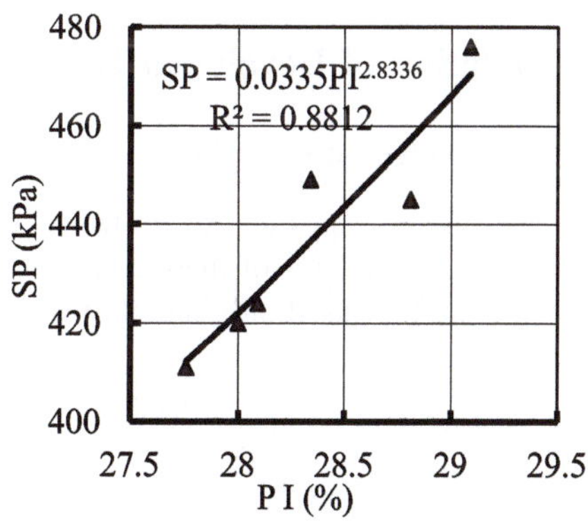

Fig. 16. Plasticity index vs. swelling pressure

5.7 Swell Potential and Plasticity Index

The relationship between swell potential and plasticity index is shown in Fig. 17. From this figure it can be seen that there is a tendency to increment of the swell potential as the plasticity index increases and manifested in a linear relation. A suggested equation was developed as, (R^2 = 0.8741):-

$$S\ (\%) = 0.0001 PI^{3.365} \tag{11}$$

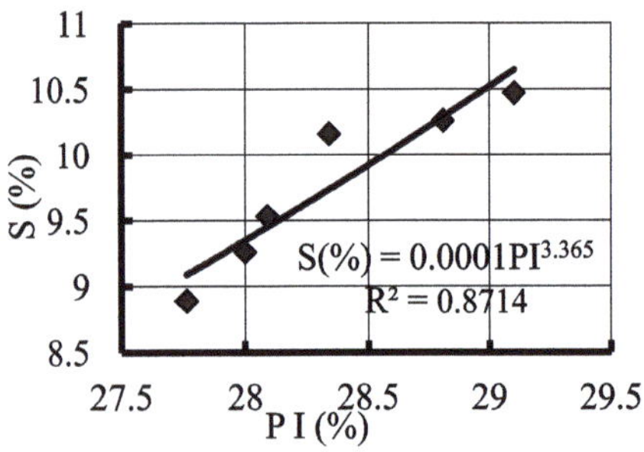

Fig. 17. Plasticity index vs. swell potential

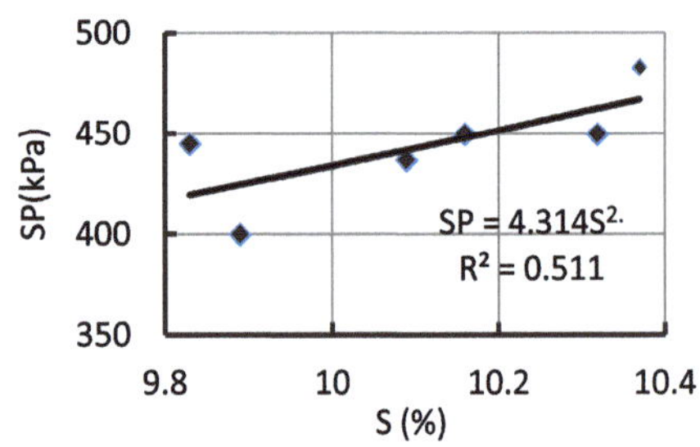

Fig. 18. Swell potential vs. swelling pressure

5.8 Swelling Pressure vs. Swell Potential

The relationship between swelling pressure and swell potential is shown in Fig. 18. From this figure it can be seen that there is a tendency of increment of swelling pressure as the swell potential increases and manifested in a linear relation. A suggested equation was developed as, (R^2 = 0.511):-

$$SP\ (kPa) = 4.31 S^2 \tag{12}$$

6 Validation of Developed Relationship

6.1 Swell Potential Equations

There are different equations suggested by different researches to determine indirectly the swell potential, some of these equations are listed below

Seed et al. [18] used only the plastic index to predict the swell potential

$$S\ (\%) = 0.00216\, PI^{2.44} \tag{13}$$

The above equation is applied only to soils with clay contents between 8 and 65%.

Nayak and Christensen equation [20]

$$S = 0.0229PI^{1.45}\frac{c}{w} + 6.38 \quad (14)$$

Where:-
c = clay content
w = initial water content

This equation was derived from analysis of test data from 18 artificial soil samples compacted at optimum moisture content by the standard Proctor method and allowed to swell under a surcharge load of 7 kPa.

Schneider and Poor equation [20]

$$S = 0.66 \times 10^{\left(0.9.\frac{PI}{w} - 1.19\right)} \quad (15)$$

Chen [16]

$$S\ (\%) = 0.2558e^{0.0838PI} \quad (16)$$

Saleem equation [22]

$$S\ (\%) = 2.7993e^{0.0341PI} \quad (17)$$

Abdallah equation [23]

$$S\ (\%) = 2.2 + 0.1(PI) \quad (18)$$

Weston equation [24]

$$S = 0.000195LL^{17.17}w^{-2.33} \quad (19)$$

Data for this equation are collected from undisturbed residual black clay soil samples with very wide ranges of engineering indices and this method therefore may be proved to be very useful for predicting the swelling for both natural and compacted soils, Samples were tested under a surcharge load of 1 kPa.

Figure 19 shows the relationship between swell potential and plasticity index for studied soils in comparison with those predicted from the equations proposed by different authors, [15, 16, 19, 20]. From this figure, it can be seen that the swell potential value obtained from suggest equation in this study is in a good agreement with that obtained from equation proposed by Seed et al. [18].

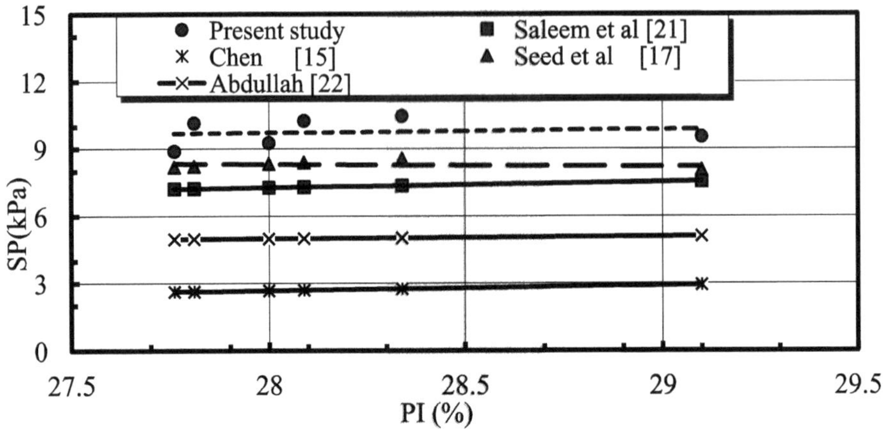

Fig. 19. Plasticity index vs. swelling pressure for different models

6.2 Swelling Pressure Equations

The swelling pressure is calculated according to the empirical equation that is presented below by Rashid [25].

$$SP\ (kPa) = 32.47\,S \tag{20}$$

Figure 20 shows the relationship between swelling pressure and swelling potential for studied soils in comparison with this predicted from Eq. (20). From this figure, it can be seen that the swelling pressure values obtained from suggest equation in this study gives overestimated values than that given by Eq. (20), this may be attributed to the different conditions and formations of soil.

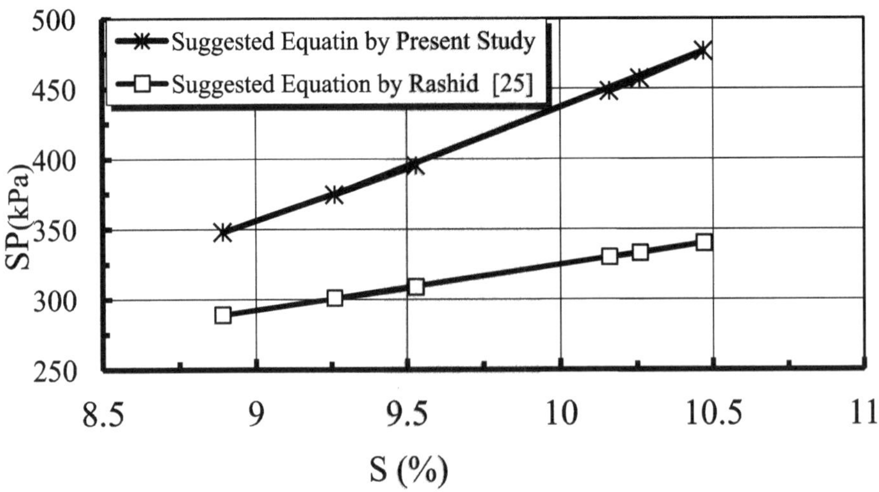

Fig. 20. Swell potential vs. swelling pressure

7 Conclusions

Analysis and interpretation of the laboratory results and the field observations led to the following findings:-

1. The soil in El-Mahrowsa contains the clay mineral montmorillonite predominantly.
2. The mineralogical and chemical compositions of the shallow marine clays are very important factors affecting their geotechnical characteristics.
3. The studied soil was geotechnical classified as high expansive soil (CH) according to USCS.
4. The regression analysis showed that there is a relationship between index properties and swelling characteristics of studied expansive soils.
5. Evaluation of the previously developed equations with the present study showed that the necessity of formulation of specific equations for specific areas.
6. The newly developed equations can be used for estimation of swelling characteristic of the studied soil.
7. It became clear from conducted tests, as liquid and plastic limit increases, the swell potential also increases.
8. We recommend using both swelling potential and swelling pressure for determining swelling index to classify the expansivity of expansive soil in practice.

Acknowledgements. Greatly thanks to all members in the Mining and Civil Engineering Department, Faculty of Engineering, Al-Azhar University, for their valuable cooperation that was highly needed during the conduction of this study.

References

1. Grim, R.E.: Clay Mineralogy, 2nd edn, pp. 166–174. McGraw-Hill, New York (1968)
2. Badry, M.M.: Geological and geotechnical studies on some types of sedimentary rocks in Egypt. M.Sc. thesis. Faculty of Science, South Valley University, Qena, Egypt (2004)
3. Basher, A.A.: Application of geophysical techniques at new Qena city, Qena, Egypt. M.Sc. thesis. South Valley University, Qena, Egypt (2003)
4. Bosy, A., Ahmed M., Tawfic, M.: Mapping of Pliocene-Pleistocene rock units using enhanced thematic mapper plus ETM+: case study, Wadi Qasab Area, South East Sohag, Egypt. In: The Seventh International Conference on the Geology of Africa, Assiut, Egypt, pp. VIII-1–VIII-13 (2013)
5. Ismaiel, H.A.H., Ibrahim, A.M.: Lime chemical stabilization of Nile silt soil exposed along Nile Banks at Taramsa Village, Qena Region, Egypt. In: The Second Symposium on the Geological Resources in the Tethys Realm, Aswan, Egypt (2013)
6. Badry, M.M.: Geotechnical and sediment logical properties of the sediments and the rocks exposed along the new upper Egypt-Red Sea Road, Eastern Desert, Egypt. Ph.D. thesis. Faculty of Science, South Valley University, Qena, Egypt (2011)
7. Mahrous, A.M., Tantawi, M.M., Hassan, E.: Evaluation of the engineering properties of some Egyptian limestones as construction materials for highway pavements. Constr. Build. Mater. **24**, 2598–2603 (2010)

8. Ahmed, A.: Surveying and geotechnical studies for the evaluation of deformation along Qena-Safaga road. M.Sc. thesis. Faculty of Engineering, Minia University, Egypt (2012)
9. Egyptian Code of Soil Mechanics and Foundations design and Construction, ECP 202/5-2001 laboratory tests, Cairo, National Center for Housing & Construction Research (2009)
10. Holtz, W.G., Gibbs, H.J.: Engineering properties of expansive clays. Transactions, ASCE **121**, 641–677 (1956)
11. Egyptian Code of Soil Mechanics & Foundations design & Construction, ECP 202/5-issue 2001, Foundations on problematic soils, Part 5, Cairo. National Center for Housing & Construction Research (2009)
12. Mitchell, J.K.: Fundamental of Soil Behaviour. Wiley, New York (1976)
13. Getaneh, B.: Investigation into index properties and swelling characteristics of expansive soils of Bahir Dar. Unpublished M.Sc thesis. AAU (2005)
14. Skempton, A.W.: The colloidial activity of clays. In: Proceeding 3rd International Conference on Soil Mechanics and Foundation Engineering, Switzerland, vol. 1, pp. 57–61 (1953)
15. Carter, M., Bentley, S.P.: Correlation of Soil Properties. Pentech Press, London (1991)
16. Chen, F.H.: Foundations on Expansive Soils. Elsevier Scientific Publishing Company, Amsterdam (1988)
17. Altmeyer, W.T.: Discussion on engineering properties of expansive clays. Proc. Am. Soc. Civ. Eng. **81**(658), 17–19 (1955)
18. Seed, H.B., Woodward, R.J., Lundgren, R.: Predication of swelling potential for compacted clays. J. Soil Mech. Found. Eng. Div. ASCE **88**(SM4), 53–87 (1962)
19. Geraid, J.: Review of expansive soils. ASCD J. Geotech. Eng. Div. **100**(6), 667–687 (1974)
20. Nayak, N.V., Christensen, R.W.: Swelling characteristics of compacted expansive soil. Clays Miner. **19**, 251–261 (1971)
21. Scheneider, P.: Research report. TR-9-74-University of Texas, Arlington (1974)
22. Saleem, Abdel, R.: R.: Study properties of soil expansive in Daraa Governorate. Damascus Univ. J. **9**(2), 11–25 (2003)
23. Abdullah, A.: Geotechnical properties of expansive clay shale in Tabuk, Saudi Arabia. J. Asian Earth Sci. **5**, 747–757 (2005)
24. Weston, D. J.: Expansive roadbed treatment for Sothern Africa. In: Proceedings of the Fourth International Conference on Expansive Soils, ASCE, Denver, Co., vol. 1, pp. 339–360 (1980)
25. Rashid, I., Farooq, K., Mujtaba, H., Khan, A.H.: Treatment of expansive clays through compaction control. Pak. J. Sci. **65**(1), 103–107 (2013)

Contribution of Zeolite to Deterioration of Pelitic Rocks from Asu-River Group and Variation of the Deterioration with Water Sorption

Tochukwu A.S. Ugwoke(✉)

Department of Geology, Michael Okpara University of Agriculture, Umudike, Nigeria
tcugwoke@yahoo.com

Abstract. The contributions of water-absorbent minerals to the deterioration of some pelitic rocks have been assessed. The assessment followed subjecting nine pelitic rock samples collected from five Albian pelitic quarry units to aggregate impact value (AIV) test, x-ray diffraction (XRD) and degradability test. The AIV was to ensure that the rocks have the acceptable mechanical competencies of rock aggregates; the XRD was to determine the water-absorbent mineralogy of the rocks while the degradability test was to determine the susceptibility of the rocks to deterioration caused by inter-seasonal wetting and drying and also to determine their water absorption potentials. Results of the degradability and XRD tests were further subjected to regression analysis to infer the water-absorbent mineral species that contribute most to deterioration of the rocks. The AIV test revealed that all the nine samples are mechanically competent. Results of the XRD revealed that eight samples contain appreciable amount (35%) of water-absorbent minerals notably: clay-minerals, zeolite, oxides and hydroxides. Results of the degradability test revealed that five samples showed significant (>1%) deterioration. It also revealed that pelitic rocks susceptible to deterioration do not absorb the same amount of water in every wetting process. The regression analysis revealed that amongst the identified water-absorbent minerals, zeolite contributes most to deterioration of the studied pelitic rocks. This work has shown that individual water-absorbent mineral species contribute more to pelitic rock deterioration, caused by inter-seasonal wetting and drying, than summed water-absorbent minerals and that deterioration of pelitic rock is followed by an increased water-absorption.

Keywords: Deterioration · Pelitic rock · Asu-River Group · Zeolite · Water sorption

1 Introduction

Rock deterioration is the physical depreciation or disintegration of rock when exposed to weathering agents like water, heat or organisms. The extent to which different rocks deteriorate is a check to their relative durability. Rock deterioration is dependent on rock's inherent properties like mineralogy and effective porosity and also on environmental

J. Wasowski et al. (eds.), *Engineering Geology and Geological Engineering for Sustainable Use of the Earth's Resources, Urbanization and Infrastructure Protection from Geohazards*, Sustainable Civil Infrastructures, DOI 10.1007/978-3-319-61648-3_4

factors like rainfall (wetting) and temperature. Generally, rocks occurring in the tropical regions are more susceptible to deterioration than those that occur in arid regions. This is because tropical regions are characterized by intense weathering agents like heavy rainfall (water) than arid regions that is characterized by very little or no rainfall. Works by Fitzner and Kalde (1991) and Weiss (1992) had shown that the hygroscopic properties (water-absorbing potential) of the minerals contained in a rock control the rate at which the rock deteriorates. Rocks rich in water-absorbent minerals like clay-minerals, zeolites, feldspar and calcite are susceptible to deterioration. According to Hudec (1980), Christaras (1991) and Ruedrich et al. (2010), such rocks deteriorate due to repeated wetting and drying; a process (deterioration) that manifest as change in colour and subsequent progressive disintegration into fragments starting from the rock surface to the inner portion.

Works by Cobanoglu et al. (2003) and Garcia-del-cura et al. (2011) reveal that the extent to which rocks absorb water and deteriorate is also controlled by their effective porosity (connected pores). Rocks rich in water-absorbent minerals but have low effective porosity will not be able to absorb enough water and thus will not show appreciable deterioration in comparison with those rich in water-absorbent minerals and have high effective porosity. Thus, mineralogy and effective porosity are two inherent properties that control the deterioration of rocks. These assessments of rock deterioration has been based on their summed water-absorbent minerals with little or no interest on the possible effect of the individual water-absorbent minerals and the extent to which each of them absorb water during the deterioration periods.

This work assesses the possible contribution of individual water-absorbent minerals to deterioration of some pelitic rocks, using samples from south-eastern Nigeria as a case study and also compares the amount of water absorbed during deterioration period(s) with the amount absorbed during non-deterioration periods. Findings of the present work shall be helpful in assessing the durability of pelitic rocks occurring in tropical regions of the world especially those rocks that contain multiple water-absorbent minerals.

2 Regional Geology

The study area, located in South-eastern Nigeria, is underlain by the Albian Asu-River Group, which is the oldest geologic unit in the Benue Trough (Nwajide 2013). According to Grant (1971) and Burke et al. (1971), Benue Trough is the third failed arm of a triple rift system which formed due to separation of South American plate from African plate in the Neocomian/early Gallic Epoch. The basin (Benue Trough) subsequently received Asu-River sediments from the bordering Basement complex in the Albian Stage. Asu-River Group is composed mostly of shales, limestones and lenses of sandstone (Ojo 1992; Obiora 2002; Obiora and Umeji 2004). Nwachukwu (1972) and Ofoegbu (1983) reported that the basin experienced tectonic events before (Turonian Stage) and after (Santonian Stage) receiving its first phase sediments (Asu-River). The tectonic events were characterized by volcanic eruptions which resulted to baking and/or hydrothermal alteration of the Asu-River sediments resulting to pelitic rocks in some locations within the group.

3 Climate of South-Eastern Nigeria

According to Koppen's climatic classification system (Kottek et al. 2006; Peel et al. 2007), south-eastern Nigeria is located in tropical savanna climatic region characterized by wet and dry seasons. The wet season occurs from April to October with monthly average precipitation (rain and dew) varying from 90 mm to 250 mm and atmospheric temperature varying from 23 °C to 32 °C. Dry season occurs from November to March with monthly average precipitation (mostly dew) varying from 15 mm to 50 mm and atmospheric temperature varying from 22 °C to 32 °C. Most often, a short dry season that lasts for about two weeks, characterized by no rainfall and the same atmospheric temperature as the long dry season, occurs in August or September. In other words, the region is characterized by long and short wet seasons and long and short dry seasons. During the wet seasons, rocks experience saturation due to high precipitation while during dry seasons; they (rocks) are completely dried due to prolonged high temperature and very low precipitation.

4 Materials and Methods

4.1 Sampling

The rock quarry outcrops, from where nine pelitic rock (baked shales) samples were collected, lie between latitudes 6°11′41″N and 6°22′23″N and between longitudes 8°02′24″E and 8°08′28″E of South-eastern Nigeria (Fig. 1). Each of the samples was divided into three portions.

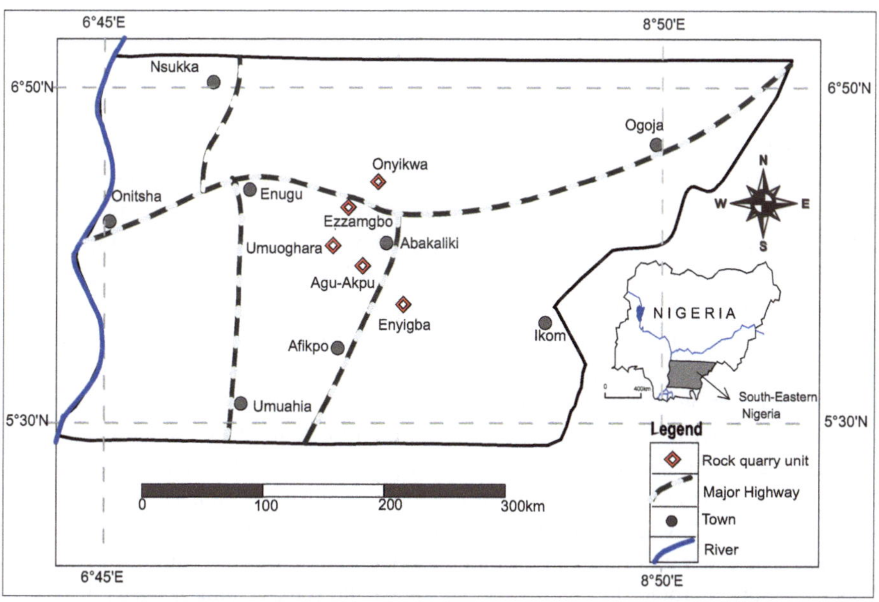

Fig. 1. Map of South-eastern Nigeria showing location of the rock quarry outcrops from where samples were collected

4.2 Testing

The first portion was subjected to x-ray diffraction (XRD) test to identify the dominant and water-absorbent minerals present in the rocks while the second portion was subjected to aggregate impact value (AIV) test to ascertain their mechanical stability. The AIV test was done according to Indian Standard (IS: 5640 1970). The third portion was subjected to long soaking and complete drying degradability test, similar to methods described by Ugwoke (2014) and Okogbue and Ugwoke (2015), to simulate the wetting and drying that occur over complete cycle of wet and dry season in south-eastern Nigeria. Results of the XRD and degradability tests were further subjected to regression analysis to determine the water-absorbent mineral(s) that contribute most to deterioration of the rocks.

4.3 X-Ray Diffraction Test

About 5 g pulverized dry rock sample was subjected to X-ray diffraction test using Schimadzu X-Ray diffractometer (XRD-6000) for a scan range of 0 and 70 2θ to generate the diffractogram. Peaks of the diffractogram denote minerals present in the analyzed sample. The minerals were identified and labelled by matching the peaks with software mineral cards installed in the diffractometer. The peak matching was done two times. Firstly, was to identify the dominant minerals contained in the rock and secondly, to identify the water-absorbent minerals contained in the rock. Relative abundances of the identified minerals were calculated from heights and population of the peaks.

4.4 Degradability Test

Batches of dry lump rock test samples, each weighing between 120 g and 180 g, were washed with water and finger pressure to remove loose/dust particles adhering to their surfaces, surface-dried with water-absorbent cloth to dry the water adhering to the surface and air-dried for 24 h to achieve natural dryness. The weight range 120 g to 180 g, which is triple the slake durability sample weight range of 40 g to 60 g (American Society for Testing and Materials; ASTM D4644-87 1998), was used in order to observe a bigger picture of the samples' deterioration as the use of the 40 g to 60 g weight range in trial tests resulted to near total breakdown of some samples and hence difficulty in assessing their actual degree of deterioration.

The test sample was weighed (W_{dry1st}), placed in a non-corrodible can and filled with potable water of natural climatic temperature (23 °C to 26 °C) determined with a mercury glass-tube thermometer inserted in the water. Potable water was used (instead of distilled water) in order to achieve similar water chemistry as rain. The set-up was left for 6 days (144 h), which is triple of the standard ASTM-C97 (1990) rock saturation duration. The water was carefully decanted and the sample washed with fresh water and finger pressure. Particles that got detached during the soaking and/or washing process were carefully picked, oven-dried at 105 °C for 24 h to achieve complete drying, cooled to ambient temperature and thereafter weighed (W_p). The intact rock samples were surface-dried with cloth, weighed (W_{wet}), oven-dried at 105 °C for 24 h to achieve complete drying and re-weighed (W_{dry}). This process was repeated for 6 cycles simulating a 3 years tropical wet and dry seasons, each year made up of 2 wet seasons and 2 dry seasons earlier explained to occur in south-eastern Nigeria. At the end of each cycle, the completely dried intact rock was used as the next test sample (W_{dry}). The percentage mass of water absorbed during each soaking (WtAb) and cumulative percentage of mass lost (Ma_{Lost}) were calculated using Eqs. (1) and (2), respectively.

$$\mathrm{WtAb} = \left[\frac{W_{wet} - W_{dry}}{W_{dry1st}}\right] \times 100\% \quad (1)$$

$$\mathrm{Ma}_{Lost} = \left[\frac{\left(\sum_{i=1}^{n=6} W_p\right)}{W_{dry1st}}\right] \times 100\% \quad (2)$$

Where
W_{wet} = Weight of wet intact rock (after soaking)
W_{dry} = Weight of dry intact rock (after complete drying)
W_{dry1st} = Weight of dry intact rock at the beginning of the test
W_p = Weight of dry detached particle(s)
$\sum_{i=1}^{n=6} W_p$ = sum of W_p

4.5 Regression Analysis

The cumulative percentages of mass lost (Ma_{Lost}) and the identified water-absorbent mineral groups were subjected to regression analysis using Statistical Program for Social Sciences (SPSS 16.0) to assess the mineral group(s) that contributes most to deterioration of the samples. The analysis was done following stepwise linear regression method using Ma_{Lost} as the dependent variable while the amounts (in % by mass) of water-absorbent mineral groups were used as the predictor variables.

5 Results and Discussion

Examples of the labelled XRD diffractogram are shown in Fig. 2 while the relative abundances of the minerals contained in the rocks are shown in Table 1. The aggregate impact value (AIV) and cumulative deterioration of the samples due to the degradability test are shown in Fig. 3. The deterioration history curves for samples that showed significant (>1% by mass) deterioration in the degradability test are shown in Fig. 4 while their water absorption history curves are shown in Fig. 5. Benchmark of 1% has been taken as significant deterioration in order to be as conservative as possible as samples that showed deterioration less than 1% were regarded as being non-degradable.

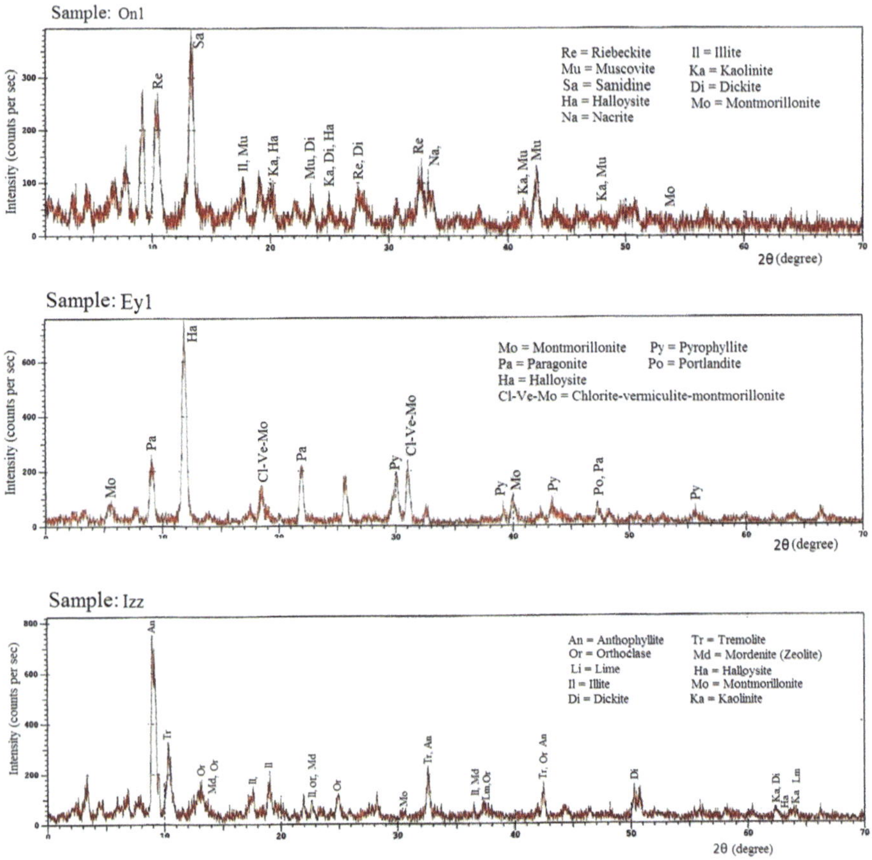

Fig. 2. Examples of the labelled diffractogram

Summary of the regression model of the cumulative percentage of mass lost due to the degradability test and water-absorbent mineral groups is shown in Table 2.

Table 1 reveals that amongst all water-absorbent minerals (clay-minerals, zeolites, oxides, hydroxides and feldspar) identified in the analyzed rocks, clay-minerals have the highest amount. Then, amongst the identified clay-minerals, Illite is the commonest phase. Illite occurs in 8 samples while smectite and kaolinite occur in 6 and 4 samples respectively. It implies that illite forms/crystallizes easier than kaolinite and smectite. Einsele (2000) and Nichols (2009) had earlier noted that illite has higher stability to diagenetic and metamorphic processes than other clay mineral types (phases).

Following comparison with the Indian Road Congress (IRC 1970) specifications, (see Fig. 3), which is a check on the studied rocks' mechanical stability, it is shown that all the analyzed pelitic rocks have AIV indicating that they are good rock aggregates. Works by Chang (1973), Kazi and Al-Molki (1982), Turk and Dearman (1988), Irfan (1994) and Al-Harthi (2001) had shown that AIV has strong relationship with other rock mechanical properties like strength, Los Angeles abrasion value (LAAV) and

Table 1. Relative abundance of minerals contained in the analyzed rock samples

Sample	Mineralogy (the mineral groups and their relative abundance are indicated in parenthesis)			Summed water-absorbent minerals (%)
	Dominant minerals	Water-absorbent minerals		
		Clay minerals	Other water-absorbent minerals	
On1	Muscovite (mc, 23.37), Riebeckite (am, 18.89)	Kaolinite (k-cl, 14.43), Dickite (k-cl, 10.14), Illite (i-cl, 9.40), Halloysite (k-cl, 7.87), Nacrite (k-cl, 6.24), Montmorillonite (s-cl, 0.74)	Sanidine (fd, 8.78)	57.60
On3	Osumillite (mi, 27.62), Muscovite (mc, 27.62)	Dickite (k-cl, 14.80), Nacrite (k-cl, 10.63) Montmorillonite (s-cl, 10.21), Kaolinite (k-cl, 7.18), Illite (i-cl, 1.55)		44.37
Izz	Anthophyllite (am, 34.45), Tremolite (am, 28.47)	Illite (i-cl, 7.99), Dickite (k-cl, 3.08), Kaolinite (k-cl, 1.95), Montmorillonite (s-cl, 0.83), Halloysite (k-cl, 0.55)	Orthoclase (fd, 12.45) Mordenite (ze, 8.13), Lime (ox, 2.09)	37.07
Um1	Truscottite (tr, 25.08), Paragonite (mc, 23.28)	Illite (i-cl, 12.68), Montmorillonite (s-cl, 7.31), chlorite-vermiculite-montmorillonite (x-cl, 6.11), Beidellite (s-cl, 2.08)	Faujasite (ze, 23.43)	51.61
Um2	Lizzardite (sp, 49.84), Chrysotile (sp, 35.20)	Illite (i-cl, 14.95)		14.95
Ag1	Truscottite (tr, 41.18), Talc (Ta, 17.14)	Illite (i-cl, 11.11)	Faujasite (ze, 30.55)	41.66
Ag2	Parahopeite (ox, 32.28), Muscovite (mc, 28.34)	Kaolinite (k-cl, 9.70), Illite (i-cl, 7.79), Nacrite (k-cl, 7.72), Dickite (k-cl, 2.41)	Mordenite (ze, 10.00), Periclase (ox, 1.73)	39.35
Ey1	Paragonite (mc, 22.72), Pyrophyllite (am, 20.45)	Halloysite (k-cl, 22.44), Chlorite-vermiculite-montmorillonite (x-cl, 14.20), Montmorillonite (s-cl, 11.92)	Portlandite (oh, 8.24)	56.80
Ey2	Antigorite (sp, 41.84), Tremolite (am, 19.80)	Illite (i-cl, 18.80), corrensite (s-cl, 6.42)	Mordenite (ze, 13.12)	38.34

mc = mica, am = amphibole, sp = serpentine, mi = milarite, tr = truscottite, ox = oxide, ta = talc, oh = hydroxide, k-cl = kaolinite clay, cl = other clay-mineral, i-cl = illite clay, s-cl = smectite clay, x-cl = mixed clay, fd = feldspar, ze = zeolite aggregate crushing value (ACV). It implies that, mechanically, the studied rocks can serve well as rock aggregates. On the other hand, the second aspect of Fig. 3 (mass lost), which is a check on the materials' (rocks) relative durability, reveals that 5 of the samples showed significant (>1%) deterioration. This implies that the other 4 samples, that did not show significant deterioration, are durable. Comparing the two aspect of Fig. 3, therefore, indicates that durability of the pelitic rocks does not relate to their

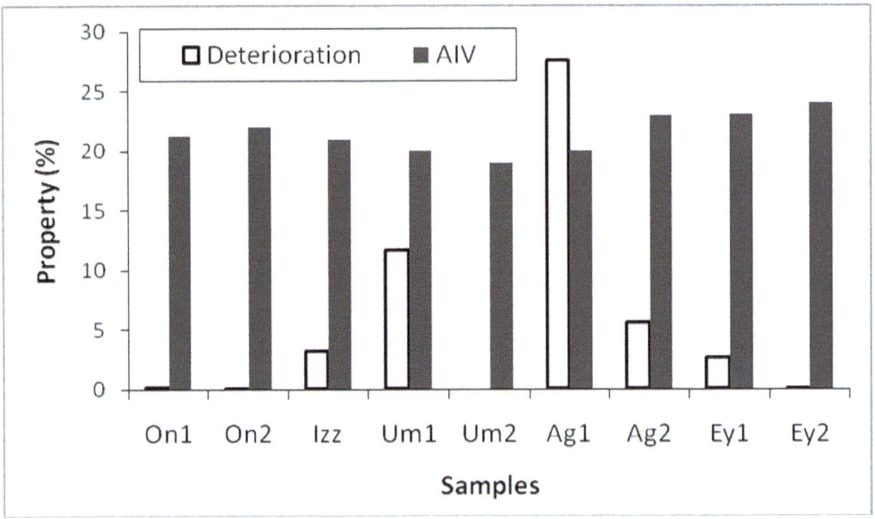

Fig. 3. Aggregate impact values and cumulative deterioration of the samples

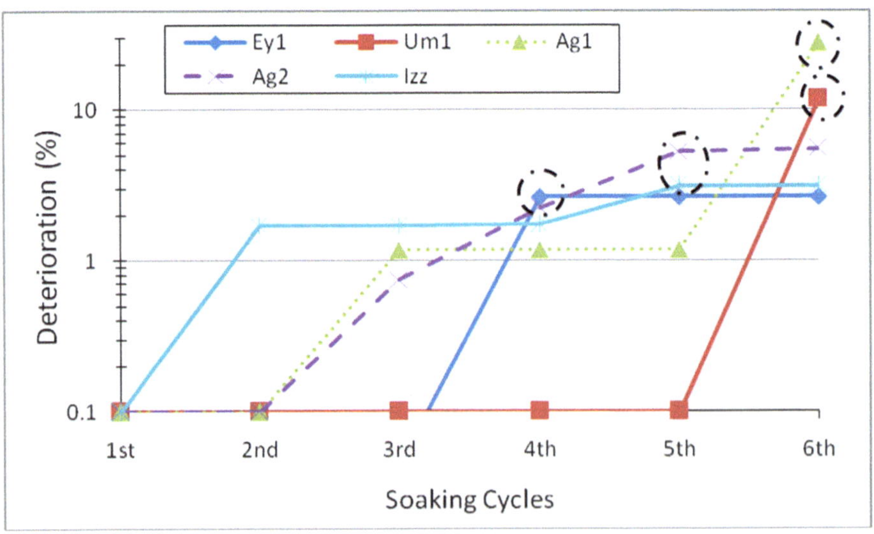

Fig. 4. Deterioration history curves of samples that showed significant deterioration

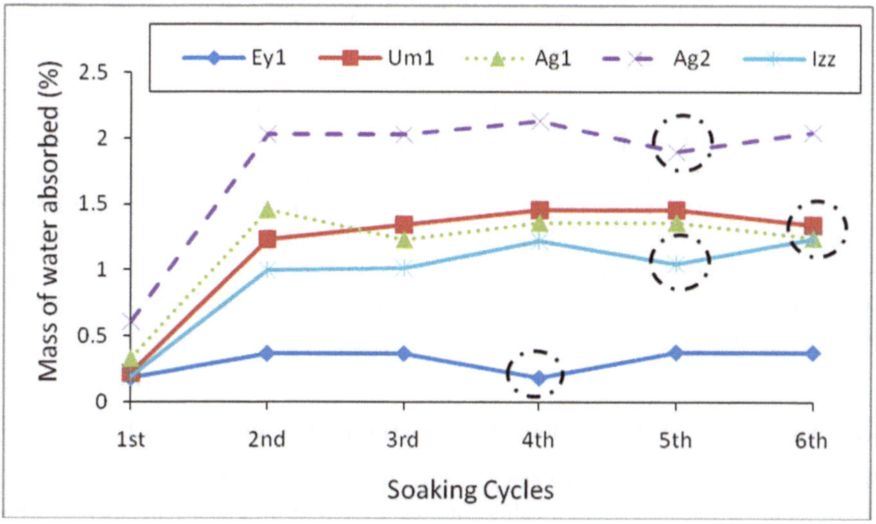

Fig. 5. Water absorption history curves of the samples that showed significant deterioration

mechanical stability. In other words, it is either that mechanical stability of a rock is not a guarantee to its durability or that factors controlling mechanical stability are different from those controlling durability.

5.1 Contributions of Zeolite to Deterioration of the Studied Rocks

It can be seen from Fig. 3 and Table 1 that the degree of deterioration is not proportional with either the clay mineral content or summed water-absorbent minerals (wam_{sum}). Sample Ag1 that showed the highest deterioration (27.6%) contains 11.1% of clay minerals and 41.6% wam_{sum} while sample On1, which did not show significant deterioration, contains 48.8% clay minerals and 57.6% wam_{sum}. These findings suggest that neither the clay minerals nor the summed water-absorbent minerals (wam_{sum}) contribute significantly to deterioration of the rocks. Table 2 reveals that amongst all the water-absorbent minerals (including wam_{sum}) used as predictor variables to rock deterioration, only zeolite emerged as the significant predictor.

From Table 2, the relationship existing between cumulative percentage of mass lost (Ma_{lost}) in the degradability test and amount zeolite is given by Eq. 3:

$$Ma_{lost} = -0.89 + (0.70 * \text{Zeolite}) \quad (R = 0.88) \tag{3}$$

Where
Ma_{lost} = cumulative percentage of mass lost
Zeolite = amount of zeolite contained in the rock (%)
R = correlation coefficient

Zeolite contribution to deterioration of the rocks is also evident in Table 1. The samples containing zeolite are those that showed significant deterioration. It follows that zeolite contributes most to deterioration of the studied pelitic rocks in comparison to other water-absorbent minerals. It implies that, although it is a widely accepted fact that mineralogy of rock contributes to rock deterioration, it may not be the summed water-absorbent minerals contained in a rock rather the individual water-absorbent mineral species. In the present case, zeolite contributes more than other water-absorbent minerals contained in the rocks. Remedy to deterioration of these rocks can be achieved by mixing them with binding medium like in making concrete cake and pavements. Such binding media (e.g. mortar and bitumen) will envelop the rocks from water thereby preventing deterioration.

5.2 Variation of Water Sorption with Deterioration of the Studied Rocks

Figure 4 shows that none of the samples that showed significant deterioration started deteriorating in the first cycle. Each started deteriorating in a subsequent cycle. This agrees with work of Ruedrich et al. (2010) that deterioration of the rocks is caused by cyclic wetting and drying not by either wetting or drying. Figure 4 also shows that one of the samples (Um1) showed its significant deterioration in the last (6th) cycle. Termination of this test in duration less than 6 cycles would have resulted to the loss of deterioration information of some samples.

Comparing Figs. 4 and 5, it can be seen that the deterioration cycles are accompanied by graphical 'apparent' decrease in water absorption (circled points of Figs. 4 and 5). Contrary to expectation, the rocks absorbed more water in cycle(s) following the deterioration cycle. This is due to water sorption and deterioration processes associated with wetting and drying of the rocks. During wetting process, the outer portions ordinarily become wet and saturated before the inner portions due to water permeation through possible deterioration pathways (connected pores, micro-fractures and/or water-absorbent minerals) of the out parts (Koch et al. 2008). Deterioration (detachment) of such outer portion is accompanied by two sequential water sorption processes. Firstly, there is a decrease in amount of water absorbed by the intact rock. Secondly, new outer portion is exposed, which results to permeation of more water through deeper deterioration pathways and thus an increase in the amount of water absorbed during the next wetting process. The first process explains the graphical decrease in amount of water absorbed during deterioration cycle while the second process explains the increase in the amount of water absorbed in the following cycle. It implies that, in service, excavated/quarried pelitic rocks exposed to inter-seasonal wetting/drying of south-eastern Nigeria absorb more water in periods subsequent to their deterioration period than they absorb in their deterioration period.

Table 2. Relationship of the rocks' deterioration and water-absorbent minerals

Predictor variables		Relationship of the water-absorbent minerals with long soaking deterioration		
Variables	Minerals	Correlation coefficient (R)	Significance (sig)	Regression coefficient (Beta)
Variables entered (predictors)	Zeolite	0.882	0.001	0.701 (constant = −0.890)
Variables removed (excluded)	Illite	0.099	0.393	NA
	Smectite	−0.287	0.211	NA
	Kaolinite	−0.424	0.111	NA
	Other clay	−0.380	0.139	NA
	All clay	−0.522	0.061	NA
	Feldspar	−0.201	0.289	NA
	Oxide & Hydroxide	−0.115	0.376	NA
	wam_{sum}	0.068	0.426	NA

wam_{sum} – sum of water-absorbent minerals
NA = not applicable
Dependent variable - Mass lost in the degradability test
Regression method: Stepwise (Criteria: Enter if sig. $\leq$ 0.050, Remove if sig. > 0.050)

6 Conclusions

The following conclusions are drawn from this work:

1. The individual water-absorbent mineral species contributes more to deterioration of pelitic rocks than the summed water-absorbent minerals. Zeolite contributes more to deterioration of the studied pelitic rocks than other water-absorbent minerals.
2. Non-durable pelitic rocks do not absorb the same amount of water in all complete wetting process. Pelitic rocks absorb more water in cycles following their deterioration periods caused by inter-seasonal wetting and drying than in the deterioration periods.
3. The studied pelitic rocks, from Asu-River Group, are richer in illite clay mineral phase than kaolinite and smectite phases.
4. Mechanical stability of rocks does not have relationship to their durability. Rocks can be mechanically stable but still not durable.

References

Al-Harthi, A.A.: A field index to determine the strength characteristics of crushed aggregate. Bull. Eng. Geol. Environ. **60**, 193–200 (2001)

ASTM (American Society for Testing and Materials): Soil and Rock, Building Stones: Annual book of ASTM standards, vol. 04.08. ASTM-C97, Philadelphia, Pennyslvania (1990)

ASTM (American Society for Testing and Materials): Standard test method for Slake durability of shale and similar weak rocks. ASTM D4644-87, West Conshohocken (1998)

Burke, K., Dessauvagie, T.F.J., Whiteman, A.J.: Opening of the gulf of Guinea and geological history of the Benue depression and Niger Delta. Nat. (Phys. Sci.) **233**, 51–55 (1971)

Chang, F.K.: A study of the aggregate impact test in relation to aggregate in Thailand. M.Sc. thesis. Asian Institute of Technology, Bangkok (1973)

Christaras, B.: Durability of building stones and weathering of antiquities in Creta/Greece. Bull. Eng. Geol. **44**, 17–25 (1991)

Cobanoglu, I., Ozpinar, Y., Ozbek, A.: Engineering properties of tuffs in the Sandikli region (Afyon-Turkey) and their possible use as concrete aggregates. Bull. Eng. Geol. Env. **62**, 369–378 (2003)

Einsele, G.: Sedimentary Basins, Evolution, Facies and Sediment Budget, 2nd edn. Springer, Berlin (2000)

Fitzner, B., Kalde, M.: Simulation of frost-thaw cycle and salt weathering – nature adapted material tests. In: Auger, F. (ed.) La deterioration des matériaux de construction. 12–14 June, Colloque International, La Rochelle, pp. 103–114 (1991)

Garcia-del-Cura, M.A., Benavente, D., Martinez-Martinez, J., Cueto, N.: Sedimentary structures and physical properties in travertine and tufa building stones. Constr. Build. Mater. **28**(1), 456–467 (2011)

Grant, N.K.: The South Atlantic, Benue trough and Gulf of Guinea Cretaceous triple junction. Bull. Geol. Soc. Am. **82**, 2295–2298 (1971)

Hudec, P.P.: Durability of carbonate rocks as function of their thermal expansion, water sorption and mineralogy. ASTM Tech. Publ. **691**, 497–508 (1980)

Indian Roads Congress, IRC:15: Tentative specification (for various types of construction methods) (1970)

Indian Standard, IS 5640: Method of test for determining aggregate impact value of soft coarse aggregates (1970)

Irfan, T.Y.: Aggregate properties and resources of granitic rocks for use in concrete in Hong Kong. Q. J. Eng. Geol. **27**, 25–28 (1994)

Kazi, A., Al-Molki, M.E.: Empirical relationship between Los Angeles abrasion and aggregate impact value test. In: proceedings of 4th International Congress, International Association of Engineering Geology, vol. 6, pp. 293–299 (1982)

Koch, R., Racatalanu, C.P., Bucur, I.I.: Examples of weathering and deterioration of tertiary building stones at St. Micheal's church in Cluj Napoca (Romania). Stud. Universitutis Babes-Bolyai, Geol. **53**(2), 25–39 (2008)

Kottek, M., Grieser, J., Beck, C., Rudolf, B., Rubel, F.: World map of the Koppen-Geiger climate classification updated. Meteorol. Z. **15**(3), 259–263 (2006)

Nichols, G.: Sedimentology and Stratigraphy, 2nd edn. Blackwell/Wiley, London (2009)

Nwachukwu, S.O.: The tectonic evolution of the Southern portion of the Benue Trough, Nigeria. Geol. Mag. **109**, 411–419 (1972)

Nwajide, C.S.: Geology of Nigeria's Sedimentary Basins. CSS Bookshops Limited, Lagos (2013). 565p.

Obiora, S.C.: Evaluation of the effects of igneous bodies on the sedimentary fill of the lower Benue rift and vice versa. Ph.D. Dissertation. Department of Geology, University of Nigeria, Nsukka (2002)

Obiora, S.C., Umeji, A.C.: Petrographic evidence for regional burial metamorphism of sedimentary rocks in the Lower Benue rift. J. Afr. Earth Sci. **38**, 269–277 (2004)

Ofoegbu, C.O.: A model for the tectonic evolution of Benue Trough of Nigeria. Geol. Rundsch. **73**, 1007–1018 (1983)

Ojo, K.A.: The southern part of the Benue Trough (Nigeria) Cretaceous stratigraphy, basin analysis, paleogeography, and geodynamic evolution in the equatorial domain of the South Atlantic. NAPE Bull. **7**(2), 131–152 (1992)

Okogbue, C.O., Ugwoke, T.A.S.: Influence of petrogenesis on suitability of some pelitic rocks as construction aggregates in South-eastern Nigeria. Geotech. Geol. Eng. **33**(6), 1395–1407 (2015)

Peel, M.C., Finlayson, B.L., McMahon, T.A.: Updated world map of the Koppen-Geiger climate classification. Hydrol. Earth Syst. Sci. **11**, 1633–1644 (2007)

Ruedrich, J., Bartelsen, T., Dohrmann, R., Siegesmund, S.: Building sandstone integrity affected by the process of hygric expansion. Environ. Earth Sci. (2010). doi:10.1007/s12665-010-0767-0

Turk, N., Dearman, W.R.: An investigation into the influence of size on the mechanical properties of aggregates. Bull. IAEG. **38**, 143–149 (1988)

Ugwoke, T.A.S.: Effect of petrogenesis on suitability of some pelitic rocks, from Asu-River Group, Abakaliki metropolis, as construction aggregates in the tropics. Unpublished M.Sc. thesis. Department of Geology, University of Nigeria, Nsukka (2014). 179pp.

Weiss, G.: Die Eis- und Salzkristallisation im Porenraum von Sandsteinen und ihre Auswirkung auf das Gefüge unter besonderer Berücksichtigung gesteinsspezifischer Parameter. In: Münchner Geowissenschaftliche Abhandlungen, B9, Pfeil, München (1992)

Detection of Elastic Region Varied by Inherent Anisotropy of Reconstituted Toyoura Sand

Bao Ngoc Le, Hirofumi Toyota(✉), and Susumu Takada

Nagaoka University of Technology, Nagaoka, Niigata, Japan
s155068@stn.nagaokaut.ac.jp,
toyota@vos.nagaokaut.ac.jp,
stakada@konomi.nagaokaut.ac.jp

Abstract. Inherent anisotropy affects deformation and strength characteristics of sand deposits. Accurate deformation modulus and shear strength are important parameters to design and construct soil structures such as retaining wall, foundation and tunneling. In this study, specimens having various angles of depositional plane (sedimentation angle) were artificially made using an inclined container to estimate soil inherent anisotropy. The container was designed to prepare easily a specimen with different inclined depositional plane by changing positions of its side walls. Specimens are made using Toyoura sand by air pluviation method to attain the same relative densities regardless of the depositional angle. The strength and deformation characteristic were evaluated using both bender elements (BE) method and local small strain (LSS) measurements. To compare the accuracy of acquired results, the both tests were conducted in the same specimens under drained condition of triaxial tests. Results from the experiments indicate that initial shear modulus has tendency to slightly increase with an increase of the depositional angle. However, shear strength has the obvious inverse relationship with changing of the depositional angles. Concerning elastic region, the elastic region tends to shrink with an increase of the depositional angle in spite of increasing of the initial shear modulus with the depositional angle. In comparison between BE and LSS tests, the initial shear moduli obtained from LSS tests are slightly greater than those obtained from BE tests.

1 Introduction

Casagrande and Corrilo (1944) pointed out two types of anisotropy in soils, namely inherent and induced anisotropy. The inherent anisotropy was defined as soil particles' fabric made during sedimentation. On the other side, the induced anisotropy was considered as anisotropy caused when stresses and/or strains applied to soils. In natural sedimentation process, different depositional angles of soil grain can cause by many factors. The folding (uplift and subsidence) was created by the tectonics movement. Sedimentation by water current and wind blowing will form not only different grading but also different depositional angles in soil layers.

J. Wasowski et al. (eds.), *Engineering Geology and Geological Engineering for Sustainable Use of the Earth's Resources, Urbanization and Infrastructure Protection from Geohazards*, Sustainable Civil Infrastructures, DOI 10.1007/978-3-319-61648-3_5

Profound understanding about the anisotropy of soils can be used to settle many geotechnical engineering problems such as designing and constructing the underground structures, and prediction of ground deformation.

The inherent anisotropy was mainly concentrated in this study. Soil response at small strain level, where the deformation of soils is assumed linearly and elastically, was investigated. Many previous studies proved that anisotropy, related not only to flat and elongated shape of grains but also to depositional manner, affected both strength and deformation characteristics of sands. Oda (1972) demonstrated that the inherent anisotropy had relations with both shape of constituting grains and depositional manner. Arthur and Menzies (1972) investigated the inherent anisotropy was caused from deposition which has a certain sedimentation angle in specimen. Yamada and Ishihara (1979) also conducted true triaxial test on cubical loose sand specimen and found that inherent anisotropy affected strongly deformation characteristics of sand in range of small shear strain. Guo (2008) demonstrated that particle shape and direction of orientation dominated the degree of anisotropy by using results of modified direct shear test with different angles of sedimentation. Tong et al. (2014) studied the inherent fabric anisotropy to investigate effects of the bedding plain on the peak friction angle using direct shear test. The other papers (Oda et al. 1978; Ochiai and Lade 1983; Wong and Arthur 1985) have also indicated that the inherent anisotropy has the significant effects to strength and stress - strain relations of sands.

Concerning clay's inherent anisotropy, Pennington et al. (1997) measured the anisotropy of stiffness using bender elements with both vertical and horizontal directions. The degree of anisotropy, defined as ratio of shear modulus between the horizontal G_{hh} and vertical G_{vh} plane, was to be found unequal to 1.0 in some studies (Jovicic and Coop 1998; Gasparre et al. 2007; Hight et al. 2007; Teng et al. 2013). Teachavorasinskun and Lukkanaprasit (2008) indicated that inherent anisotropy has stronger effect to elastic shear modulus than the induced stress.

A series of experiments were conducted using reconstituted sand specimens. The depositional angles of the specimens were controlled in various directions. The effects of the depositional angle on shear strength, deformation and elastics region were investigated in this study.

2 Testing Material and Preparation Method

Toyoura sand, which is Japanese standard sand, was used as the test material in this study. Toyoura sand is a sub-angular to angular poorly graded fine quartz rich sand. The particle size distribution and physical properties of Toyoura sand are shown in Fig. 1. 2-D scanning optical microscope image is shown in Fig. 2. Two principal dimensions (length and width) of sand particle were measured from the 2-D image to consider the shape of sand particles. The sand particles are somewhat flat and elongated because the length to width ratio L/W distributes from 1.0 to 1.6 and the average value of L/W is about 1.5 (Fig. 3).

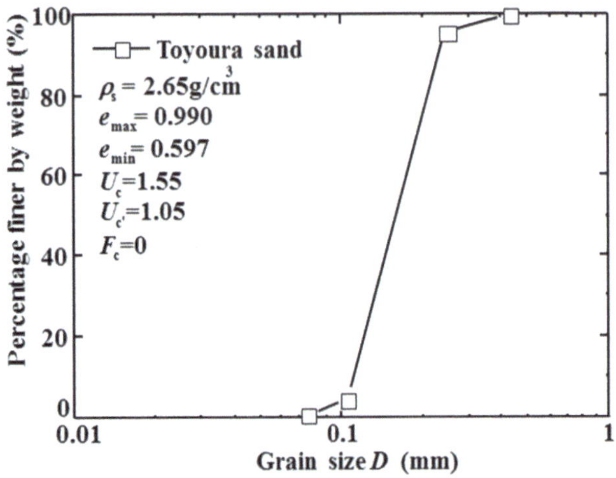

Fig. 1. Particle size distribution of Toyoura sand

Fig. 2. Scanning optical microscope image of Toyoura sand

Figure 4 shows a rectangular container to make samples with different depositional angles by changing the inclination and the position of its walls. Angle α of 0°, 22.5°, 45°, 67.5°, and 90° (Fig. 5) were chosen as the depositional angles in the study.

The inherent anisotropy is related to depositional process during sedimentation of soil particles. Therefore, specimen preparation method plays a crucial role for mechanical properties of soil with the inherent anisotropy. "Air pluviation" method was used to make specimen with different depositional angles. Toyoura sand was poured into the container through a sieve with a mesh size of 450 μm as shown in Fig. 6. The falling

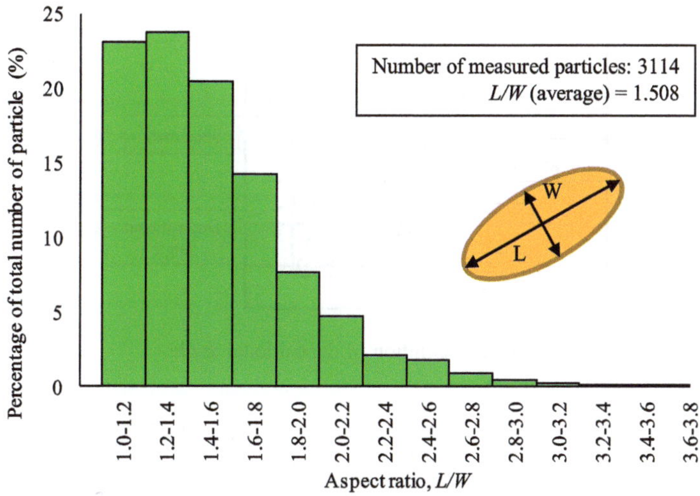

Fig. 3. Grain shape ratio distributions of Toyoura sand

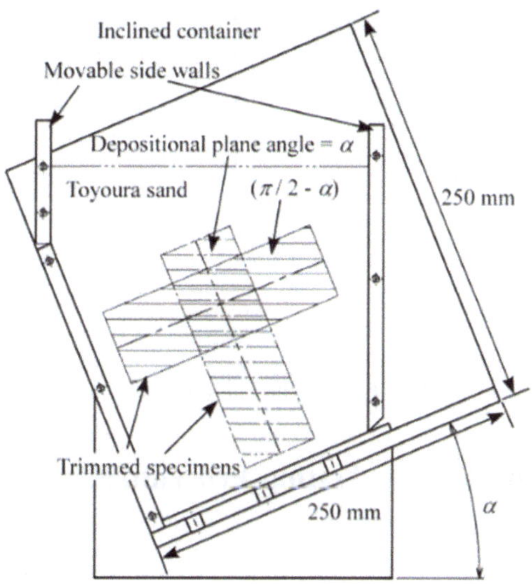

Fig. 4. Inclined container with movable wall to make specimen

height of sand was kept to be constant at about 425 mm using a hanging weight. After the sedimentation of dry sand, the container with full of sand was immersed in water for 2 h. Then, water was removed and remained water in the sand sample was pumped out by using a vacuum pump at −10 kPa. The sand sample made by this method can stand by itself. Eventually, the cylindrical specimen, d = 50 mm in diameter and h = 125 mm in height, was trimmed carefully from the soil block. This method created the dense specimens with almost constant relative densities D_r of more than 88% (as shown in Fig. 7) corresponding to void ratio less than 0.644.

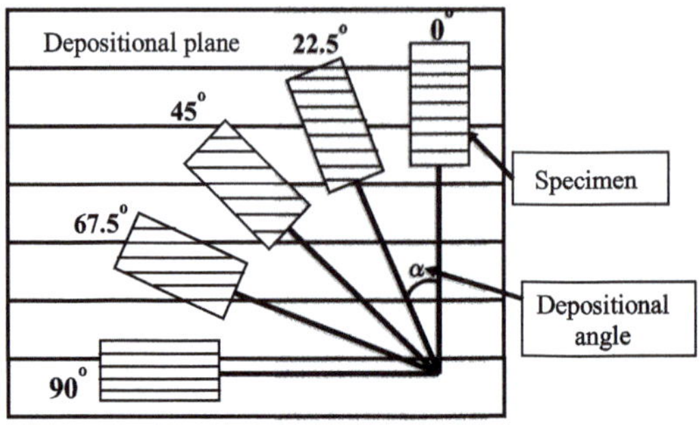

Fig. 5. Definition of depositional angle

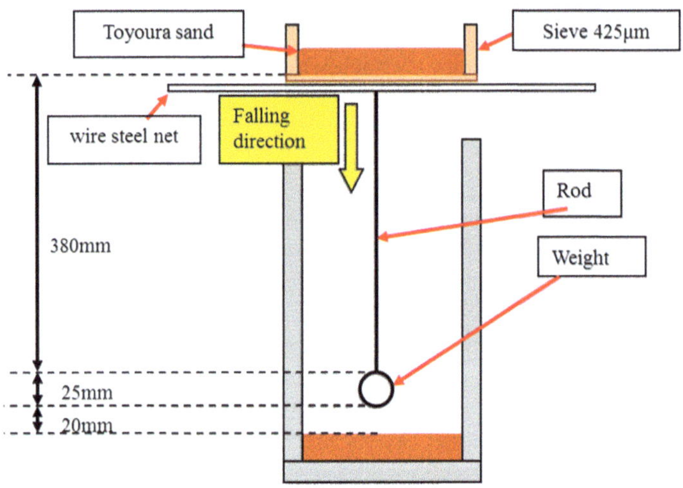

Fig. 6. Air pluviation of sand

3 Testing Apparatus and Testing Procedure

Testing Apparatus
A series of experiments using the triaxial apparatus were carried out to estimate the effects of the inherent anisotropy on shear strength and deformation characteristics.

Bender Element Test
The BE test was equipped in the apparatus to estimate the initial shear modulus G_0 from the shear wave velocity. Figure 8 depicts the schematic diagram of BE mounted on the top cap (works as the transmitter) and on the pedestal (works as the receiver). The dimensions of BE are 2.5 mm in length, 12 mm in width and 1 mm in thickness.

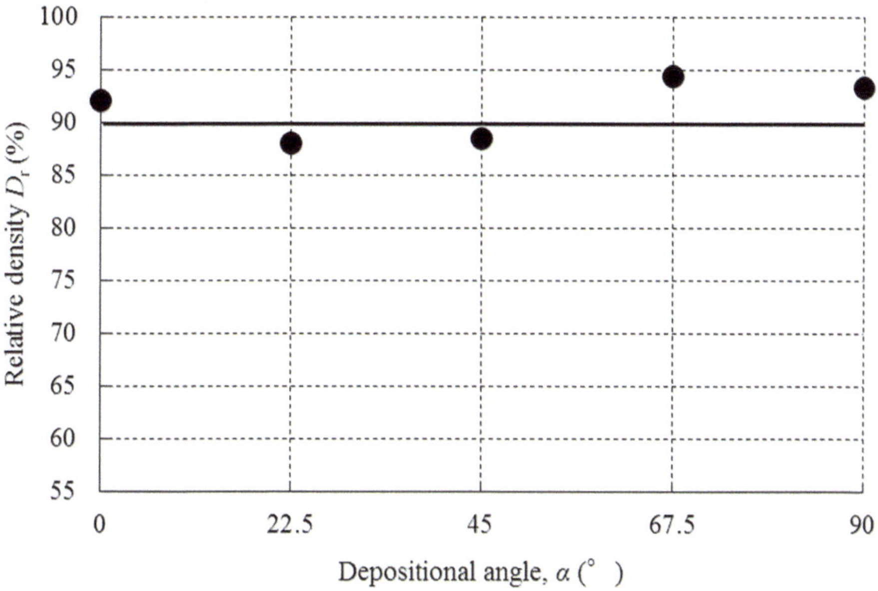

Fig. 7. Relative density in different depositional angles

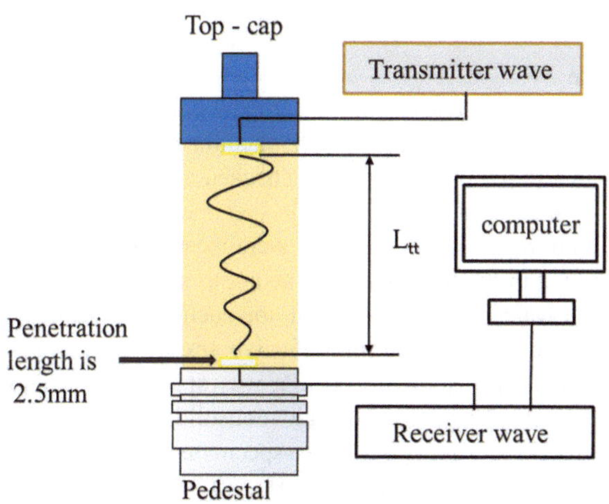

Fig. 8. Schematic of BE testing system

Local Small Strain Test

Local small strain technique was also incorporated in the apparatus to measure the accurate small strain which exhibits elastic region. Figure 9(a) describes how to set up the proximity transducers whose capacity is 2 mm in the triaxial apparatus. Two separated targets were glued directly on membrane to measure for vertical axial displacement. A couple of proximity transducers were fastened on two columns near specimen so as not to contact the targets. The distances between the targets and the proximity transducers can be adjusted from outside of the triaxial cell and the distance between 2 targets was about 80 mm (Fig. 9(b)). For measurement of radial strain, the proximity transducer was installed in the middle part of the specimen using a clamping device (Fig. 9(c)).

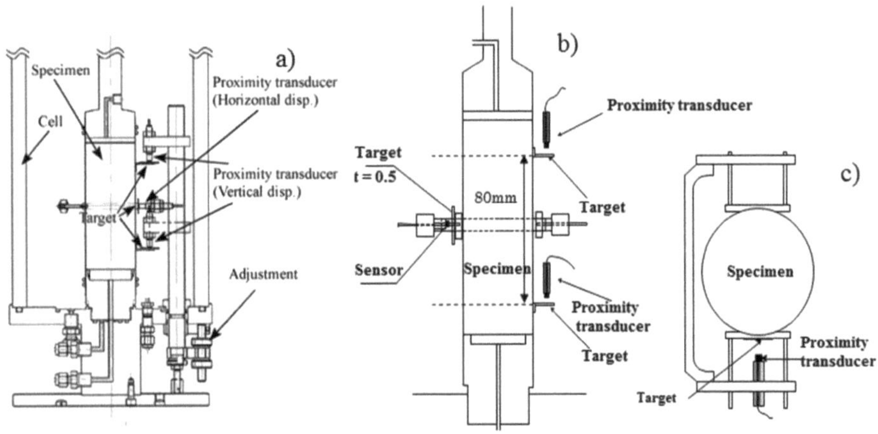

Fig. 9. Schematic of local small strain apparatus

Testing Procedure

The prepared specimen was placed into the triaxial cell. Initially, suction of −20 kPa was applied to the specimen and the double vacuum method was used for saturation process. The B value (Skempton 1954) was checked and made sure that it was greater than 0.98. Keeping suction of −20 kPa inside specimen, cell water was removed from the triaxial cell and dimensions of specimen were measured carefully. The transducers for LSS were installed in the triaxial cell so as not to affect the condition of specimen. After supplying cell water in the triaxial cell, the suction was decreased while the cell pressure was increased so that isotropic effective stress was always kept constant at 20 kPa. Then, the back pressure of 200 kPa was applied to increase the degree of saturation of specimen. Next, in isotropic consolidation process, effective stress $\sigma'_1 = \sigma'_2 = \sigma'_3 = 150$ kPa, was applied to the specimen under drainage condition. Then, BE test was carried out at frequency of 15 kHz, 20 kHz and 30 kHz to determine the shear wave velocity. Finally, monotonic compression was conducted under drain condition at constant cell pressure of 350 kPa and constant axial strain rate of 0.0025%/min for LSS tests and 0.1%/min for tests determined shear strength. The transducers for LSS did not applied for shear strength tests.

4 Testing Result and Discussion

Bender Element Test

The value of the initial shear modulus G_0 can be estimated based on elastics continuum mechanics:

$$G_0 = \rho_t V_s^2 = \rho_t \frac{L_{tt}^2}{\Delta t^2}$$

where shear wave velocity V_s can be found via the propagation distance L_{tt} and propagation time Δt. Start to start method was used as the picking of travel time Δt. ρ_t is bulk density of soil.

Relationships of the initial shear modulus G_0 and depositional angles α are indicated in Fig. 10 where the G_0 increases with an increase of α. Schematic view of sedimentation using elliptic particles is illustrated in Fig. 11. The number of contact points between particles become smaller in vertical direction as the depositional angle increases. The contact points among soil particles lead to the discontinuity and are weaker than the rigid of the particles. As a consequence, traveling time from transmitter source to receiver source changes by the depositional angle. This might be the reason why G_0 increases with an increase of the depositional angle.

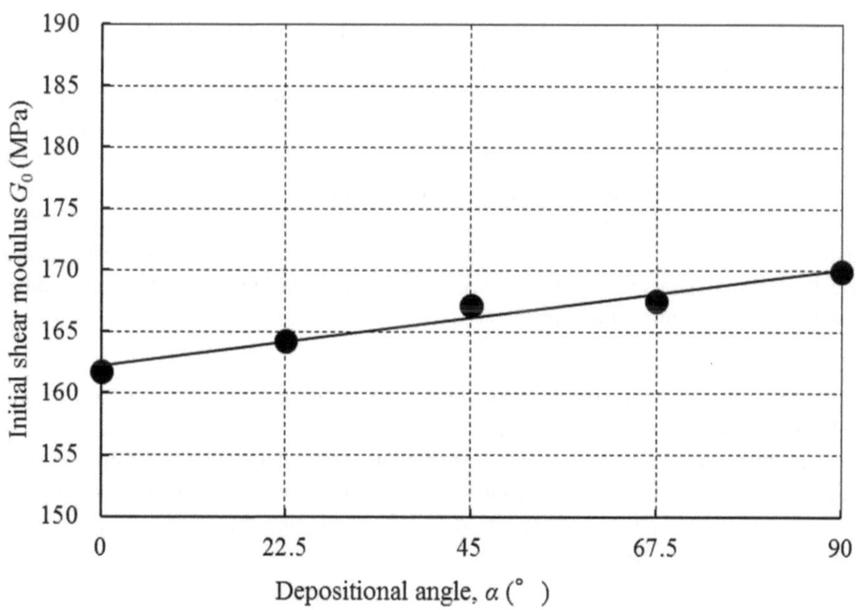

Fig. 10. Effect of depositional angle on initial shear modulus in BE test

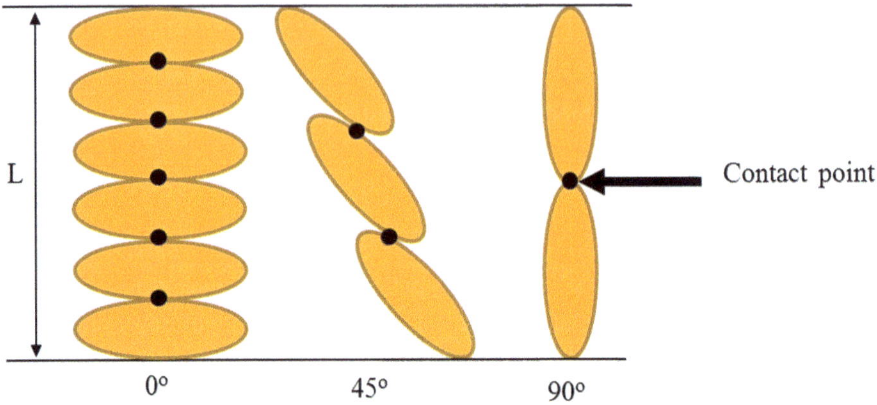

Fig. 11. Sedimentation of particles

Local Small Strain Test

The results of the triaxial tests were interpreted in terms of the deviator stress $q = \sigma'_a - \sigma'_r$, the mean effective stress $p' = (\sigma'_a + \sigma'_r)/3$ and shear strain $\varepsilon_s = 2/3(\varepsilon_a - \varepsilon_r)$. σ'_a and σ'_r are axial and radial effective stresses, respectively. Secant shear modulus G_{sec} is determined by the following equation; $G_{sec} = q/3\varepsilon_s$.

Shear Strength

As mentioned above, the preparation method used in the study could create dense specimens with the similar relative densities. Dense sand exhibits clearly peak strength and strain softening because of dilatancy induced by interlocking of sand particles. Relationships between the shear strength q_{max} and the depositional angles are summarized in Fig. 12. The depositional angle α clearly affects the shear strength. The shear strength q_{max} becomes smaller when α changes from horizontal to vertical depositional angle. However, according to the plane strain tests (Oda et al. 1978; Tatsuoka et al. 1986), results are not coinciding with those of the triaxial tests. That is lowest strength appears in the direction of the "potential" strain localization plane based on the Mohr-Coulomb failure criterion.

Deformation Characteristic

Figure 13 shows that the secant shear modulus G_{sec} remains almost constant in very small strain of less than approximately 10^{-3}%. In this range of strain, the greater secant shear modulus is obtained in $\alpha = 90^{\circ}$ than that obtained in $\alpha = 0^{\circ}$. From the shear strain of 10^{-3}% to 10^{-1}%, the secant shear moduli have a downward tendency and reach the opposite trend at shear strain of 10^{-1}% comparing the secant shear moduli at shear strain less than 10^{-3}%.

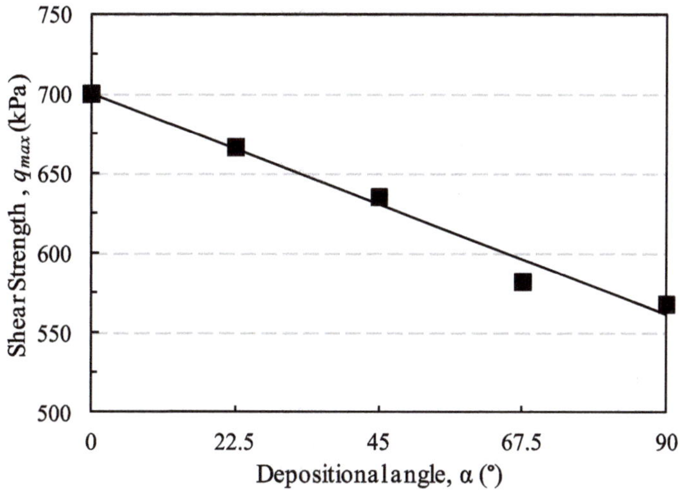

Fig. 12. Effect of depositional angle on shear strength

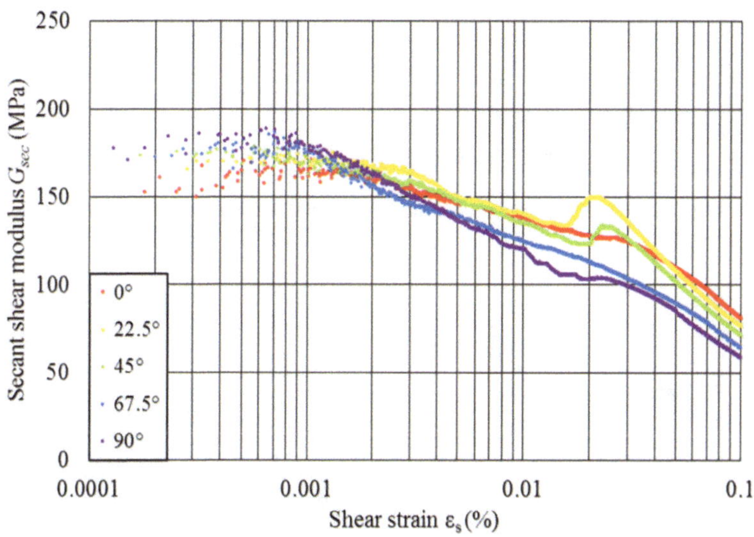

Fig. 13. Effect of depositional angle on secant shear modulus

Elastic Behavior of Reconstituted Sand

In small shear strain less than 10^{-3}%, soil is assumed that it behaves linearly and elastically. Figure 14 shows the relations between the initial shear modulus G_0 obtained from LSS tests and the depositional angles α. The same tendency with the BE tests (Fig. 10) was obtained from the LSS tests. The reason is unclear, but the inclination of the line in Fig. 16 obtained from LSS tests is slightly larger than that obtained in BE tests (Fig. 10).

The values of the secant shear moduli were reversed between initial (less than $10^{-3}\%$) and larger shear strain level (Fig. 13). Therefore, elastic regions, where G_0 remains constant, were estimated from Fig. 13. Figure 15 indicates that though the data is scattering, the elastics region has the tendency to shrink with an increase of the depositional angle. While, as mentioned before, the initial shear modulus increase with an increase of the depositional angle.

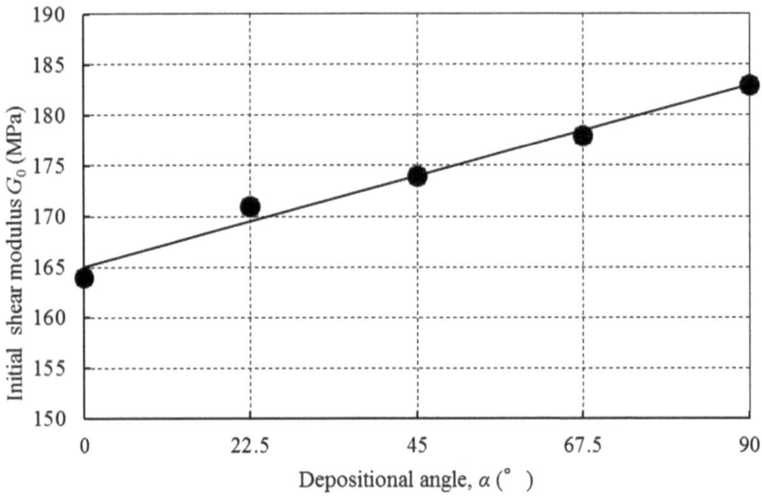

Fig. 14. Effect of depositional angle on initial shear modulus in LSS test

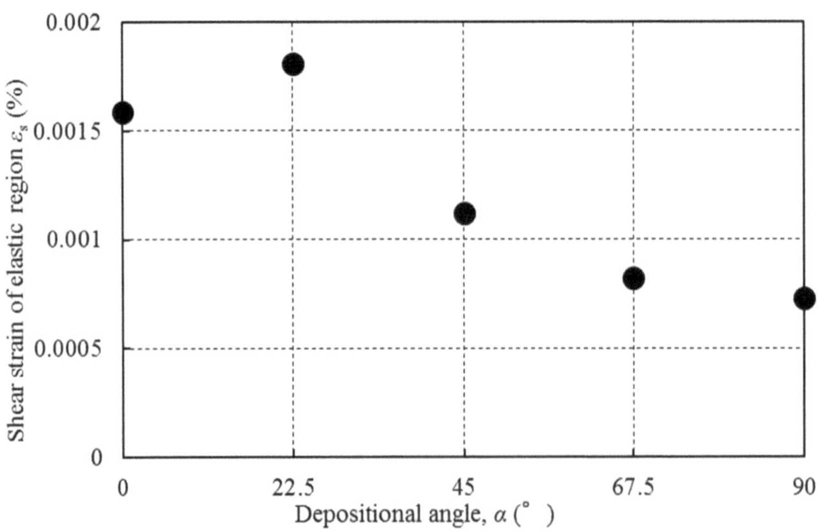

Fig. 15. Effect of depositional angle on elastic regions

The reason for the difference of the elastic regions can be explained by the difference of sand particles' skeleton (structure). In the case of $\alpha = 90^{\circ}$, as shown in Fig. 11, the number of contact points is small and rigid (particle) area is large in vertical direction. The contact points are weaker than the rigid of the particles. Then, waves can travel faster in the case of $\alpha = 90^{\circ}$. However, due to the flat and elongated shape of the grain, the soil skeleton is easy to "collapse" and easy to deform during compression comparing with the case of $\alpha = 0^{\circ}$, which has stable contact of particles against deformation and strength. As a consequence, the elastic region and the shear strength become smaller even the initial shear modulus is great.

Comparison Between BE and LSS Tests

Figure 16 shows comparison of the initial shear modulus G_0 estimated from both BE and LSS tests. Although the tendency of both results are the same, the initial shear moduli of LSS tests are slightly greater than those of BE tests. Furthermore, this difference increases with an increase of the depositional angle.

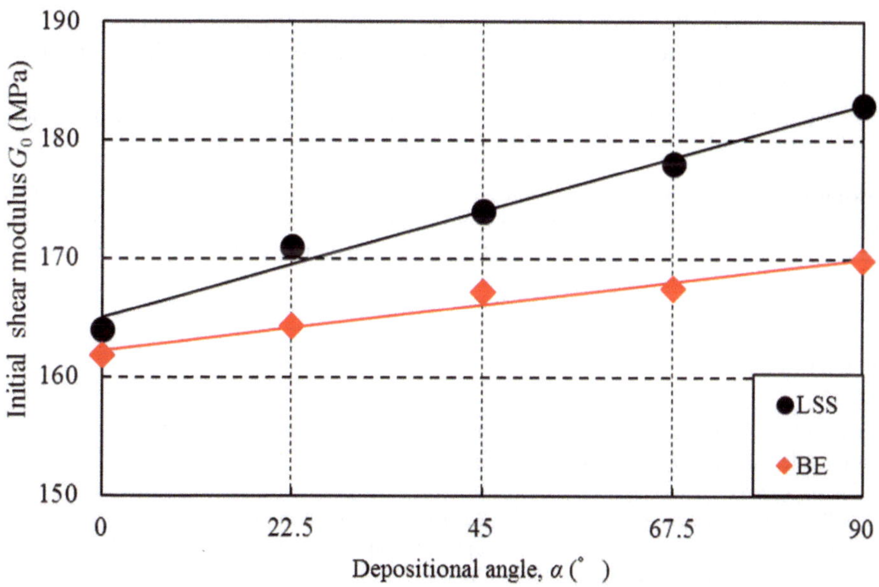

Fig. 16. Comparison of initial shear moduli obtained from BE and LSS tests

The obtained results in this research have the opposite trend compared to the data from other researchers. Chaudhary et al. (2003) studied the shear stiffness of dense Toyoura sand in hollow cylinder apparatus (HCA) employed proximity transducers and triaxial apparatus with BE tests. This data shows that the G_0 values from HCA are slightly smaller than those from BE test in isotropic stress condition and almost the same in anisotropic stress condition. The similar tendencies are obtained by Tsutsumi et al. (2006) and Wicaksono et al. (2008).

G_0 was estimated at smaller strain of 0.001% in this study. However, when G_0 was estimated at 0.001% strain on the study, both results from BE and LSS tests are almost the same. Therefore, further experiments are necessary to verify this kind of difference that induced by the different measuring methods.

5 Conclusions

The inherent anisotropy was investigated using the specimens of Toyoura sand with different depositional angles in terms of shear strength, deformation characteristics and elastic region. From the experimental results, the following conclusions are obtained:

(1) The shear strength clearly decreases with an increase of the depositional angle. However, the initial shear modulus slightly increases with an increase of the depositional angle. The maximum change of the initial shear modulus induced by anisotropy is about 10%.
(2) Although the initial shear modulus increases with an increase of the depositional angle, the elastic region reduces with an increase of the depositional angle. As a result, contradictory relation is generated between the shear strength and the initial shear modulus.
(3) The initial shear moduli in LSS tests are slightly greater than those in BE tests.

Acknowledgments. This study presented in this paper was performed with the support of current and former graduated students at Nagaoka University of Technology, Japan. The authors are very thankful with their great helps and high cooperation to rise the quality of this paper.

References

Arthur, J.R.F., Menzies, B.K.: Inherent anisotropic in sand. Géotechnique **22**(1), 115–125 (1972)
Casagrande, A., Carillo, N.: Shear failure of anisotropic material. J. Boston Soc. Civ. Eng. **31**(4), 74–87 (1944)
Chaudhary, S.K., Kuwano, J., Hayano, Y.: Measurement of quasi-elastic stiffness parameters of dense Toyoura sand in hollow cylinder apparatus and triaxial apparatus with bender elements. Geotech. Test. J. ASTM **27**(1), 23–35 (2003)
Gasparre, A., Nishimura, S., Minh, N.A., Coop, M.R., Jardine, R.J.: The stiffness of natural London clay. Géotechnique **57**(1), 33–47 (2007)
Guo, P.: Modified direct shear test for anisotropy strength in of sand. J. Geotech. Geoenviron. Eng. **134**(9), 1311–1318 (2008)
Hight, D.W., Gasparre, A., Nishimura, S., Minh, N.A., Jardine, R.J., Coop, M.R.: Characteristics of the London Clay from the Terminal 5 site at Heathrow Airport. Géotechnique **57**(1), 3–18 (2007)
Jovičić, V., Coop, M.R.: The measurement of stiffness anisotropy in clays with bender element tests in the triaxial apparatus. Geotech. Test. J. ASTM **21**(4), 3–10 (1998)
Ochiai, H., Lade, P.V.: Three-dimensional behavior of sand with anisotropic fabric. J. Geotechn. Eng. **109**(10), 1313–1328 (1983)

Oda, M.: Initial fabrics and the relations to mechanical properties of granular material. Soils Found. **12**(1), 17–37 (1972)

Oda, M., Koishikawa, I., Higuchi, T.: Experimental study of anisotropic shear strength of sand by plane strain test. Soils Found. **18**(1), 25–38 (1978)

Pennington, D.S., Nash, D.F.T., Lings, M.L.: Anisotropy of G_0 shear stiffness in Gault Clay. Géotechnique **47**(3), 391–398 (1997)

Skempton, A.W.: The pore pressure coefficient A and B. Géotechnique **4**(4), 143–147 (1954)

Tatsuoka, F., Sakamoto, M., Kawamura, T., Fukushima, S.: Strength and deformation characteristics of sand in plane strain compression at extremely low pressure. Soils Found. **26**(1), 65–84 (1986)

Teachavorasinskun, S., Lukkanaprasit, P.: Stress induced and inherent anisotropy on elastic stiffness of soft clays. Soils Found. **48**(1), 127–132 (2008)

Teng, F., Ou, C., Hsieh, P.: Measurements and numerical simulation of inherent stiffness anisotropy in soft Taipei clay. J. Geotech. Geoenviron. Eng. **140**(1), 237–250 (2013)

Tong, Z., Zhou, S., Fu, F., Dafalias, F.Y.: Experimental investigation of shear strength of sands with inherent fabric anisotropy. Acta Geotechnica **9**(2), 257–275 (2014)

Tsutsumi, T., Builes, M., Sato, T., Koseki, J.: Dynamic and static measurement of small strain moduli of Toyoura sand. Bulletin of Earthquake Resistance Structure Research Center, University of Tokyo, No. 39, pp. 91–103 (2006)

Wicaksono, R.I., Tsutsumi, T., Sato. Y., Koseki, J., Kuwano, R.: Laboratory wave measurements on Toyoura sand and Hime gravel. Bulletin of Earthquake Resistance Structure Research Center, University of Tokyo, No. 41, pp. 91–103 (2008)

Wong, R.K.S., Arthur, J.R.F.: Induced and inherent anisotropy in sand. Géotechnique **35**(4), 471–481 (1985)

Yamada, Y., Ishihara, K.: Anisotropic deformation characteristic of sand under three dimensional stress conditions. Soils Found. **19**(2), 79–94 (1979)

Wireless MEMS-Based In-Place Inclinometer-Accelerometer Array for Real-Time Geotechnical Instrumentation

Victoria Bennett[1](✉), Tarek Abdoun[1], Kathleen O'Meara[2], Matthew Barendse[3], and Thomas Zimmie[1]

[1] Rensselaer Polytechnic Institute, Troy, USA
{bennev,abdout,zimmit}@rpi.edu
[2] CB&I, The Woodlands, USA
kathleen.omearall@gmail.com
[3] New York State Department of Transportation, Geotechnical Engineering Bureau, Albany, USA
mbarendse@dot.state.ny.us

Abstract. The continued functionality of distributed levee systems worldwide is critical for the millions of people who live behind these structures and the further millions that depend on clean drinking water supplies and agriculture protected by flood-control infrastructure. Maintaining these systems has been an increased concern in the wake of natural disasters such as Hurricane Katrina in 2005. The variability of properties within geotechnical systems makes predictions of soil behavior extremely difficult, especially when soil models are not calibrated with field measurements. Accordingly, a need has arisen for a system capable of in situ, real-time monitoring of levees, embankments, and other earthen structures. The work presented herein highlights the development of new sensing technologies for use in a framework to monitor, manage and ensure the safety of geotechnical infrastructure. MEMS (Micro-Electro-Mechanical Systems)-based in-place inclinometer system, Measurand's ShapeAccelArray (SAA), is now established as a sensing tool for simultaneous measurement of 3D soil acceleration and 3D ground deformation up to a depth of one hundred meters, with an accuracy of ±1.5 mm per 30 m. Each sensor array is connected to a wireless sensor node to enable real-time monitoring as well as remote sensor configuration. This system can also include digitally integrated pore pressure measurement in the form of vibrating wire piezometers equipped with microprocessors (called SAAPs). The SAAPs convert vibrating wire data to digital data downhole, and they integrate easily into the SAA system. In situ testing was conducted in several levee sections in New Orleans, LA, USA. The sites were also instrumented with other sensors from Geokon, providing values for comparison. The preliminary testing indicates the suitability of this new multi-parameter system for inclusion in a multi-scale monitoring and health assessment framework, which will be implemented in New Orleans, LA in the near future.

Keywords: Real-Time monitoring · Geotechnical asset management · Levees

J. Wasowski et al. (eds.), *Engineering Geology and Geological Engineering for Sustainable Use of the Earth's Resources, Urbanization and Infrastructure Protection from Geohazards*, Sustainable Civil Infrastructures, DOI 10.1007/978-3-319-61648-3_6

1 Introduction

The health and state of the aging and overburdened civil infrastructure in the United States has been subjected to renewed scrutiny over the last few years. The American Society of Civil Engineers reports that this state threatens the economy and quality of life in every state, city and town in the nation. As one example, the United States Army Corps of Engineers noted in early 2007 that nearly 150 United States levees pose an unacceptable risk of failing during a major flood (ASCE 2008). Losses associated with failures of soil systems continue to grow in the United States and elsewhere in view of increased development in hazard-prone areas. In order to overcome these problems, the performance of these systems needs to be reliably predicted, and then such predictions can be used to improve design and develop efficient remediation measures. The use of advanced in situ monitoring devices of soil systems, such as the Shape Acceleration Array (SAA) system described in this paper, and the development of effective system identification and model calibration is essential to achieve these goals.

The answer to this challenge partly resides in the development of tools for short- and long-term health monitoring of existing civil infrastructure along with data reduction tools of systems identification and inverse problems. The knowledge gained from this monitoring and analysis would aid in planning for maintenance and rehabilitation of these infrastructure systems and will improve the design, construction, operation and longevity. This paper presents an autonomous sensor array that is capable of measuring in situ deformations and accelerations up to a depth of one hundred meters. The frequency and spatial abundance of data made available by this sensor array enables tools for the continuous health monitoring effort of critical infrastructure under a broad range of static and dynamic loading conditions.

The following sections present: (1) a brief description of the SAA technology, and case histories of SAA applications at (2) a bridge replacement site in New York, and (3) New Orleans levees. The developed SAA, SAA with pore-pressure measurement (SAAP), and the real-time data they provide constitute a major step in the direction of establishing long-term monitoring and analysis tools capable of providing a realistic picture of large deformation response and pending failure of soil and soil-structure systems.

2 Sensor Description

The concept of the presented MEMS-based, in-place inclinometer-accelerometer instrumentation system is centered on measurements of angles relative to gravity, using triaxial MEMS (Micro-Electro-Mechanical Systems) accelerometers, which are then used to evaluate inclinations (i.e. deformations). The same MEMS accelerometers also provide signals proportional to vibration during earthquakes or construction activities. Three accelerometers are contained in each 30 cm long rigid segment for measuring x, y, and z components of tilt and vibration. The rigid segments are connected by composite joints that are designed to prevent torsion but allow flexibility in two degrees of freedom. The system, called Shape Acceleration Array (SAA), is manufactured by Measurand, Inc and is capable of measuring three-dimensional (3D) ground deformations at 30 cm intervals and 3D acceleration at 2.4 m intervals to a depth of 100 m. The SAA system uses temperature-calibrated MEMS accelerometers within rigid segments connected by composite joints that prevent torsion but allow flexibility in two degrees of freedom. The SAAs are factory-calibrated and completely sealed, requiring no field assembly or calibration. Because each segment of the SAA contains three orthogonal sensors, arrays can be installed vertically or horizontally. The intended array orientation does not need to be specified prior to installation. Orientation can be selected in the software. The sensor arrays are transported to the jobsite on an 86 cm diameter reel and can be lowered into vertical, see Fig. 1, or pushed into horizontal, 25 mm casing. The initial shape of the installation, or the absolute deviation of the installation from a virtual vertical or horizontal line, can be immediately viewed on a computer. In the case of near-vertical installations, the vertices correspond to the joint-centers of the array in 3D. For near-horizontal installations, the vertices show vertical deformation only versus horizontal position (Abdoun et al. 2007; Danisch et al. 2004).

Fig. 1. Vertical installation of SAA at field site.

Arrays are constructed by connecting subarrays of eight segments end-to-end. Microprocessors, one per subarray, collect data from groups of sensors and transmit this digital data to the surface using just two communication wires. Because they require only two communication wires, arrays are thin enough to fit into 25 mm casing for installation and are flexible enough to be rolled up on a reel for shipping and storage. Traditional probe inclinometers require guide casing to measure ground deformations. The torsion-restrained joints of SAAs and 3D calculation method associated with the torsion constraint make this grooved casing unnecessary.

Wireless SAA data transmission is possible with the inclusion of an on-site data acquisition system, called a wireless earth station. For the miles of field arrays installed to date, this wireless data transmission has been available within 24 h of the instrument installation. Similar to traditional probe and in-place inclinometers, data from the SAA represents deviations from a starting condition or initial reading. These data are sent wirelessly, over the cellular network, to Measurand's automated server, where data are made available to users through Measurand's viewing software and an Internet connection. Automated SAAs typically collect data once or a few times a day but this collection frequency can be specified by the user and changed at any time, through the same wireless interface used to receive the data.

A full temperature calibration is done on each MEMS sensor individually prior to its inclusion in an array. Measure and has completed a study of the temperature coefficients of the MEMS accelerometers and found that the change in output of the sensor is linear with temperature. Calibration files associated with each SAA allow the automatic calibration for temperature effects in each individual sensor. A digital temperature sensor is included within the SAA near each microprocessor. Thus, each temperature sensor calibrates the MEMS sensors in the eight segments surrounding it. This configuration is deemed sufficient for typical underground applications, as the ground temperature is usually constant below 1.5 m from the surface. A denser construction of temperature sensors would be possible but only necessary if a large temperature gradient is expected across any consecutive eight segments, or if the temperature gradient is in a location where ground deformations are expected (Bennett et al. 2012).

Many geotechnical instruments now contain MEMS devices. The perceived novelty of the SAA system is not the use of this sensor but rather that, when compared to other in-place inclinometer systems, the SAA arrives ready to install and is less vulnerable to shear because of its short segment lengths and more flexible casing. The SAA casing is inexpensive 2.5 cm electrical conduit, not the grooved casing required for the manual inclinometer probe and other in-place inclinometer systems.

2.1 Stability with Time

Data corrections procedures have been used for several decades to remove any long-term drift of the gravity-sensing transducers from the calculated deformation data of probe inclinometers. The use of MEMS accelerometers virtually eliminates concerns of long-term drift in the SAA. MEMS accelerometers are manufactured from pure silicon, using photolithographic methods developed by makers of computer chips. They were originally developed for, and continue to be used in, the automotive industry for airbag deployment. Automotive sensors undergo exhaustive testing in extreme environments and must remain stable for at least 10–15 years. The internal structure of MEMS accelerometers is based on the bending of cantilevered beams of pure silicon, with dimensions less than 1 mm, due to the force of gravity. Deflections of the tiny beams are measured using electric-field technology, also built into the MEMS using photolithographic techniques. The lack of sliding parts and the near-inertness of pure silicon make for a stable sensor technology. Stability over time has been confirmed by data from three SAAs, sampled several times per day, over a period of 1.5 years. The deformations from portions of the arrays known to be in stable soil (below observed shear zones) were monitored over the entire period of measurement. Data in the stable soil were found to deviate from their initial readings by no more than ±1.5 mm (Bennett 2010), including in arrays over 32 m long, see Fig. 2.

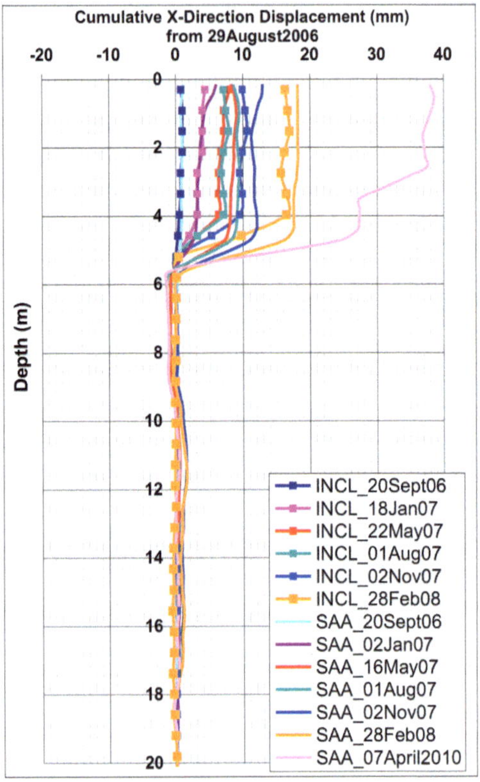

Fig. 2. Comparison of SAA (smooth lines) and traditional inclinometer (lines with markers) data for a nineteen-month monitoring period with updated (4 year) SAA reading.

2.2 Accuracy of Deformation Measurement

The accuracy of the deformation measurement of the SAA is ±1.5 mm per 30 m. This figure can be directly compared to the reported system accuracy of traditional probe inclinometers, ±7.6 mm per 30 m, because both of these specifications are referenced to a virtual straight line, or the initial reading of both instruments. The SAA system accuracy specification was derived empirically from thousands of frames of wireless data over a period of 1.5 years, from three different field locations (Dasenbrock 2011). The MEMS accelerometers mounted in pipes on a mechanical goniometer readout have an absolute accuracy similar to that of conventional inclinometers and excellent 'linearity' over a 45° range. 'Linearity' is actually the match of the arcsine of the output to the tilt in degrees since MEMS accelerometers exhibit a sinusoidal response to tilt (output voltage = sin (angle)). Accuracy, as for traditional probe inclinometers, is best near either pure vertical or pure horizontal (probe inclinometers are usually specified within ±3° of vertical). In the case of MEMS accelerometers, the sinusoidal response causes a very gradual degradation away from the pure pose, due to a decrease in slope of the sinusoid away from its "zero-crossing". At ±10° from vertical, the sine curve slope is degraded by only 1.5%, and at ±45° is degraded by only 29%. Similar numbers apply to deviation from a horizontal position, due to the use of three MEMS accelerometers per segment. Because the SAA is left in place permanently or semi-permanently, it does not have the potential of errors due to mechanical mismatch between a wheeled instrument and a grooved casing, and there is less opportunity for operator error. This advantage is most apparent when the casing becomes extremely distorted.

3 SAA Field Installation at NYSDOT Bridge Replacement Site

The SAA system was installed at a NYSDOT bridge replacement site over the Champlain Canal in upstate New York. A brief site history and description of the installation process of the NYSDOT site is provided below along with a comparison between the horizontal SAA systems and settlement plates. For more information on this site, please see Barendse (2008).

The instrumentation plan for this site included the use of two 32 m (104 ft) long SAAs. One SAA was oriented horizontally and the other vertically to monitor the settlement and the lateral displacement, respectively, of a thirty-meter deep soft clay deposit. Based on soil strength and consolidation testing performed on undisturbed boring samples, prefabricated vertical drains (PVDs) and surcharge fills were used to accelerate the consolidation and strength gain of the clay layer prior to driving piles for the bridge (Abdoun et al. 2008). A 4.5 m high, geosynthetic reinforced earth wall was constructed on the east bank of the Champlain Canal to mimic the load of the proposed bridge abutment, upon which an additional 1.5 m of fill was placed. With the surcharge in place, ground displacements began to accumulate and the lateral displacement of the foundation soils could be discerned. The horizontal SAA was installed after the PVDs had been driven, just prior to the construction of the surcharge embankment, approximately 5 m east of the west-most extent of the embankment and approximately 0.3 m west of a row of PVDs.

Figure 3 shows the settlement profile from the horizontal SAA and a row of settlement plates (SP1, SP2 and SP3). This figure includes the horizontal SAA settlement data shown as a contour plot through February 2008, at which time the array was extracted prior to the pile installation at the site. The settlement plate profile is only provided through August 2007 in Fig. 3, though it can be seen that the shape and values of the profiles from both methods of instrumentation is quite similar. The settlement plates (SP1, SP2 and SP3) experienced greater total settlement, approximately 280 mm versus 225 mm maximum observed SAA settlement. This difference is attributable to the fact that the settlement plates were located approximately 4 m east of the horizontal SAA, a location bearing more of the surcharge load.

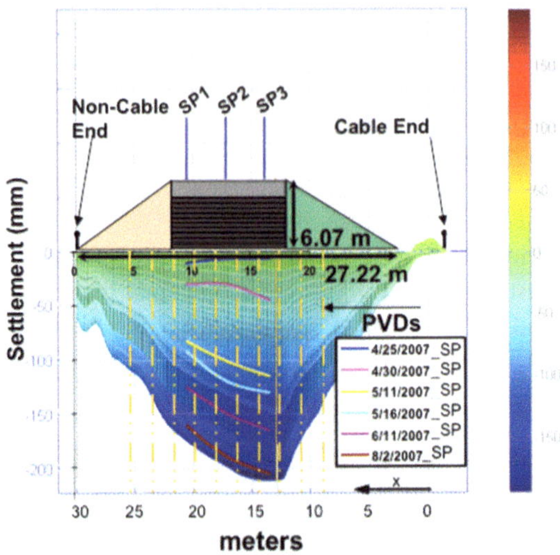

Fig. 3. Settlement profile from horizontal SAA (contour plot) and nearest set of surface settlement plates (discrete lines).

This project demonstrates the usefulness of SAAs for construction monitoring. The information provided by these two SAA systems helped NYSDOT engineers evaluate the effectiveness of the geotechnical treatments utilized at this site, namely surcharge loading and PVDs. Information from the horizontal installation, especially, helped engineers make decisions about the surcharge waiting period during construction. Specifically, the settlement profile beneath the embankment and the lateral squeeze of the underlying soft clay layer were available in real-time. Had it been necessary, the construction schedule at this site might have been accelerated based on interpretation of the real-time settlement and rate of settlement information provided by the horizontal SAA. The same methodologies applied at this site could be used for longer-term monitoring of foundation soils of permanent structures.

4 SAA Field Installation in New Orleans, LA

New Orleans, Louisiana is located in Orleans Parish along the eastern edge of the Mississippi River. This area is extremely susceptible to flooding because the majority of the city is located below sea level in swamplands and it is surrounded by different bodies of water such as Lake Pontchartrain to the north, Lake Borgne to the east which connects to the Gulf of Mexico, and the Mississippi River running through portions of the city (Britsch and Dunbar 2007).

New Orleans is a good testing location because it is a massively developed urban area inhabited by thousands of people whose safety depends on the success of the levee system throughout the city. There are also existing geotechnical subsurface investigations from the US Army Corps of Engineers who responded immediately to Hurricane Katrina. These reports allow for a knowledgeable decision to be made on where to install instruments instead of blindly placing them. This information allows us to further develop our analysis. There are hundreds of miles of natural and man-made levees native to the city of New Orleans allowing for a plethora of potential locations to install the instrumentation. Two sites were chosen for this project: London Avenue Canal and V-Line Levee.

Instrumentation will be highlighted at the London Avenue Canal site, where there are a total of four SAAs, two Shape Accel Array Piezometers, four extensometers, seven piezometers, two differential global positioning systems, one absolute global positioning system, and two InSAR reflectors.

4.1 London Avenue Canal

At this location, all instruments are installed on the east side of the canal (Fig. 4). The SAAs were installed in June 2012 and correspond to three sections on the eastern side of London Avenue Canal. The southernmost cross section of SAA includes one middle instrument; SAA-M1 installed to a depth of −42.5 ft, and is located at Station 75+00. The middle cross section includes two SAAs and two SAAPs. At the middle of the levee, there is one SAA known as SAA-M2 installed to a depth of −46.4 ft and one SAAP known as SAAP-M2 with sensors at depths of −21.3 ft and −41.3 ft. At the toe of the levee, there is one SAA known as SAA-T2 installed to a depth of 46.5 ft and one SAAP known as SAAP-T2 with sensors at depths of −21.4 ft and −41.4 ft. This cross section is at Station 76+60. The northernmost cross section of SAA includes one middle instrument; SAA-M3 installed to a depth of −45.2 ft, and is located at Station 78+40.

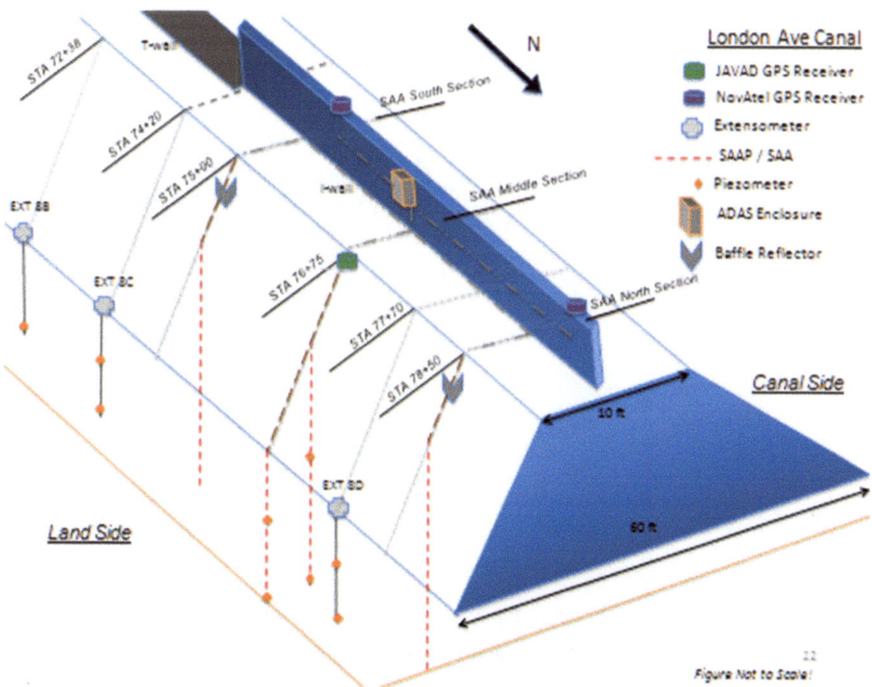

Fig. 4. Layout of instrumentation systems at London Avenue canal.

The extensometers and piezometers, installed in the same borehole, are located on both the east and west side of the canal and were installed in June 2012. There are four sets of extensometers and piezometers at London Ave. The set installed on the west side of the canal includes extensometer, EXT 8A which extends to −40.9 ft, and piezometers, P8 A1 and P8 A2 with elevations of −38.7 ft and −16.6 ft respectively, and is located at Station 72+40. Note that there are two piezometers measuring the piezometric surface elevation at different depths. On the east side of the canal, there are three sets of extensometers and piezometers. At Station 72+38 is the southernmost set, which includes extensometer, EXT 8B installed to a depth of −50.5 ft, and piezometer, P8 B1 with an elevation of −40.3 ft. The middle set is composed of extensometer, EXT 8C installed to a depth of −48.1 ft, and piezometers, P8 C1 with an elevation of −39.1 ft and P8 C2 with an elevation of −19.1 ft, located at Station 74+20. At Station 77+63 is the northernmost set includes extensometer, EXT 8D installed to a depth of −50.5 ft, and piezometers P8 D1 with an elevation of −40.6 ft and P8 D2 with an elevation of −20.6 ft.

The global positioning systems, JAVAD and NovAtel, are located at the same stations corresponding to the Shape Accel Arrays. There are a total of four global positioning systems at London Avenue Canal including one JAVAD system and three NovAtel Systems. The JAVAD unit is at Station 76+75 on the canal wall while the

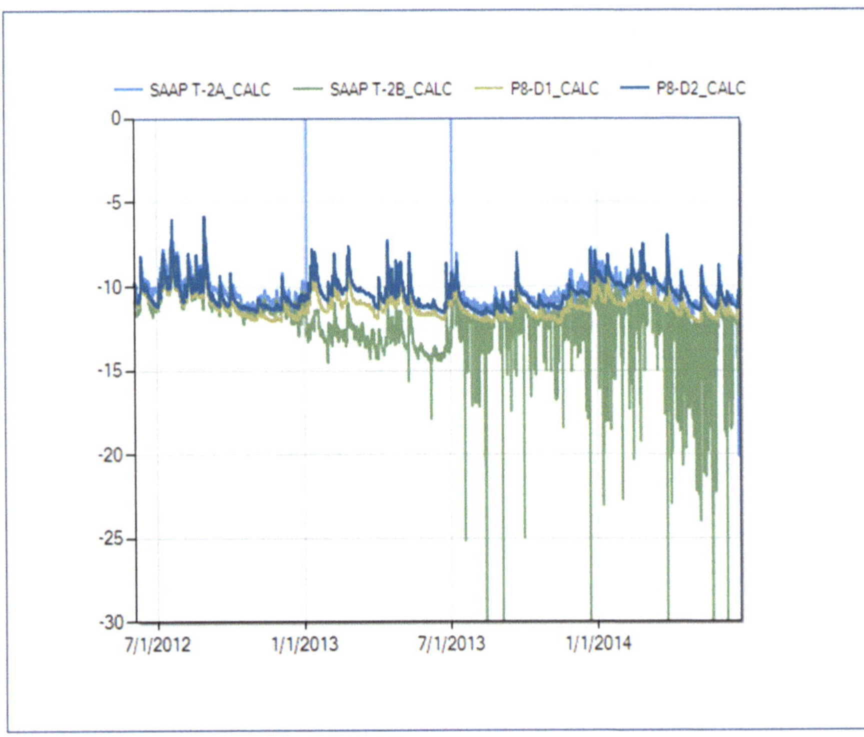

Fig. 5. SAAPZs and Piezometers at London Ave.

NovAtel units are organized in a south, center, and north pattern with the southernmost unit at Station 75+00, the center at Station 76+75, and the northernmost unit at Station 78+50. The Interferometric Synthetic Aperture Radar reflectors are installed on the upper section of the levee slope and are used to reflect the radar signal to monitor ground deformation. There are two InSAR reflectors located at Station 74+20 and Station 77+63.

The Shape Accel Array Piezometers can be compared to the traditional Geokon piezometers. The SAAPZ sensors, SAAP-T2A and SAAP-T2B, and piezometer sensors, P8D1 and P8D2, located closest at about 100 ft apart, were selected for comparison, see Fig. 5. SAAP-T2A displays a piezometric surface elevation of about −10 ft while SAAP-T2B shows an elevation between −10 and −11 ft. The piezometers P8D1 and P8D2 show almost identical measurement of a piezometric surface elevation fluctuation around −10.5 ft. The plot displays unexplained and concerning jumps and extreme fluctuations occurring in the SAAPZ sensors that disparage the quality of data. However, the general trend depicts fluctuation steadily around a depth of 10 ft. This suggests that the water table level is at a depth of 10 ft which is consistent with reports of existing conditions at London Avenue Canal. Because of this, the SAAPZ are viable instrumentation, in a general manner, for monitoring pore pressure but there are technicalities that need to be addressed to ensure the quality of data.

5 Conclusions

This paper presented successful field applications of the Shape Acceleration Array (SAA) and Shape Acceleration Array with Pore Pressure Measurement (SAAPZ) systems at an active bridge realignment site on a 30 m deposit of very soft clay and in two levee sites in New Orleans, LA, which demonstrate how these systems could be utilized for real-time health monitoring of civil infrastructure. Field data comparisons confirm that the SAA/SAAPZ instrumentation system is a viable solution for use in monitoring and assessing levees and other earthen structures. The SAAPs integrate reliable and proven vibrating wire technology developed by Geokon, Inc. into the existing SAA system for identifying accelerations, deformations, and pore pressure measurements. The additional insight into subsurface behavior provided by the new SAAPs is integral in the development of a comprehensive system for monitoring and management of civil infrastructure. The SAAP system shows promise for future use in other applications and for incorporation into early warning management systems for civil infrastructure in New Orleans, The Netherlands, and other locations. Studies are planned to capitalize on the capabilities of the SAA and identification techniques to analyze the long-term mechanisms of flood-control infrastructure.

Acknowledgments. The SAA development was partially funded by the National Science

Foundation (NSF) sensor program (Grant No. CMS-0330043) and is manufactured by Measurand, Inc. This NSF support is gratefully appreciated. The authors wish to thank Lee Danisch, Terry Patterson and Christiane Levesque of Measurand, Inc. for their contribution to these projects, including their efforts toward all the field installations. The authors would like to acknowledge the valuable collaboration with Geocomp Corporation. The authors would also like to express their gratitude to the engineers, drillers and maintenance staff who participated in these field installations, without whom this research would not have been possible.

References

ASCE: Raising the grades: small steps for big improvements in America's failing infrastructure, Report Card for America's Infrastructure, 2 January 2008. http://apps.asce.org/reportcard/2005/actionplan07.cfm

Abdoun, T., Bennett, V., Danisch, L., Barendse, M.: Real-time construction monitoring with a wireless shape-acceleration array system. In: Proceedings of GeoCongress 2008: Characterization, Monitoring, and Modeling of GeoSystems (GSP 179), New Orleans, LA, 9–12 March 2008, pp. 533–540 (2008)

Abdoun, T., Bennett, V., Danisch, L., Shantz, T., Jang, D.: Field installation details of a wireless shape-acceleration array system for geotechnical applications. In: Proceedings of SPIE, San Diego, CA, 19–22 March 2007, vol. 6529 (2007)

Barendse, M.B.: Field evaluation of a mems-based real-time deformation monitoring system. Geotech. News Mag. **26**(1), 41–44 (2008)

Bennett, V., Abdoun, T., Zeghal, M., Koelewijn, A., Barendse, M., Dobry, R.: Real-time monitoring system and advanced characterization technique for civil infrastructure health monitoring. Adv. Civil Eng. J. **2012** (2011). Article ID 870383, Advances in Instrumentation and Monitoring in Geotechnical Engineering Special Issue, Hindawi Publish

Bennett, V.G.: Development of a MEMS-based in-place inclinometer-accelerometer array for monitoring and evaluation of geotechnical systems, Ph.D. dissertation, Rensselaer Polytechnic Institute, Troy, NY (2010)

Britsch III, L.D., Dunbar, J.B.: Geology of the New Orleans area and the canal levee failures. Technical Report from American Society of Civil Engineers. Journal of Geotechnical and Geoenvironmental Engineering, Special Issue: Performance of Geo-Systems during Hurricane Katrina (2007)

Danisch, L.A., Lowery-Simpson, M.S., Abdoun, T.: Shape-Acceleration measurement Device and Method, Patent Application (2004)

Dasenbrock, D., Abdoun, T., Bennett, V.: Real-time structural health monitoring of landslides and geotechnical assets with shape accelarrays. In: Proceedings of GeoFrontiers 2011 – Annual Geo-Institute Conference, Dallas, TX (2011)

Study of Bench Stability in the Phosphate Mine (Algeria)

Mohamed Fredj[1,2(✉)], Abdallah Hafsaoui[1,3], Youcef Khadri[1,4], Boukarm Riadh[2], Radouane Nakache[1,3], Abderrazak Saadoun[1,2], and Kamel Menacer[1,3]

[1] Laboratory: Mineral Resources and Planning, University of Annaba, Annaba, Algeria
fredj_khiero@yahoo.com
[2] Mining and Geology Department, Faculty of Technology, University of Bejaia, Béjaïa, Algeria
[3] Mining Department, Faculty of Earth Science, University of Annaba, BP 12, Annaba, Algeria
[4] Mechanical Department, Faculty of Technology, University of Annaba, Annaba, Algeria

Abstract. The results of analyses performed and aimed to the investigation of the stability of a rock slope in open pit mine in Tebessa (East Algeria). Processes of sliding, toppling and rolling of large blocks, affect this slope. Sloped bench faces that compose the overall slope in the studied case in this paper are evaluated from the stability prospective by applying a variety of stability analysis methods. The adopted approach is based on laboratory tests on understudied samples to designate the mechanical parameters and numerical modeling by implementing the finite element method. The latter, a vital tool to the quantitative determination of deformation mechanisms in large slope instabilities, is used to unravel the uncertainty of mechanical homogeneity properties of the involved materials at the level of discrete meshes in numerical computations that type of application proposes a procedural combination of an assortment of calculation stability methods through three steps. The first step is to determine the mechanical parameters of Kef-Essnoun rock mass by using slope mass rating (SMR) classification scheme, developed by Romana, to depict the strength of an individual rock slope. This system is based on the rock mass rating (RMR) geo-mechanical classification system of rocks, developed by Bieniawski, who refurbished that system with quantitative guidelines to get the rate of influence of adverse joint orientations. The second step is to uses the abacus method to estimate the stability of open-pit mines. The last step is to use a numerical modeling by applying Plaxis 8.2 calculation code. The analyses indicate that the slip problem is the presence of fragile phosphate-marl interfaces at the base of the phosphate layer.

1 Introduction

Slope stability analysis forms an integral part of the opencast mining operations during the life cycle of the project.

J. Wasowski et al. (eds.), *Engineering Geology and Geological Engineering for Sustainable Use of the Earth's Resources, Urbanization and Infrastructure Protection from Geohazards*, Sustainable Civil Infrastructures, DOI 10.1007/978-3-319-61648-3_7

Most of slope movements can be categorized into four categories: planar, wedge, toppling, and circular failures based on their geometrical and mechanical nature of the discontinuity and the conditions of the rock masses (Wang et al. 2013). Planar failures occur along prevalent joints dipping toward the slope with strike near parallel to the slope face (Alemdag et al. 2014). Wedge failures occur along two joints from different families whose intersection dips toward the slope (Jhanwar 2011; Budetta and De Luca 2015). Toppling failures occur along a prevalent family of joints, which dip against the slope with a strike near parallel to the slope face (Krautblatter et al. 2012; Jhanwar 2013). Circular failures occur along slip paths (Hadji et al. 2013; John et al. 2014). All these methods are characterized by the uncertainty of quantitative data within the studied massif. All discontinuities and their properties are generally defined according to their structural analysis of rock surface outcrop. These data are fundamental for stability calculations; however, the interpolation of these observations from the surface to the inside of the rock massive remains problematic.

Rock slope can be classified pursuant to several parameters to measure such as predictors of rock behavior (Abd-Allah et al. 2013). Other systems include Deere's rock quality designation (RQD) system (Deere et al. 1969), Franklin's shale rating system and size-strength system (Franklin 1986) were developed to deduce rock masses in the mining and tunneling industries. Until now, the most commonly used rock classifications are the slope mass rating (SMR) (Romana 1991) and the rock mass rating (RMR) (Bieniawski 1984). These systems include a description of rock mass quality by a quantitative classification, based on a formal set of parameters that are universally adopted. The limitation of these systems is referred to their consideration only geological factors, and these systems are essentially classifying only risk values (Pantelidis 2009; Paul et al. 2012; Jhanwar 2012).

Kinematic analysis is based on the orientation of the combination of discontinuities, the slope face, the upper slope surface, etc. to determine whether certain modes of failure can possibly occur.

Bieniawski geomechanics classification and numerical validation of the related results (Sarkar et al. 2012) are selected to analyze quarries edges stability in Kef-Essnoun open-pit phosphate mine. Since instability expresses a set of factors, leading to transformation of mines geometries during the excavation, in the case of unsupported excavations, two fundamental mechanisms of instability can be meted (Hoek and Brown 1980). A geometrical approach, using the method of graphs, or geomechanical approach, using SMR classification, would have been sufficient to avoid this problem. The main purpose of this work is to provide the engineering operator with a simple methodology to ensure better safety conditions in mines and open pits.

2 Geological Setting and Operating Method

2.1 Geological Setting

Kef-Essnoun phosphate mine is located on the southern flank of Jebel Onk Cretaceous anticline (34.726784 E, 7.895978 N). It is about 7 km in the southeast of

Bir el Ater City in the south of Tebessa province in northeast of Algeria and about 21 km to the Algerian-Tunisian border (Fig. 1). The study site has an area of approximately 250 ha and belongs to the same mining basin than Metlaoui phosphate mine (SW Tunisia) (Mokadem et al. 2014). It contains approximately the half of Algeria phosphate reserves (estimated at 2 billion tons of reserves), (Dass Amiour et al. 2013).

Fig. 1. Location map of the study area

The complexity of the structural directions that characterizes Jebel Onk region is referred to two major compressive deformation phases: a post Eocene Atlases phase marked by (NW–SE) faults system and a post Villafranchian sub-meridian tightening phase (Fig. 2). The structural inversion and the quasi-vertical, relief of the southern edge of the understudied area, reflect the sub-meridian tightening conditioned by a deep accident of the base.

2.2 Operating Method

The mine of Jebel Onk aspires to be an economic hub and a center of development, exploitation, and transformation of phosphate.

The pit is excavated as benches with slope angles of 75° to 85°, 30 m in height, and 10 m in width (Fig. 3). The depth of the base of mine is 70 m. The thickness of the phosphate layer is about 40 m; the barren covering consists of a series of Y Persian dolomite limestone with silex, locally overcome by lutetium limestone, Miocene sands, and quaternary alluvium.

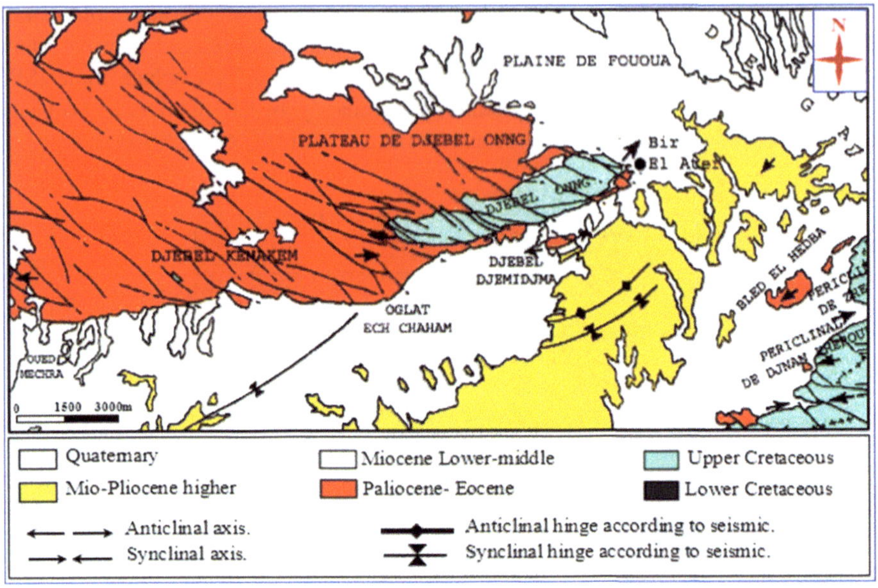

Fig. 2. Geological map of the study area

Fig. 3. Quarry of Kef Essnoun before sliding

3 Instability of Kef-Essnoun Quarry

The high-rise slope, carved in rock masses with a developed geological structure may display severe rotational instabilities (e.g., Kef-Essnoun open-pit mine, located in the south of Tebessa province, NE Algeria). The instability mechanism transpires along the structural plans of rock mass. This instability problem, which has caused the sliding of the quarry edge, is due to the lack of geotechnical studies, which could avert this disaster. On September 8, 2007 at 5:00 a.m., upon sliding of the northeastern side of Kef-Essnoun quarry, a large mass of rock got detached from the massif and fell completely. Depending on the shape of the failure surface, the slip appears at first as a circular sliding (Fig. 4). However, according to the geological cross section (bedding aspect) and the position of the sliding surface (along the bedding planes), the slip is classified as a structural sliding (planar sliding) (Martin 2006). The circular shape of the failure surface is rather due to the straightened layers in the northern part of the quarry profile. This is certainly due to a tectonic activity. According to their physico mechanical properties, phosphates of Jebel Onk are considered as resistant materials. These hard rocks overcome soft-marl rocks, where the joint between these two formations presents a favorable surface for land sliding (Colas and Pilot 1976).

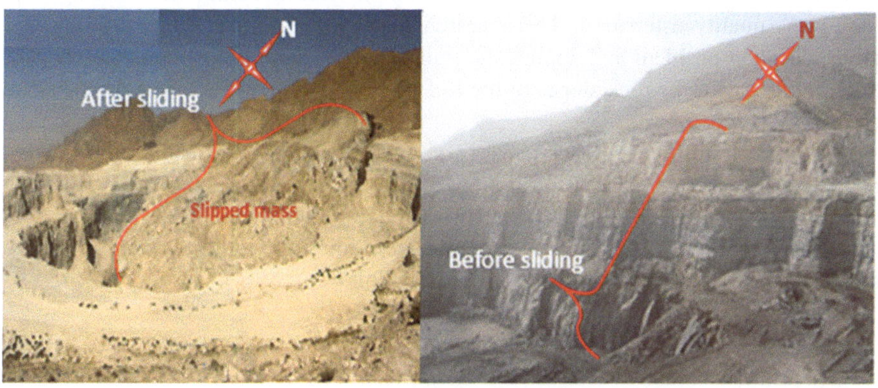

Fig. 4. Geological section of the studied slope (Kef Essnoun) before and after sliding

The mobilized rock mass consists of the phosphate layer (35 m average) mounted on the overburden (35 m on average composed of calcareous marl, dolomite and limestone benches flint Ypresian and Lutetian).

The ensemble slipped on the interface between phosphate and marl schitifiées underlying, almost completely filling the quarry.

This masse waste was attributed to some geological and anthropogenic factors as lithology and exploitation mode.

The geological factors (unrelated to the exploitation mode) are behind the slide:

- The presence of a phosphate marl interface at the base of the phosphate layer. (Plasticity and low coefficient of friction of the training). “A friction angle $\varphi < 15°$ when α, slip plane is equal to the friction angle φ, sliding is possible.
- The quasi- vertical turnaround of this layer.
- In the end the highly tectonized character of the massif at the origin of intense fracturing (North/South fracture, North-West/South – East fractures).

Factors related to mining method are as follows:

- The opening of a front to a height of 70–80 m.
- The step angle of the slope of a can box (70° to 85°).
- The reduction of the safety berm (10 m).
- Vibrations related to the dynamic effect of blasting (unstable equilibrium by mass).

4 Methodology

The main objective of this paper is to present a systematic procedure that combines the rock mass classifications, abacus method, and numerical modeling for assessing slope rock mass quality estimation. This research treats with the rock slope stability as a group decision problem and applies the finite element method for numerical calculation. In this paper, the rock slopes of the Kef Essnoun open-cast mine is selected as the case study.

The first part of the approach consists of a geomechanical estimation of the studied quarries edges stability, based on the SMR classification (Romana 1985). This classification is obtained by applying the RMR system (Bieniawski 1989) and using the adjustment factors that rely on the relation between the slope and the joints, yet another factor depends on the method of excavation as follows:

$$\mathrm{SMR} = \mathrm{RMR}_{\mathrm{basic}} + (\mathrm{F1} \cdot \mathrm{F2} \cdot \mathrm{F3}) + \mathrm{F4}, \tag{1}$$

With

F1: depends on the angle between the direction of the joints and the slope: $\mathrm{F1} = (1 - \sin \mathrm{A})^2$.

F2: depends on the dip angle of joints in the plane failure mode;

F3: is related to the angle between the slope and the dip of the joints (Bieniawski 1976);

F4: is an adjustment factor depending on the slope excavation method.

A SMR index provides representative values of the EM behavior (with the exception of comments shown above). Several empirical relations can be found in the literature providing mechanical characteristics of the EM depending on RMR indices.

The RMR system uses five parameters (their ratings are added to obtain a value of basic RMR). These parameters are as follows:

$$RMR_{basic} = A_1 + A_2 + A_3 + A_4 + A_5 \tag{2}$$

A1: The compressive strength of the intact rock;
A2: The rock quality designation (RQD);
A3: Joint spacing and/or discontinuities;
A4: State of the joints;
A5: Groundwater conditions.

In this research, these characteristics are named equivalent characteristics.

The second part of the approach contains an analytic calculation (abacus method) of the stability of the edges of Jebel Onk phosphate mine. The geometric parameters of Kef Essnoun quarry before slipping happened were as follows: Height of total steps of 70 m; slope angle of 70°; safety berms of 10 m.

The last part of the approach consists of the use of finite element method for calculating quarries stability. Numerical modeling is performed within the calculation code Plaxis 8.2 using an equivalent milieu (EM), assumed to reproduce the behavior of the rock masses and contained low extension discontinuity (discontinuity in the catchment were explicitly taken into account).

The modeling retains the expressions (3), (4), and (5) of the Young modulus, cohesion, and angle of internal friction of Mohr-Coulomb equivalents (Serafim and Pereira 1983; Bieniawski 1979; Trunk and Hönisch 1989).

$$E_{eq}(GPa) = 10^{(RMR-10)/40} \tag{3}$$

$$C_{eq}(kPa) = 5 * RMR \tag{4}$$

$$\Phi_{eq}(deg) = 0,5 * RMR + 8,3 \pm 7,2 \tag{5}$$

5 Results and Discussions

The results of RMR_{basic}, we deduce that the facies Ypresian-Lutetian limestones, limestone phosphate, phosphates and Danian-Montien limestone have a good quality and more resistant. However, the facies of marl is average quality (Table 1).

According to the rock mass classification (SMR), Kef Essnoun quarry is encountering a high probability of sliding risk. A planar failure can occur unexpectedly at the phosphate marl interface (Table 2).

Table 1. Classification of Kef Essnoun rock formation (RMR system)

RMR classification parameters	Layer rating				
	Limestone Ypresian-Lutetian	Limestone phosphate	Phosphate	Marl	Danian-Montien limestone
Compressive strength	7	4	2	2	4
RQD	13	13	17	8	17
Spacing of discontinuities rating	25	25	25	20	25
Condition of discontinuities rating	15	15	15	15	15
Groundwater rating	15	15	15	15	15
RMR_{basic}	75	72	74	60	76
Rock classification	II	II	II	III	II
Description of the rocky Massif	Good rock	Good rock	Good rock	Average rock	Good rock

Table 2. Classification of kef Essnoun rock formation (SMR system)

SMR classification parameters	Layer rating				
	Limestone Ypresian-Lutetian	Limestone phosphate	Phosphate	Marl	Danian-montien limestone
F1	0, 7	0, 7		0, 7	0, 7
F2	1	1	1	1	1
F3	−25	−25	−25	−25	−25
F4	−8	−8	−8	−8	−8
(F1.F2.F3) + F4	−25, 5	−25, 5	−25, 5	−25, 5	−25, 5
SMR	49, 5	46, 5	48, 5	34, 5	50, 5
Classe	II	II	II	III	II
Description	Good	Good	Good	Average	Good
	Description: possible rupture				
	Stability: unstable				
	Rupture: possible break at the layer of marl				
	Method of reinforcement: reprofiling				

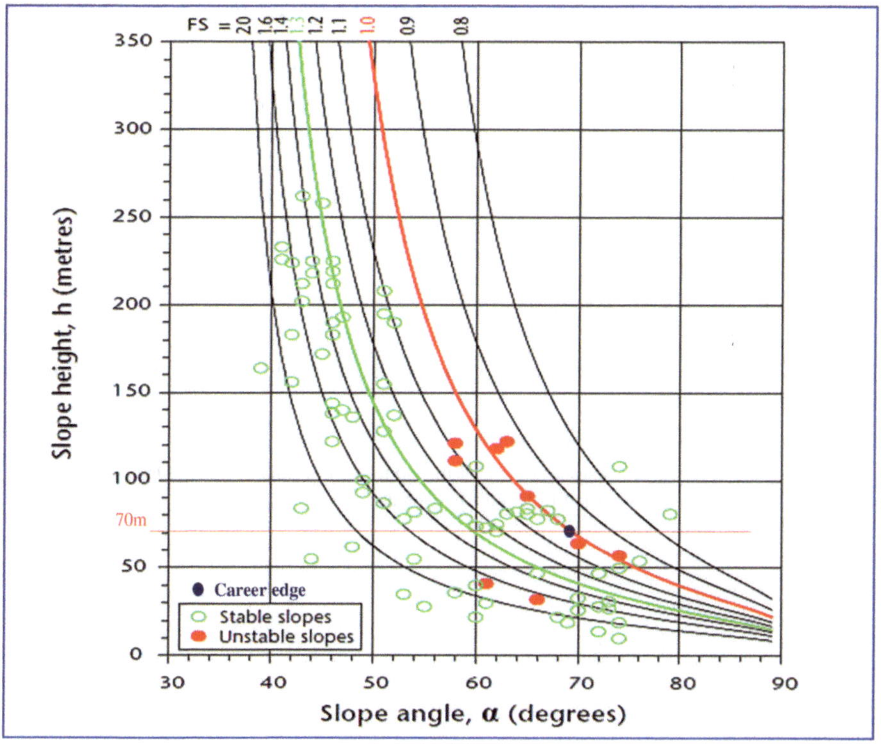

Fig. 5. Safety factor for dry and saturated slope with different slope angles

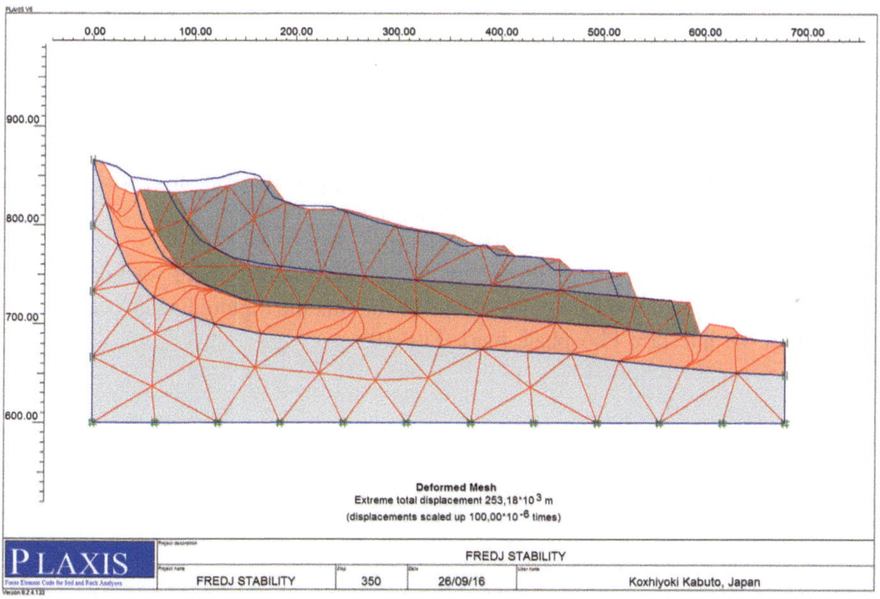

Fig. 6. Model and deformed mesh

The profile of Kef Essnoun quarry (before sliding) shows berms with 10 m of width, an edge of about 70 m of height, height of bench is 30 m and a slope angle of nearly 69°. This profile affects negatively on the stability by increasing the angle of the quarry edge. The use of the abacus (Merrien-Soukatchoff et al. 2001) shows that the quarry edge has a critical stability with a safety factor close to one (represented by a blue diamond in Fig. 5).

The model (Fig. 6), obtained by the numerical simulation with the Finite Element Code for Soil and Rock Analyses (Table 3), Plaxis 8.2, shows the development of a plastic zone at the foot of the slope (Fig. 7).

Table 3. The physical and mechanical properties of the different layers of along the north side of Kef Essnoun mining

Properties	Symbols units	Rock formations				
		Limestone Ypresian-Lutetian	Limestone phosphate	Phosphate	Marl	Danio limestone-montien
Compressive strength	R_c MPa	60	/	25	8	40
Density unsaturated	ρ_{insat} kn/m^3	27	26	21	23	27
Saturated density	ρ_{sat} kn/m^3	27.46	26.5	24.81	24	27.46
Cohesion	c kn/m^2	5400	2300	2300	160	3600
Internal friction angle	φ (°)	37	37	37	16	37
Dilatant	ψ (°)	7	7	7	0	7
Young's modulus	e MPa	27000	24000	24000	1000	27000
Poisson's ratio	ν	0.250	0.28	0.28	0.25	0.25

The resulted shear values 1.77 * 10^3 kN/m^2 (Fig. 9) with a total increment of 6.86 * 10^3 m (Fig. 8) show that this layer cannot support the weight of the overlying formations (Fig. 10). The total displacements are of maximum values at the upper layers, reaching values equal to 253.10 * 10^3 m.

Depending on the result of the safety factor, the finite element method (Plaxis) shows that the slope of Kef Essnoun is unstable and that the rupture line is located more precisely at the interface Phosphate marl (Fig. 10).

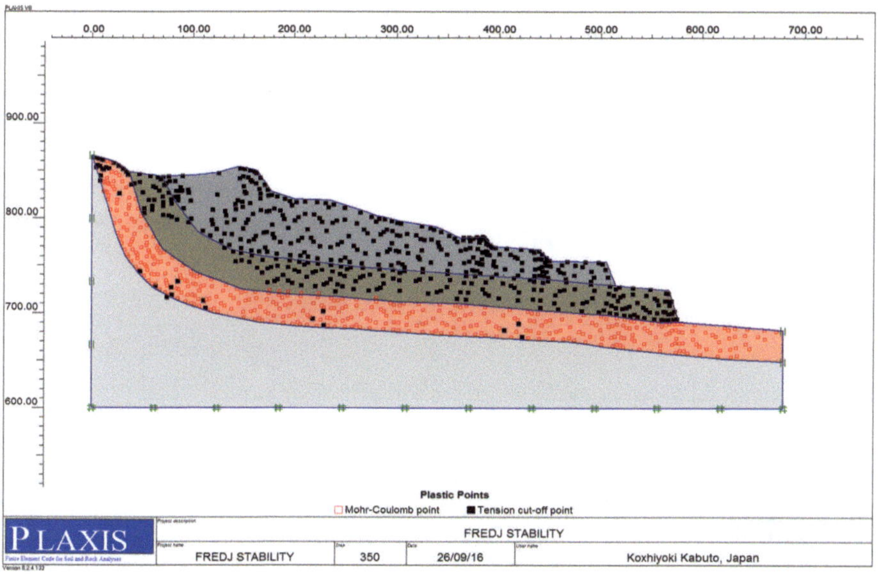

Fig. 7. Plastic points in Kef Essnoun quarry.

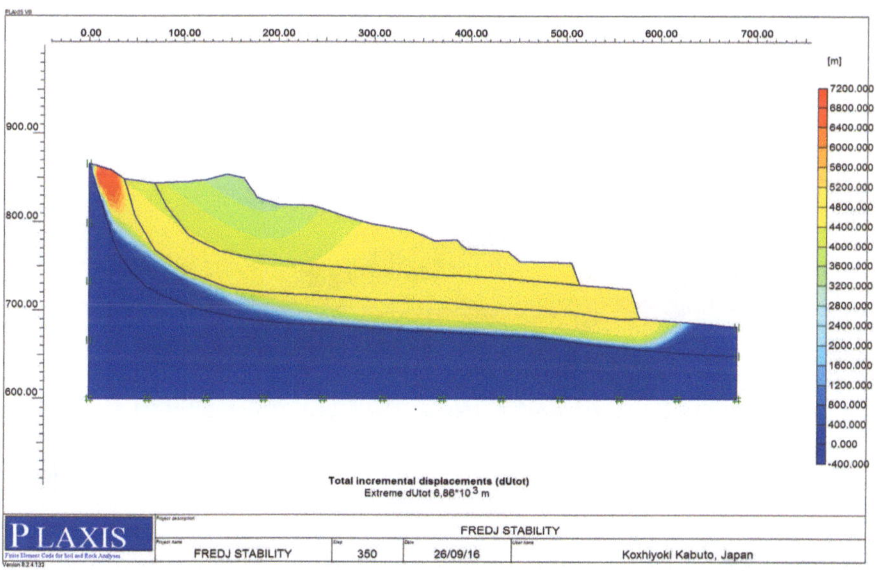

Fig. 8. Total incremental displacements of Kef Essnoun quarry.

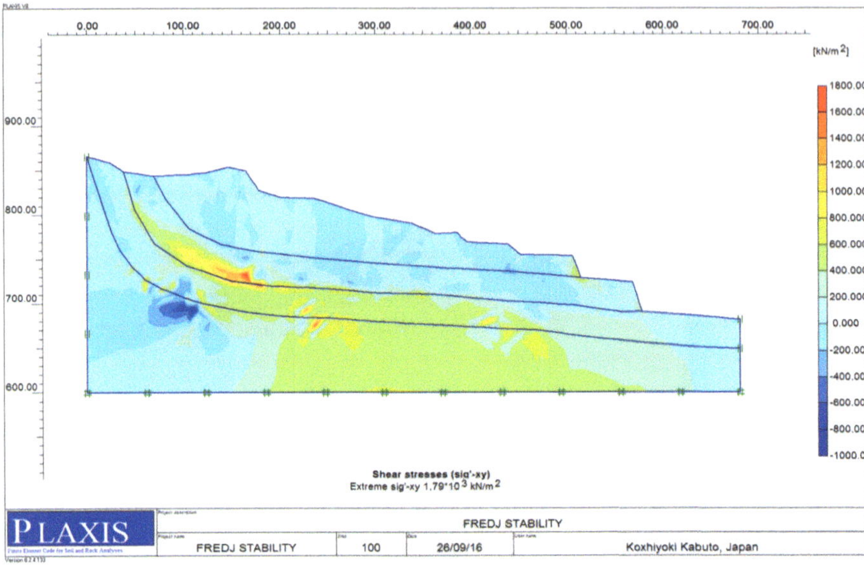

Fig. 9. Shear stresses of Kef Essnoun quarry.

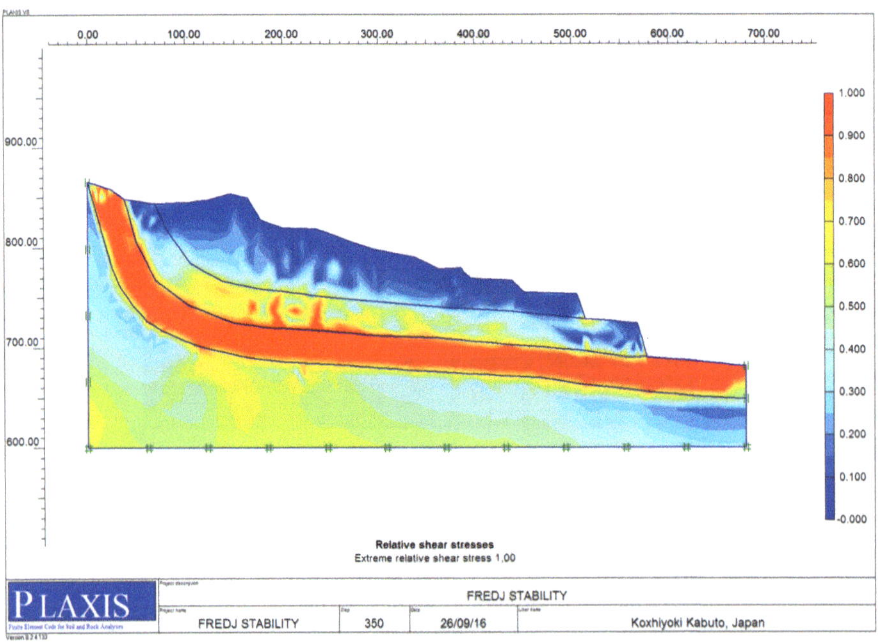

Fig. 10. Relative shear stresses of Kef Essnoun quarry

6 Conclusions and Recommendations

The use of the rock masses classifications techniques (RMR, SMR) has allowed the proposition of a methodology of estimating physic mechanical characteristics at a massive slope scale. Moreover, the numerical modeling is based on these parameters.

The empirical approach can be used only to pre-dimension as a commencement point or a mechanical study. The combined use of several analysis approaches allows reliable results.

Analysis Plaxis software confirmed, on the one hand, the results obtained by the deterministic method, on the other hand, it was possible to see that the displacement (deformation) of the slope Kef Essnoun is important. The presence of a low resistance layer at the base of the overlying layers may present landslide risk, even if the geometric conditions are favorable for the stability of the studied structures.

The results of semi-empirical and analytical methods have shown that the slope of Kef Essnoun is unstable due to the presence of potential slip planes in the marl layer. In these circumstances, it is advisable to leave an abutment at the base of the operating pit (formed by a more resistant rock) to prevent planar sliding. The control of geometric parameters of quarries is an absolute necessity to manage the problems of opencast mines stability.

Acknowledgements. The author would like to thank the staff of the mine laboratory of Annaba University (Algeria). In addition, a special thanks to **A. HAFSAOUI** (Annaba University - Algeria) for her objective comments and corrections.

References

Abd-Allah, A.M.A., El-Sawy, E.K., El-Sayed, S.A.S., El-Fakharani, A., Amin, M.S.M.: Rock slope stability and design in Arafat–Muzdalifa area, Saudi Arabia. Arab J. Geosci. (2013). doi:10.1007/s12517-013-1030-2

Alemdag, S., Akgun, A., Kaya, A., Gokceoglu, C.: A large and rapid planar failure: causes, mechanism, and consequences (Mordut, Gumushane, Turkey). Arab. J. Geosci. **7**(3), 1205–1221 (2014)

Bieniawski, Z.T.: Rock mass classification of jointed rock engineering. In: Proceedings Symposium on Exploration for Rock Engineering, Balkema, Johannesbourg, pp. 97–106 (1976)

Bieniawski, Z.T.: The geomechanics classification in rock engineering applications. In: Proceedings of the 4th Congress International Symposium of Rock Mechanics, Montreux, vol. 2, pp. 41–48 (1979)

Bieniawski, Z.T.: Rock Mechanics Design in Mining and Tunneling, p. 272. Rotterdam, A.A. Balkema (1984)

Bieniawski, Z.T.: Engineering Rock Mass Classifications, p. 251. John Wiley & Sons, New York (1989)

Budetta, P., De Luca, C.: Wedge failure hazard assessment by means of a probabilistic approach for an unstable sea-cliffin. Nat. Hazards **76**, 1219–1239 (2015)

Colas, G., Pilot, G.: Description et classification des glissements de terrain. Bulletin de liaison des laboratoires des ponts et chaussées, Numéro spécial **2**, 21–30 (1976)

Dass Amiour, M., Mezghache, H., Elouadi, B.: The use of three physico-chemical methods in the study of the organic matter associated with the sedimentary phosphorites in Jebel Onk Basin, Algeria. Arab. J. Geosci. **6**(2), 529–541 (2013)

Deere, D.U., Merritt, A.H., Coon, R.F.: Engineering classification of in situ rock: air force systems command. Kirtland Air Force Base, report AFWL-64-144 (1969)
Franklin, J.A.: Size-strength system for rock characterization: proceedings of the symposium of rock characterization to mine design. In: American Society of Mining Engineers Annual Meeting, pp. 11–16 (1986)
Hadji, R., Boumazbeur, A., Chouabi, A.: Geologic, topographic and climatic controls in landslide hazard assessment-using GIS modeling a case study of Souk Ahras region NE Algeria. Quat. Int. **302**, 224–237 (2013)
Hoek, E., Brown, E.T.: Empirical strength criterion for rock masses. J. Geotech. Eng. Div. ASCE **106**(GT9), 1013–1035 (1980)
Jhanwar, J.C.: Slope failures in the open-cast coal mines of Wardha valley coalfield in central India: a study. Rock Mech. Rock Eng. **44**(D5), 635–639 (2011). ISSN 0723-2632
Jhanwar, J.C.: A classification system for the slope stability assessment of opencast coal mines in central India. Rock Mech. Rock Eng. **45**(D4), 631–637 (2012). ISSN 0723-2632
Jhanwar, J.C.: Slope stability issues in the open-cast limestone and iron ore mines in India—case studies. Indian Min. Eng. J. **52**(D7), 8–12 (2013). ISSN 0019-5944
John, L.P., Srikrishnan, S., Verma, C.P., Jhanwar, J.C., Roy, P.P.: Slope stability assessment approach for multiple seams highwall mining extraction. Int. J. Rock Mech. Min. Sci. **70**, 444–449 (2014). ISSN 1365-1609
Krautblatter, M., Moser, M., Schrott, L., Wolf, J., Morche, D.: Significance of rock fall magnitude and carbonate dissolution of rock slope erosion and geomorphologic work on alpine limestone cliffs (Reintal, German Alp). Geomorphism 167,168, 21–34 (2012)
Martin, P.D.: Ces risques que l'on dit naturels, 506 p. Editions Eyrolles, Paris (2006)
Merrien-Soukatchoff, V., Quenot, X., Guglielmi, Y.: Apports de méthodes géomécaniques quantitatives à l'investigation de grands versants instables: application au glissement de la Clapière (Saint- Etienne, Alpes Maritimes) (2001)
Mokadem, N., Hamed, Y., Saad, A.B., Gargouri, I.: Atmospheric pollution in North Africa (ecosystems-atmosphere interactions): a case study in the mining basin of El Guettar-M'Dilla (Southwestern Tunisia). Arab. J. Geosci. **7**(5), 2071–2079 (2014)
Pantelidis, L.: Rock slope stability assessment through rock mass classification systems. Int. J. Rock Mech. Min. Sci. **46**(2), 315–325 (2009)
Paul, A., Singh, A.P., John-Loui, P., Singh, A.: Validation of RMR-based support design using roof bolts by numerical modeling for underground coal mine of Monnet Ispat, Raigarh, India—a case study. Arab. J. Geosci. **5**(6), 1435–1448 (2012)
Romana, M.: New adjustment rating for application of the Bieniawski classification to slopes. In: Proceedings of International Symposium on Rock Mechanics, Mining Civ Works, ISRM, Zacatecas, Mexico, pp. 59–63 (1985)
Romana, M.: SMR classification. In: Wittke, W. (ed.) Proceedings of 7th Congress on Rock Mechanics, vol. 2. ISRM, Aachen (1991)
Sarkar, K., Singh, T.N., Verma, A.K.: A numerical simulation of landslide-prone slope in Himalayan region. Arab. J. Geosci. **5**, 73–81 (2012)
Serafim, J.L., Pereira, J.P.: Consideration of the geomechanics classification of Bieniawski. In: Proceedings of International Symposium on Engineering Geology and Underground Constructions, pp. 1133–1144 (1983)
Trunk, U., Hönisch, K.: Cited at rock mechanics design in mining and tunneling, Bieniawski, p. 183 (1989)
Wang, G., Wu, F., Ye, W.: Stability analysis for toppling failure of unstable rock in three gorges reservoir area, China. Rock Charact. Model Eng. Des. Methods, 431–435 (2013). doi:10.1201/b14917-77

Cracking Mechanism Along the North Batter of Maddingley Brown Coal Open Pit Mine, Victoria, Australia

Lei Zhao and Greg You(✉)

School of Engineering and Information Technology,
Federation University Australia,
University Drive, Mt Helen, VIC 3353, Australia
g.you@federation.edu.au

Abstract. Cracks manifested on the north batter at Maddingley Brown Coal Open Pit Mine, Victoria, Australia in November 2013. The crack opened varies from trace to approximately 150 mm wide was located at about 20 m back from the coal face and extended for approximately 50 m on the eastern side and terminated 10 m away from the access road. Site investigations and remedy measures were implemented immediately after the cracking. This study involves in a three dimensional modeling on the cracking mechanism using finite element method (FEM) encoded in Plaxis 3D software program. From the study, it was found that the initial model based on north batter being stable for many decades tended to lead a circular critical path while the model after overburden removal showed a trend of block sliding. The safety factor of initial north batter was 1.38 through safety analysis while it was decreased to 1.17 for the coal batter after overburden removal. But the simulated shear and tensile strains indicated that the coal batter after overburden removal in fact experienced block failure. Furthermore, the simulated location of cracking was in good agreement with the actual location, and the simulated heave of the coal seam was in good agreement with the experience in Victoria brown coal open pit mining. The observed vertical crack would be a combined action of the overburden removal and the groundwater flow in the unconfined aquifer.

Keywords: Batter stability analysis · Block sliding · Brown coal · FEM · Ground crack · Open pit mining

1 Introduction

Victoria is well-known for its huge quantity of reserved brown coal (approximately 430 billion tones) according to the Department of Economic Development, Jobs, Transport and Resources, Victoria, Australia (2016), which represents 22.6% of the world's recoverable brown coal reported by Australian Atlas of Mineral Resources, Mines, and Processing Centres (2012). The distribution of brown coal in Victoria is shown in Fig. 1. As the main fuel for generating electricity in Victoria, brown coal has been exploited by open cut mining since the early 1920s in the Latrobe Valley (Hutchings et al. 1977). Accompanied with numerous activities of open pit mining in Victoria,

J. Wasowski et al. (eds.), *Engineering Geology and Geological Engineering for Sustainable Use of the Earth's Resources, Urbanization and Infrastructure Protection from Geohazards*, Sustainable Civil Infrastructures, DOI 10.1007/978-3-319-61648-3_8

slope failure has become an inevitable problem that can cause considerable environmental damage and loss of lives and properties. In past decades some failures were observed in this region, such as two slips along clay seams below the coal occurred at Yallourn North Open Cut Mine in 1950 and 1957 (Learmonth 1985); tension cracks appeared in the southern wall of Maddingley Brown Coal Open Pit in 1994 reported by Golder Associates (2011); a failure occurred in the northeast batter at Yallourn East Field Mine on 14th November 2007 (Mining Warden 2008); an embankment constructed to divert the Morwell River across the Yallourn mine failed during an extreme rainfall in 2012 (Hepburn 2014). Learmonth (1985) reported the overburden had a rotational circular slip while block and wedge failures formed in brown coal due to opened joints at Morwell Open Cut Mine.

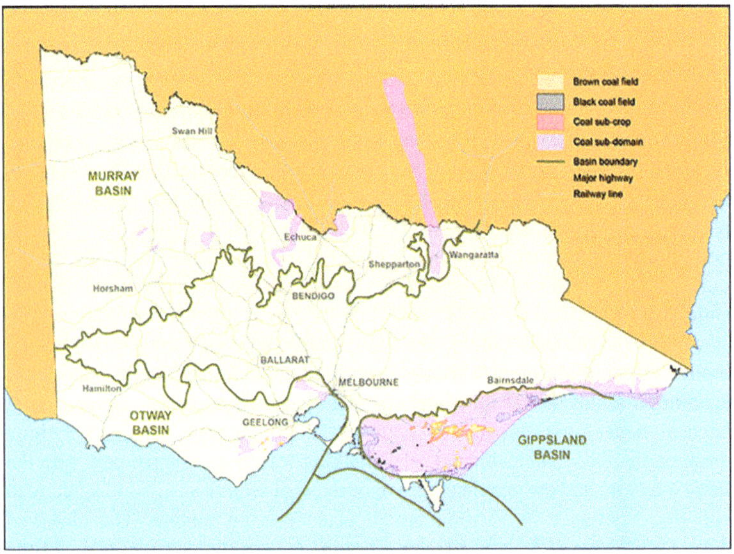

Fig. 1. Brown Coal Distribution in Victoria, Australia (Department of Economic Development, Jobs, Transport and Resources, Victoria, Australia 2016)

Making clear the slope failure mechanism is significant to predict and avoid the future ground movement in open pit mining. However, it is a considerably complicated geotechnical issue as a mass of factors need to be taken into calculation. Nowadays numerical simulation has been widely used for solving such a complex issue, considering its powerful data processing capacity and user friendly interface. In literatures, limit equilibrium method (LEM) and finite element method (FEM) are often adopted to conduct slope stability analyses (Chang and Huang 2015; Chen et al. 2014; Maiorano et al. 2014; Djerbal and Melbouci 2015; Jamsawang et al. 2015; Usluogullari et al. 2015). Compared to the more advanced FEM built in strength reduction method, the LEM has the disadvantages such as time consuming and assumption-dependence (Ozbay and Cabalar 2015). When it comes to the development of geo-engineering model, 2D models are used more than 3D models due to its simplicity. However 2D model was regarded as conservative by Duncan (1996) compared to the gradually mature application of 3D simulation. In most situations a 2D model can only be used to make horizontal and vertical sections but three dimensional model of slope stability analysis can give more realistic result and the sliding mechanism analysis can be more comprehensive, especially when the case with non-uniform resistance distribution (Chang and Huang 2015). In addition, 3D model enables 3D visualization and details such as estimated sliding volume. In this paper, the latest version of Plaxis 3D finite element software was employed to investigate the formation mechanism of cracks emerged on the north batter of Maddingley open cut mine, Victoria, Australia.

2 Mine Site and Geological Background

The Maddingley Brown Coal (MBC) open pit is a small scale mine and is located 60 km northwest of Melbourne, Victoria, Australia (Fig. 2). It has been mined as an open pit since 1946 and the mined pit has been used as a landfill since 1978. The Maddingley coal seam at Bacchus Marsh is a stratum of the Werribee Formation. The coal seam is generally underlain by the gravel layer or undifferentiated sediments and is conformably overlain by fluviatile clays, sands and minor gravels, marine silts, and by Quaternary basalts to the south and south-west. The coal seam is typically 35 km in length, 10 km to 15 km in width and from 35 m to 60 m in thickness. The coal is rarely fractured and is predominantly a dark brown earthy variety of lignite with little impurity known as a small percentage of the matrix maybe silty or sandy (Golder Associates 2006).

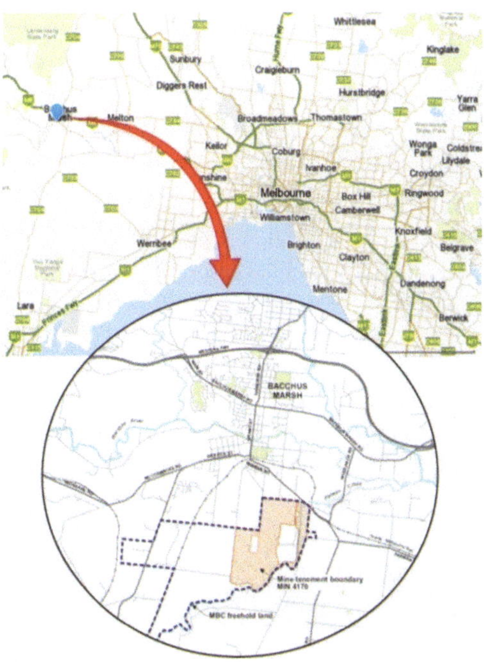

Fig. 2. Relative location of Maddingley brown coal site (URC Australia 2013)

There are three main hydrostratigraphic units underlying mine site (Golder Associates 2014a, b). The most upper aquifer, Fyansford Formation, consisting mainly of silts, sands and clays, with a thickness of 5–20 m, is considered as an unconfined aquifer and includes overlying Quaternary sediments. The Maddingley Coal Seam (Upper Werribee Formation) is regarded as an aquitard separating the upper unconfined Fyansford Formation from the lower confined Werribee Formation, due to its high clay content and low hydraulic conductivity (10^{-8} m/s). The lower Werribee formation consists of mainly silty soils and sands and it is known as a confined or semi-confined aquifer.

3 Crack Investigation

The north batter, which is in the north of the site, had been stable for many decades without any mining activities. Overburden stripping activities commenced in early 2013. The overburden is approximately from 10 m to 20 m in thickness, consists of

Fig. 3. North batter coal face (pre-crack, overburden in place) with locations of monitoring system established immediately after the observation of crack (URS Australia 2013a, b)

Fig. 4. Crack on the coal seam (URS Australia 2013a, b)

sand, sandstone and clay and is stripped ahead of coal mining. The mining advances in a northerly direction. The northern coal batter is approximately 25 m high in a single bench, at a slope angle > 80°. At the western end of the coal bench, there was a 5 m high toe buttress to provide support to the bench (Fig. 3) (URS Australia 2013a, b).

In November 2013, an east-west striking crack was observed on the top of coal seam approximately 20 m from the crest of the coal face (Department of State Development, Business and Innovation 2013). The aperture varies from trace to 150 mm wide (Fig. 4) extended 50 m and terminated 10 m away from the access road on the eastern side (URS Australia 2013a, b). The aperture is wider in the east (bottom in Fig. 4) than in the west (top in Fig. 4). The less horizontal movement in the west would benefit from the toe buttress (Fig. 3) (Department of State Development, Business and Innovation 2013). A number of survey markers were installed on north batter on 19 November 2013 (in Fig. 3) to monitor the displacements of the batter moving upwards, eastwards and northwards. The monitored result has been reporting weekly (Golder Associates 2014a, b).

4 Numerical Analysis and Results

4.1 Establishment of 3D Geological Model

A 3D geological model was developed based on an aerial survey map with 1 m interval of contour made in June 2012. The 3D model was initially created using AutoCAD Civil 3D and then exported to Plaxis 3D (Table 1). The dimensions of model are 200 m long,

Table 1. Soil properties adopted in Plaxis 3D simulations

Layer	Unit	Fyansford Formation	Werribee Formation	Intact MBC	Broken MBC	Fill
Drainage type		Drained	Drained	Undrained A	Drained	Drained
Unsaturated unit weight	kN/m^3	16	17	5	5	15
Saturated unit weight	kN/m^3	19	20	11.5	11.5	18
Poisson's ratio	-	0.3	0.3	0.27	0.27	0.3
Cohesion	kN/m^2	9	5	150	20	5
Friction angle	°	27	30	30	30	30
Young's modulus	kN/m^2	130	95	37	37	15
Dilatancy angle	°	0	0	0	0	0
Horizontal permeability k_x	m/d	0.0086	0.0086	0.00086	0.1	0.1
Horizontal permeability k_y	m/d	0.0086	0.0086	0.00086	0.1	0.1
Vertical permeability k_z	m/d	0.00086	0.00086	0.00086	0.1	0.1

100 m wide and 109 m high, covering the crack in the northern batter. The model consists of Fyansford Formation (overburden, 87 m–109 m), Intact Maddingley Brown Coal (50–87 m), Werribee Formation (0–50 m), Fill (55–60 m) and Broken coal (50–55 m) (Fig. 5a).

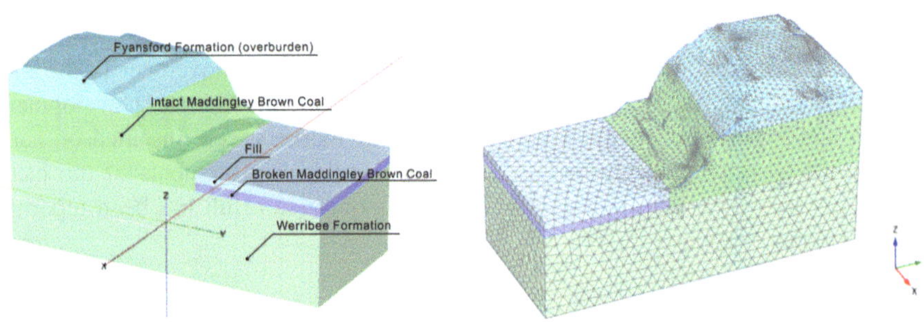

(a) **(b)**

Fig. 5. 3D Model: (a) Geological model, (b) Plaxis 3D FEM mesh (in very fine mesh)

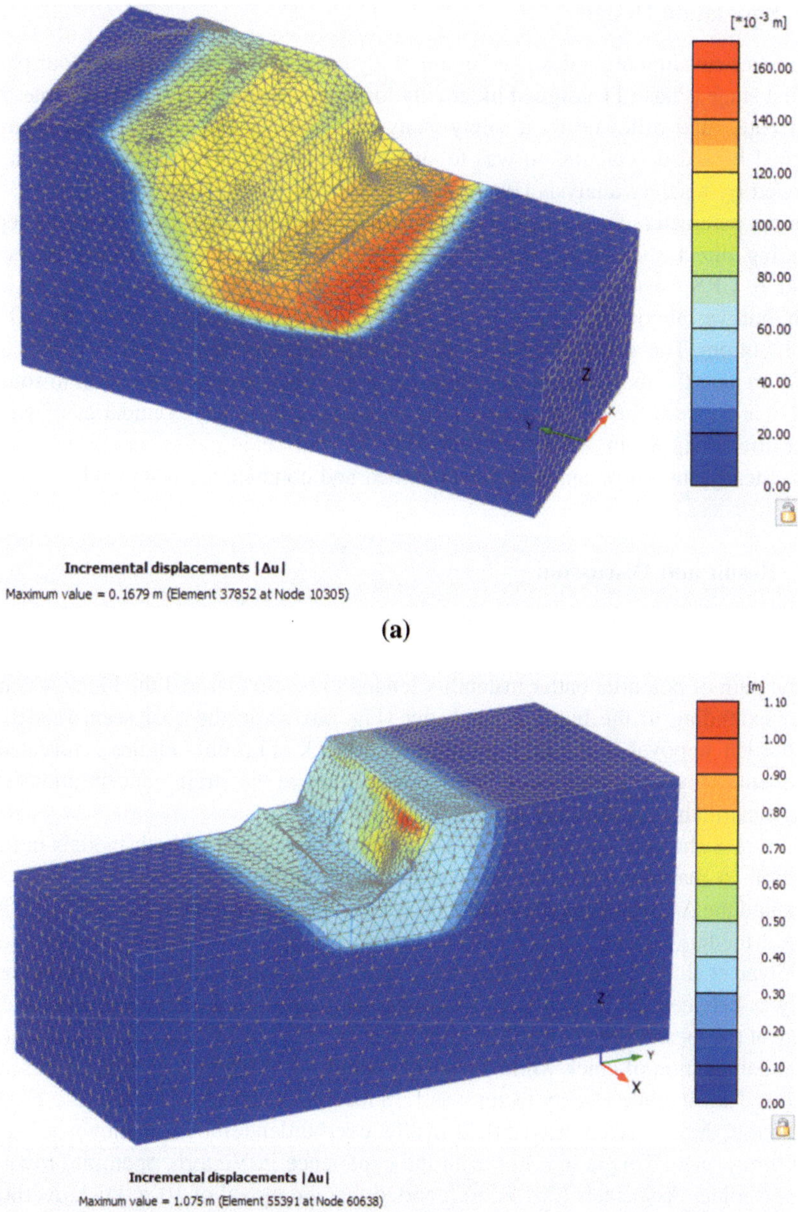

(a)

(b)

Fig. 6. Incremental displacement resulted from safety analysis: (a) Before overburden removal; (b) After overburden removal

4.2 Simulation Design

The numerical simulation designed in this study consists of two stages and four phases: the first stage (Phase 1) assigned by gravity loading was to define the initial state of the north batter and followed by a safety analysis (Phase 2); the second stage (Phase 3) assigned by plastic calculation was to simulate the effect of overburden removal and followed by a safety analysis (Phase 4).

Input parameters for the numerical simulations were adopted from technical reports including direct shear tests, triaxial tests and permeability tests (Golder Associates 2014a, b; URS Australia 2014). The groundwater table in the north batter was set at RL 91 m that was about 4 m above the top surface of coal seam while it was RL 60 m at the pit bottom. The assumed groundwater flow surface was from RL 91 m in the coal to RL 59 m near the toe of the batter; the water flow direction was from north to south, in other words, from the north batter to the pit bottom. Thus the boundaries of groundwater flow along Xmin, Xmax and Zmin (Fig. 6) were set as closed and the others were open. Steady state flow analysis was assumed and conducted in this study.

4.3 Result and Discussion

Figure 6 shows the incremental displacements of Stage 1 and 2, which were the initial model and the model after overburden removal, respectively. In the initial model, the critical path of potential batter instability tended to be circular and the block was much larger extending to the back of the batter (Fig. 6a) while the coal seam model after overburden removal was likely to slide as a block (Fig. 6b). Figure 7 revealed the incremental deviatoric strains of Stage 1 and 2, where the strain concentrations along the critical paths are standing out.

Figure 8 showed the incremental Cartesian normal strain of both models in the yy direction. A maximum $\Delta\varepsilon_{yy} = 0.34$ was mobilized on the exposed vertical face of the block and the $\Delta\varepsilon_{yy}$ between 0.04 and 0.06 was on top and bottom of the coal seam in the Stage 2 model (Fig. 8b), in contrast, the maximum $\Delta\varepsilon_{yy}$ was much lower for Stage 1. Tolooiyan et al. (2014) reported that the range of tensile strain limit for Victorian brown coal was between 0.005 and 0.01. This suggests that the coal seam experienced tensile failure at the location of about 20 m from the crest in Stage 2 that was in agreement with the actual location of crack while the tensile failure would not occur in the coal seam in Stage 1. The simulated safety factor was 1.38 for Stage 1 and 1.17 for Stage 2. From the simulation, the coal seam heaved 0.22 m after overburden removal as shown in Fig. 9 in a 50 times scale. This is in a line with the experience in Victoria open pit brown coal mines (Golder Associates (2014a, b) reported that a removal of 10 m thick overburden could generate a heave of about 0.15 m and probably 70% of which has occurred during the process of overburden removal and the remaining is expected to happen over a few years).

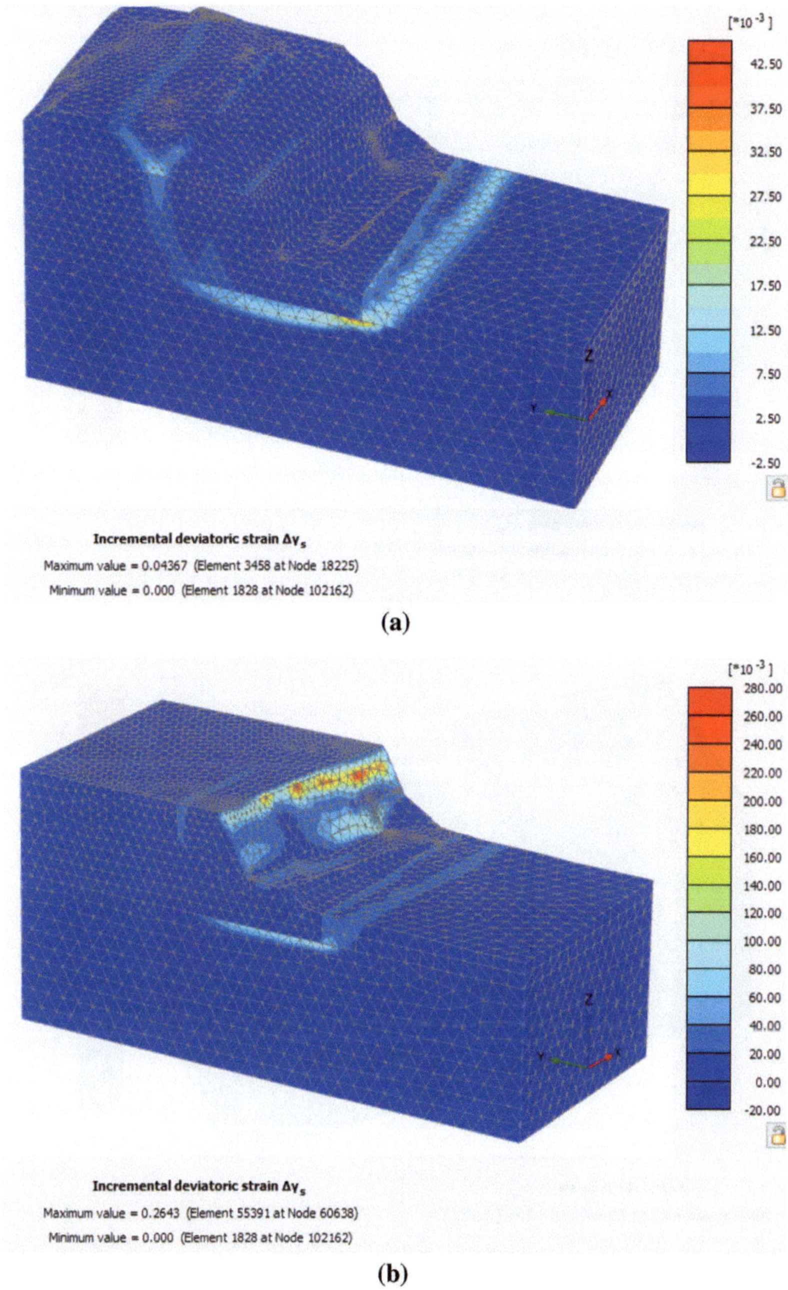

Fig. 7. Incremental deviatoric strains resulted from safety analysis: (a) Before overburden removal; (b) After overburden removal

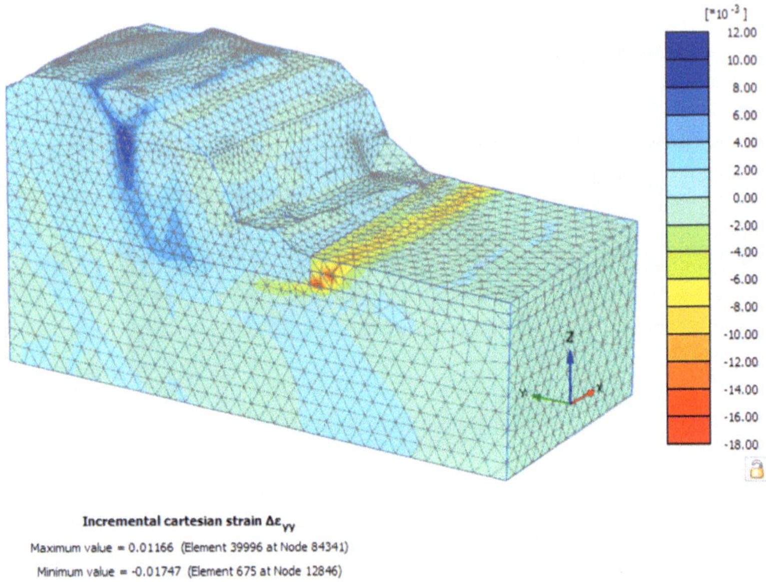

(a)

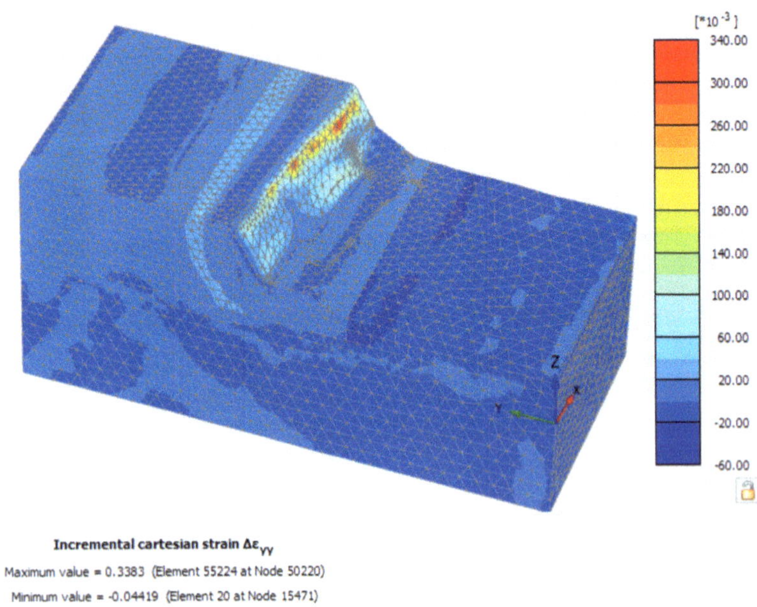

(b)

Fig. 8. Incremental cartesian tensile strains resulted from safety analysis: (a) Before overburden removal; (b) After overburden removal

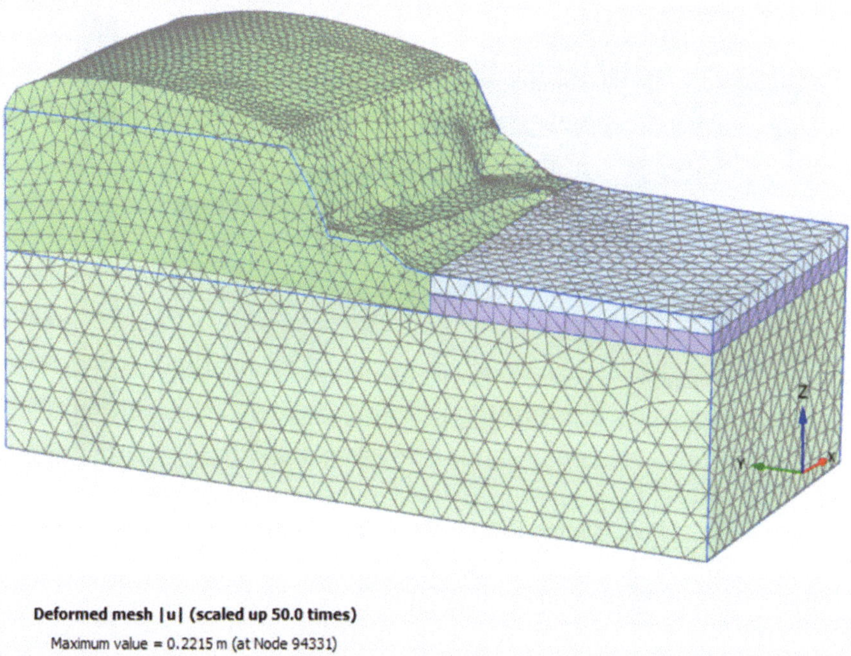

Fig. 9. Heave of coal seam, emerging after overburden removal (showing in a 50 times scale)

From the FEM analyses, the initiation of the vertical crack on top of the coal seam was induced by overburden removal, and a block failure of the north batter could eventuate along the critical path passing the crack (Figs. 6b and 7b). In addition, groundwater played an important role to aggravate the movement and instability of the block when the crack was filled with water, which would provide an additional driving force towards the pit. The steady state pore pressure resulted from steady-state flow analysis is shown in Fig. 10. Figure 11 demonstrated the plastic points and the tensile failures of north batter after overburden removal. A fact that cannot be neglected is the possible existing joints in the coal seam. Steeply dipping tight joints filled with white clay were observed (Golder Associates 2011), and a prominent joint was observed at the south eastern coal face. However, this study did not consider the effect of pre-existing joints and clay within coal seam of the north batter.

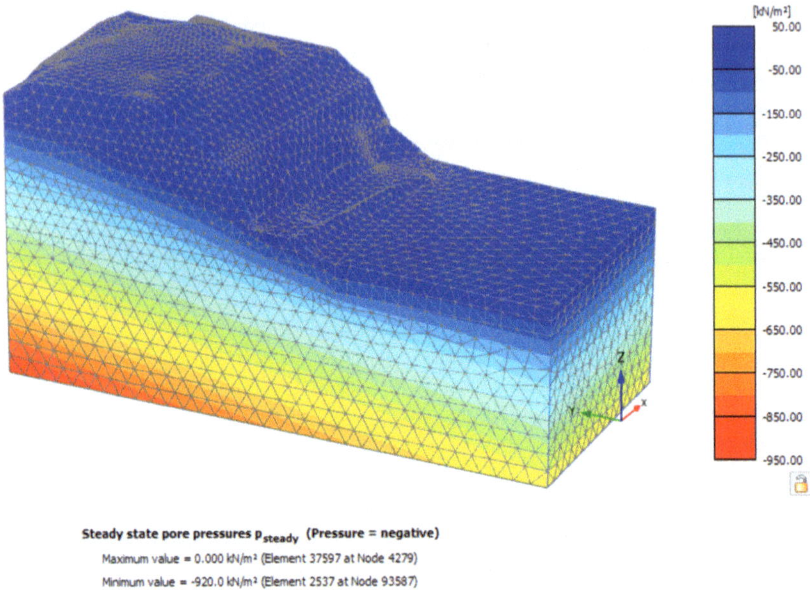

(a)

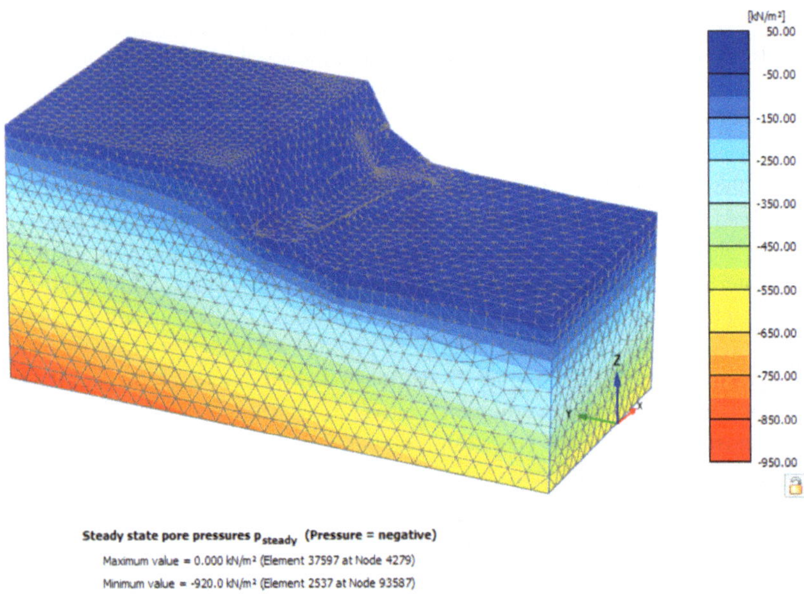

(b)

Fig. 10. Steady state pore pressure resulted from steady-state flow analysis: (a) Before overburden removal; (b) After overburden removal

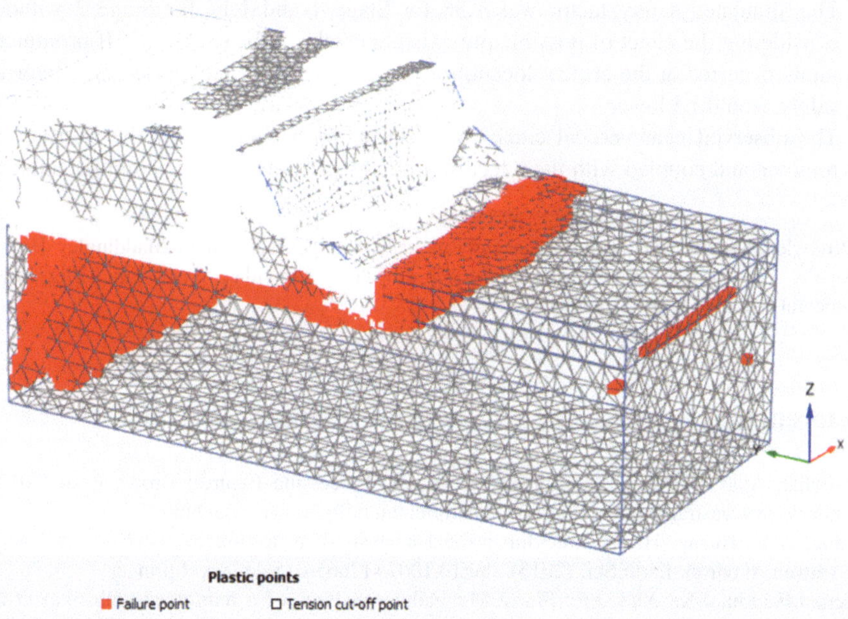

Fig. 11. Plastic points and the tensile failures of north batter after overburden removal related to plastic calculation of Phase 2

5 Conclusion

A 3D numerical model was established using Plaxis 3D FEM program to investigate the formation of the cracks on the north batter at Maddingley Brown Coal Open Pit Mine, Victoria, Australia. From the study, the following conclusions can be drawn.

1. The Stage 1 model of the north batter, which was stable for many decades, tended to lead a circular critical path while the Stage 2 model after the overburden removal showed a critical path of block sliding. The batter displacement of Stage 2 is far larger than the displacement of Stage 1 that means the excavation activity has a noticeable effect on the batter displacement.
2. The simulated potential location of tension crack in the coal seam in the Stage 2 model was about 20 m from the crest that was in agreement with the actual location. The simulated heave of the coal seam after overburden removal was in good agreement with the experience in Victoria brown coal open pit mining.
3. In Stage 2 the high tensile strain in the horizontal yy direction occurred in the top of the coal seam and at the bottom of the coal block along the critical path, where also coupled with high shear strain concentrations. But both tensile strain and shear strain are quite small in Stage 1 model.

4. The simulated safety factor was 1.38 for Stage 1 and 1.17 for Stage 2 without considering the effect of possible pre-existing cracks in the coal seam. If prominent joints occurred at the critical location in the coal seam, it is expected the factor of safety would be lower.
5. The observed near vertical crack was believed to be induced by the overburden removal and coupled with the effect of groundwater flow in the unconfined aquifer.

Acknowledgements. The authors sincerely express their appreciation to Maddingley Brown Coal Pty. Ltd. for their support of this research project, in particular, to Mr. Tim Tillig, Environmental, Quality & Safety Officer.

References

Australian Atlas of Mineral Resources, Mines, and Processing Centres: Brown Coal (2012). http://www.australianminesatlas.gov.au/aimr/commodity/brown_coal.html

Chang, K.T., Huang, H.C.: Three-dimensional analysis of a deep-seated landslide in central Taiwan. Environ. Earth Sci. (2015). doi:10.1007/s12665-015-4128-x. Springer

Chen, J.F., Liu, J.X., Xue, J.F., Shi, Z.M.: Failure analyses of a reinforced embankment by strength reduction and limit equilibrium methods considering hardening of soft clay. KSCE J. Civ. Eng. (2014). doi:10.1007/s12205-014-0288-6. Springer

Department of State Development, Business and Innovation: Geotechnical comments on batter movement (2013)

Department of Economic Development, Jobs, Transport and Resources, Victoria, Australia: Lignit/Brown Coal (2016). http://www.energyandresources.vic.gov.au/earth-resources/victorias-earth-resources/coal

Djerbal, L., Melbouci, B.: Numerical modeling of the climate effect on the evolution of the landslide of Ain El Hammam (Algeria). Eng. Geol. Soc. Territory (2015). doi:10.1007/978-3-319-09057-3_220. Springer

Duncan, J.M.: State of the art: limit equilibrium and finite element analysis of slopes. J. Geotech. Eng. **122**(7), 577–596 (1996)

Golder Associates Pty. Ltd.: Mine Work and Operations Plan Revision 2. Prepared for Maddingley Brown Coal (2006)

Golder Associates Pty. Ltd.: Mine Risk Issues Assessment. Prepared for Maddingley Brown Coal (2011)

Golder Associates Pty. Ltd.: Geotechnical Assessment Northern Coal Batter. Prepared for Maddingley Brown Coal (2014a)

Golder Associates Pty. Ltd.: Maddingley Brown Coal Numerical Groundwater Model. Prepared for Maddingley Brown Coal (2014b)

Hepburn, S.: Accidents or bad regulation? Why Victoria's coal mines keep failing. The Australian (2014). http://www.theaustralian.com.au/business/business-spectator/accidents-or-bad-regulation-why-victorias-coal-mines-keep-failing/news-story/eaaba87b0ea5119a28c97393fd1dbd73

Hutchings, R., Fajdiga, M., Raisbeck, D.: The effects of large ground movements resulting from brown coal open cut excavations in the Latrobe Valley, Victoria. Development and Planning Division, Fuel Department, Safety Electricity Commission of Victoria (1977)

Jamsawang, P., Boathong, P., Mairaing, W., Jongpradist, P.: Undrained creep failure of a drainage canal slope stabilized with deep cement mixing columns. Landslides (2015). doi:10.1007/s10346-015-0651-9. Springer

Learmonth, A.P.: Geomechanics Working in the Power Industry. The National Engineering Conference, Melbourne (1985)

Maiorano, R.M.S., Russo, G., Viggiani, C.: A landslide in stiff, intact clay. Acta Geotech. (2014). doi:10.1007/s11440-013-0249-0

Mining Warden-Yallourn Mine Batter Failure Inquiry (2008). http://eagcg.org/common/pdf/Yallourn.pdf

Ozbay, A., Cabalar, A.F.: FEM and LEM stability analyses of the fatal landslides at Çöllolar open-cast lignite mine in Elbistan. Landslides (2015). doi:10.1007/s10346-014-0537-2. Springer, Turkey

Tolooiyan, A., Mackay, R., Xue, J.: Measurement of the tensile strength of organic soft rock. Geotech. Test. J. (2014). doi:10.1520/GTJ20140028

URC Australia Pty. Ltd.: Mine Work Plan Variation, MIN 4701. Prepared for Maddingley Brown Coal (2013)

URS Australia Pty. Ltd.: 14 November 2013 Inspection. Prepared for Maddingley Brown Coal (2013a)

URS Australia Pty. Ltd.: Maddingley Brown Coal-Emergency Buttress. Prepared for Maddingley Brown Coal (2013b)

URS Australia Pty. Ltd.: Hydrogeological Risk Assessment Maddingley Brown Coal Landfill. Prepared for Maddingley Brown Coal (2014)

Usluogullari, O.F., Temugan, A., Duman, E.S.: Comparison of slope stabilization methods by three-dimensional finite element analysis. Nat. Hazards (2015). doi:10.1007/s11069-015-2118-7. Springer

Research Progress on Comprehensive Control Technologies for Abandoned Coal Mine Hidden Disasters in China

Wen Li[1,2,3](✉)

[1] Mine Safety Technology Branch of China Coal Research Institute, Beijing 100013, China
18611184568@163.com
[2] State Key Laboratory of Coal Mining and Clean Utilization, China Coal Research Institute, Beijing 100013, China
[3] Beijing Mine Safety Engineering Technology Research Center, Beijing 100013, China

Abstract. The number of coal mines in China has been reduced from 87000 in 1995 to about 10000 in 2015, some of the closed coal mines have been incorporated into bigger coal corporations, and others become abandoned coal mines. Such abandoned coal mines hidden disasters as old goafs water disaster, instability disaster by large scale goafs collapse, coal spontaneous combustion and poisonous or harmful gas disaster in goafs, spontaneous combustion disaster of coal waste piles, land pollution of mining area, destruction of ecological environment have been important factors that influencing safety production and social lives. In order to promote ecological civilization construction and improve the comprehensive control technology level of abandoned coal mines hidden disasters, the paper analyzed the characteristics and comprehensive control technologies progress, systematically discussed the feasibility of hidden disasters detection, monitoring and forecasting, comprehensive control and utilization for abandoned coal mines. The results showed that the hidden disasters comprehensive control technologies were mainly including data platform construction of hidden disasters factors detection, monitoring and forecasting platform construction (goaf instability and collapse monitoring, underground water level and quality monitoring, temperature and spontaneous combustion monitoring, toxic and harmful gases monitoring, ecological environment monitoring), comprehensive control and utilization (ecological restoration effect evaluation, gas reserves prediction and drainage effect evaluation, underground space comprehensive utilization of abandoned coal mines). With the introduction of data platform construction and cloud computing, the comprehensive control of abandoned coal mines hidden disasters can be automatically, effectively and in real time.

1 Introduction

The number of coal mines in China has been reduced from 87000 in 1995 to about 10000 in 2015, some of the closed coal mines have been incorporated into bigger coal corporations, and others become abandoned coal mines. Such abandoned coal mines

J. Wasowski et al. (eds.), *Engineering Geology and Geological Engineering for Sustainable Use of the Earth's Resources, Urbanization and Infrastructure Protection from Geohazards*, Sustainable Civil Infrastructures, DOI 10.1007/978-3-319-61648-3_9

hidden disasters as old goafs water disaster, instability disaster by large scale goafs collapse, coal spontaneous combustion and poisonous or harmful gas disaster in goafs, spontaneous combustion disasters of coal waste piles, land pollution of mining area, destruction of ecological environment have been important factors that influencing safety production and social lives.

On the study of comprehensive control technologies for abandoned coal mines hidden disasters, many experts and scholars have studied such traditional aspects as the types and characteristics, ecological environmental problems and control, goaf instability evaluation and control, water disaster control and water pollution evaluation, goafs spontaneous combustion and fire area control, harmful gas prevention, spontaneous combustion disaster of coal waste piles, etc. Based on the previous studies, the paper systematically discussed the characteristics and core problems of comprehensive control for abandoned coal mines hidden disasters, analyzed the research status and progress of hidden disaster detection, monitoring and forecasting, comprehensive control and utilization for abandoned coal mines, proposed main control technologies and methods. While big data platform and cloud computing technology were introduced to realize automatic, real time and effective control for abandoned coal mines hidden disasters.

2 The Characteristics and Core Problems of Comprehensive Control for Abandoned Coal Mines Hidden Disasters in China

The abandoned coal mines hidden disasters are mainly classified into eight major disasters and such corresponding characteristics as unpredictable goafs instability (1), irregular subsidence on the surface of goafs (2), larger harmful old goafs water disaster (3), serious pollution in underground water (4), frequency goafs temperature anomaly or spontaneous combustion disasters (5), toxic or harmful gases (6), spontaneous combustion disaster of coal waste piles (7), destructed ecological environment (8), etc. The core problems of comprehensive control for abandoned coal mines hidden disasters are detection, monitoring and forecasting, comprehensive control and utilization. For the above eight major disasters, the core problems of comprehensive control for abandoned coal mines hidden disasters are shown in Table 1.

Table 1. The core problems of comprehensive control for abandoned coal mines hidden disasters in China (on behalf of '√')

Core problems	(1)	(2)	(3)	(4)	(5)	(6)	(7)	(8)
Detection	√		√		√		√	
Monitoring and forecasting	√	√	√	√	√	√	√	√
Comprehensive control and utilization		√	√		√	√	√	√

Take the disaster of the irregular subsidence on the surface of goafs for example. On 23, June, 2012, two mine earthquake were happened in the place of Shibadun coalmine of Yulin City, China, and the earthquake level reached 2.8 and 3.2 grad. Through the investigation, about 66700 m^2 area of collapse subsidence on the surface of goafs, however there is not any sign before the disaster happened. In addition, several irregular subsidence disasters had been happened in other coalmines of Yulin City China and no regular conclusion was obtained.

In the process of control, the principles are summarized and put forward, including ground survey in advance, undermine detection follow up, hazard assessment and evaluation in time, comprehensive control for disaster reduction, monitoring and supervisory in place, safety measures for guarantee. So, the hidden disasters can be controlled and thus achieve the goal of classification and gradient utilization.

3 Researches Progress on the Abandoned Coal Mines Hidden Disasters Detection

The abandoned coal mines hidden disasters are detection technologies mainly involve goafs and water disasters detection, temperature abnormality area detection where spontaneous combustion and fire area including.

Take goafs and water disasters detection for example, the surface and under-mine geological detection technologies have been widely used and developed, that is shown in Table 2. The surface geophysical detection technologies include Ground penetrating rada method, Multi-electrodes resistivity method, Time domain electromagnetic method, 2D or 3D seismic methods, Controlled-source audio magnetotellurics method and Audio magnetotellurics method, etc. While the under-mine geophysical detection technologies are including mine transient electromagnetic method, mine direct current methods (Multi-electrodes resistivity method and DC electrical penetration method), mine seismic methods (Rayleigh wave method, Reflection wave method and In-Seam Seismic method), Radio wave Penetration method (Electromagnetic wave CT), etc. The comprehensive fine detection technologies combined geophysical detection, drilling and geochemical detection have played an important role in goafs and water disasters detection, and effectively prevent occurrence of several hidden disasters.

The typical apparent resistivity comprehensive profile including goafs is shown in Fig. 1. The darker color indicates higher resistivity. Between the distance 510 m to 590 m and 650 m to 800 m, there are two higher apparent resistivity regions that old abandoned goafs were deduced. And they are verified by drilling boreholes.

No doubt that, in the aspect of geophysical detection data acquisition, processing and interpretation, there still in the stage of the independent acquisition through single or multi-channels, professional processing respectively and comprehensive interpretation. The big data could be collected from different mining area, different geophysical or drilling methods on surface or under-mine, different time, different parameter settings on the same geophysical method, etc. In order to facilitate inquire, compare and analyze the big data, how to introduce these big data into Internet and then build strata graphic information database have become one of the most important scientific requirements recently.

Table 2. The detection methods and applicable conditions of abandoned coal mines hidden disasters

Classification	Geophysical methods	Applicable conditions
Surface geophysical methods	Ground penetrating rada method Multi-electrodes resistivity method Time domain electromagnetic method 2D seismic method 3D seismic method (Controlled source) Audio magnetotellurics method	The very shallow abnormal geological bodies with depth less than 20 m are fit for Ground penetrating rada method and Multi-electrodes resistivity method The shallow abnormal geological bodies with depth between 20 m and 150 m are fit for 2D seismic method, Time domain electromagnetic method or Multi-electrodes resistivity method The medium-depth abnormal geological bodies with depth between 150 m and 400 m are fit for 2D or 3D seismic method, Time domain electromagnetic method, Controlled-source audio magnetotellurics method or Audio magnetotellurics method The deep abnormal geological bodies with depth more than 400 m are fit for 3D seismic method, Controlled-source audio magnetotellurics method, Audio magnetotellurics method or Time domain electromagnetic method
Under-mine geophysical methods	Mine transient electromagnetic method Mine direct current methods (Multi-electrodes resistivity method and DC electrical penetration method) Mine seismic methods (Rayleigh wave method, Reflection wave method and In-Seam Seismic method) Radiowave Penetration method (Electromagnetic wave CT)	The lanes head advanced detection are fit for old goafs and water distribution, abnormal geological structures such as faults, collapse columns. The applicable geophysical detection methods are Mine transient electromagnetic method, Mine direct current methods, Rayleigh wave seismic method, Reflection wave seismic method, etc. The inner working face detection are fit for abnormal geological structures and their water conductivity such as the faults, collapse columns. The applicable geophysical detection methods are DC electrical penetration method, Radiowave Penetration method (Electromagnetic wave CT), Reflection wave method and In-Seam Seismic method The roof and floor detection are fit for aquifer and resisting layer distribution, conductivity channels, limestone karst water and development situation, and fracture zone height, etc. The applicable geophysical detection methods are Mine transient electromagnetic method, Mine direct current methods, Reflection and penetration wave mine seismic methods, etc.

Remarks: two or more comprehensive geophysical detection methods should be adopted where high precision and accuracy required.

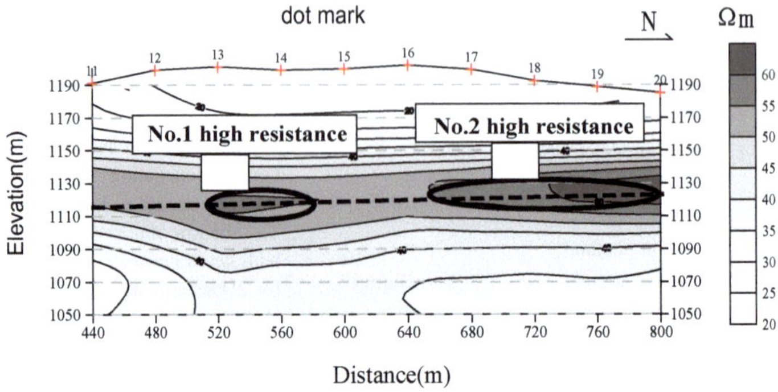

Fig. 1. The typical apparent resistivity comprehensive profile including goafs

The data platform construction of goafs and water disasters detection is based on the construction of very large scale information database, which the geological data have been digital on Internet sharing platform. Before the detection, the technique personnel can inquiry the database and analyze the data through mobile clients (for example phones or PC) to contrast the strata information and abnormal area. After the detection, the technique personnel can also input the results to the database. So the data platform can greatly improve the detection precision and ultimately achieve wisdom management and scientific decision.

4 Researches Progress on the Abandoned Coal Mines Hidden Disasters Monitoring and Forecasting

The monitoring and forecasting technologies of the abandoned coal mines hidden disasters are mainly including 5 platforms construction, that is, the goafs instability and collapse monitoring and forecasting, underground water level and quality monitoring and forecasting, temperature and spontaneous combustion monitoring and forecasting, toxic and harmful gases monitoring and forecasting, ecological environment monitoring and forecasting, see Fig. 2.

4.1 Platform Construction of Goaf Instability and Collapse Monitoring and Forecasting

Platform construction of goaf instability and collapse monitoring and forecasting aims to establish the relationship between the parameters changes and goaf instability and collapse, and thus to establish the models, through longtime monitoring the surface subsidence and deformation, the displacement variation of overlying strata. The mobile clients can be used to trace the variation of the monitoring data at any time, and then timely forecasting can be realized.

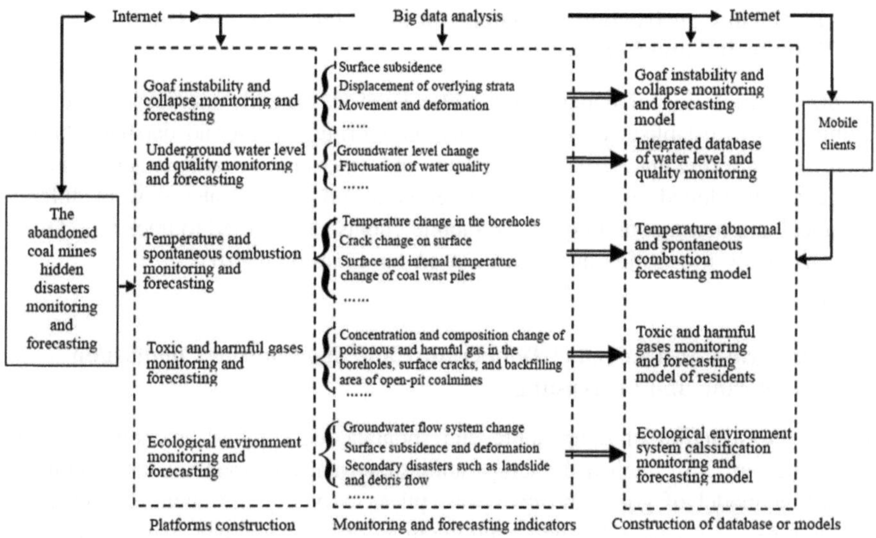

Fig. 2. Monitoring and forecasting system construction on abandoned coal mines hidden disasters

For example, the comprehensive monitoring system of goafs has been established in Shendong mining area, which involve the micro-seismic monitoring, working face pressure monitoring, boreholes stress monitoring, rock mass internal displacement monitoring, ground borehole television watching, surface displacement monitoring, etc. The comprehensive monitoring system of goafs is shown in Fig. 3.

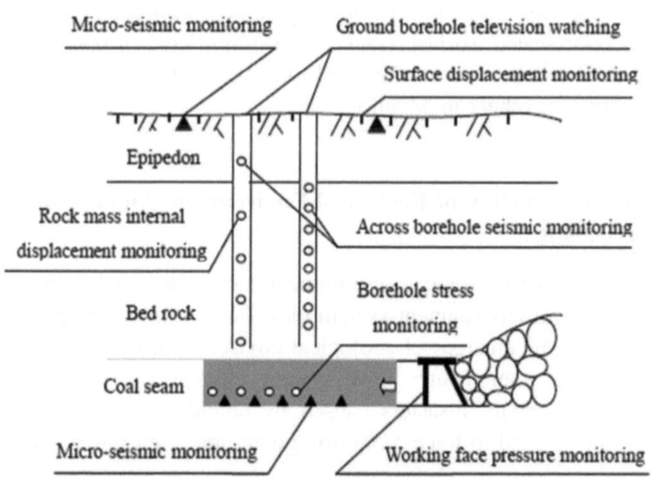

Fig. 3. The comprehensive monitoring system of goafs

4.2 Platform Construction of Underground Water Level and Quality Monitoring and Forecasting

Platform construction of underground water level and quality monitoring and forecasting aims to establish the water quality monitoring system and database through longtime monitoring underground water level and water quality variation, analyzing the change of abandoned coal mines water rebound and groundwater flow in different geological and mining conditions. The mobile clients can be used to trace monitoring, and then timely forecasting and control in advance can be realized.

4.3 Platform Construction of Temperature and Spontaneous Combustion Monitoring and Forecasting

Platform construction of temperature and spontaneous combustion monitoring and forecasting aims to establish the temperature abnormal and spontaneous combustion forecasting model of goafs and coal wast piles, carry out fire status changing and explosion risk assessment, through longtime monitoring abandoned coal mines drilling boreholes, surface cracks, surface and internal temperature abnormal area of wast piles. The mobile clients can be used to trace monitoring, and then timely forecasting and taking measures for fire prevention or closure of spontaneous combustion goafs.

4.4 Platform Construction of Toxic and Harmful Gases Monitoring and Forecasting

Platform construction of toxic and harmful gases monitoring and forecasting aims to establish the toxic and harmful gases monitoring and forecasting models of residents, carry out toxic gas leakage risk assessment, through longtime monitoring the concentration and composition changes of abandoned coal mines shafts or drilling boreholes in use or poor sealed, surface cracks, back-filling area of open pit. The mobile clients can be used to trace monitoring, and then timely forecasting and taking measures for ventilation or others in advance.

4.5 Platform Construction of Ecological Environment Monitoring and Forecasting

Platform construction of ecological environment monitoring and forecasting aims to establish the ecological environment system classification monitoring and forecasting models, and analyze the changes of ecological environment indicators, through longtime monitoring the groundwater flow system changes, surface subsidence and deformation, and secondary disasters caused by abandoned coal mining area. The mobile clients can be used to trace monitoring data, and then timely forecasting.

5 Researches Progress on the Abandoned Coal Mines Hidden Disasters Comprehensive Control and Utilization

The abandoned coal mines hidden disasters comprehensive control and utilization technologies mainly involve ecological restoration effect evaluation, gas reservation prediction and drainage effect evaluation, underground space comprehensive classification utilization of abandoned coal mines.

5.1 Ecological Restoration Effect Evaluation of Abandoned Coal Mines

Ecological restoration of abandoned coal mines is based on the ecological environment monitoring platform and such technologies as remote sensing, wireless transmission, networking, artificial intelligence, to carry out ecology partition and planning of different types of destruction through real-time monitoring the changes of soil, water, forestry, agriculture, grass, etc. The land of abandoned coal mines may carry out engineering reclamation, ecological reclamation, biological reclamation and their combination.

Ecological restoration effect evaluation may achieve real-time computing, evaluation and monitoring through different big data analysis and cloud computing. Once the effect is not good, the control programs and methods can be timely adjusted and ultimately outstanding control results can be expected.

5.2 Gas Reservation Prediction and Drainage Effect Evaluation of Abandoned Coal Mines

Gas drainage and utilization of abandoned coal mines can not only relieve the shortage of resources, but also decrease the waste gas emission thus cause air pollution to some extent in China. Now the gas drainage problem has been basically solved and the accurate prediction of gas reservation is still at the stage of research.

Gas reservation prediction and drainage effect evaluation of abandoned coal mines should analyze the relationship between the real gas drainage amount of boreholes, the corresponding gas reservation and mining or geological conditions through monitoring the gas concentration and composition. Then the gas reservation prediction models and drainage boreholes designs can be improved.

5.3 Underground Space Comprehensive Classification Utilization of Abandoned Coal Mines

In addition to the gas drainage, the underground space also be comprehensive classification utilized according to the category, location, maintainability of the abandoned coal mines. Three utilization levels are divided as follows: (1) underground space such as experiment laboratories, sightseeing, entertainment, garages. (2) underground storage or energy sites. (3) garbage back-fill stations.

The underground space comprehensive classification utilization of abandoned coal mines should be taken such measures as monitoring, evaluation and prediction of stress, fracture and displacement field according to the function classification.

6 Conclusions

The abandoned coal mines hidden disasters are mainly classified into eight major disasters and such corresponding characteristics as unpredictable goafs instability, irregular subsidence on the surface of goafs, larger harmful old goafs water disaster, serious pollution in underground water, frequency goafs temperature anomaly or spontaneous combustion disasters, toxic or harmful gases, spontaneous combustion disaster of coal waste piles, destructed ecological environment. The comprehensive control technologies for abandoned coal mines hidden disasters are detection, monitoring and forecasting, comprehensive control and utilization.

In China, there existed such three main technologies or methods for abandoned coal mines hidden disasters control as the detection platform construction in big data, platforms construction of monitoring and forecasting, comprehensive control and utilization of abandoned coal mines hidden disasters.

The platforms construction of monitoring and forecasting involved the goafs instability and collapse monitoring and forecasting, underground water level and quality monitoring and forecasting, temperature and spontaneous combustion monitoring and forecasting, toxic and harmful gases monitoring and forecasting, ecological environment monitoring and forecasting.

The comprehensive control and utilization technologies mainly involve ecological restoration effect evaluation, gas reservation prediction and drainage effect evaluation, underground space comprehensive classification utilization of abandoned coal mines.

With the introduction of big data platforms and cloud computing technology, the hidden disasters of abandoned coal mines can realize automatic, real-time and effective control.

Acknowledgments. The author gratefully acknowledges such team members as Dr. Li Hongjie, Zhang Bin, Zhang Junying, and the funding by the National Natural Science Foundation of China (51674142) and the National Science and Technology Major Project of China (2016ZX05045001-004).

Bibliography

Li, W., Li, J.: Coafs disaster characteristics of resource integration coal mines and their control countermeasures. Saf. Coal Mines **46**(7), 179–181,85 (2015)

Xu, J., Li, G., Chen, G., et al.: Dynamic changes and evaluation of land ecological quality in coal mining area. J. China Coal Soc. **38**(S1), 180–186 (2013)

Hu, Z., Long, J., Wang, X.: Self-healing, natural restoration and artificial restoration of ecological environment for coal mining. J. China Coal Soc. **39**(8), 1751–1757 (2014)

Li, W.: Risk assessment of room method goafs instability. China Saf. Sci. J. **21**(3), 95–100 (2011)

Li, W.: Mine earthquake prevention and control technology induced by old coalmine goafs based on surface isolation grouting. In: 3rd International Young Scholars' Symposium on Rock mechanics: Transit Development in Rock Mechanics, pp. 549–553. Taylor and Francis Group, London (2014)

Dong, S.: Some key scientific problems on water hazards frequently happened in China's coalmines. J. China Coal Soc. **35**(1), 66–71 (2010)

Zhou, X.: Proposals on improvement of spontaneous combustion prevention and control in mining goaf and emergency handling capacity. Coal Sci. Technol. **41**(9), 151–153 (2013)

Wu, H., Zeng, F., Yao, H., et al.: Danger evaluation and control technology of coal mine gangue spontaneous combustion. Coal Sci. Technol. **41**(4), 119–123 (2013)

Cheng, J., Pan, D., Li, W., et al.: Study on the detecting of hazard abandoned workings by ground penetrating radar on strong electromagnetic interference area. J. China Coal Soc. **35**(2), 227–231 (2010)

Cheng, J., Sun, H., Zhao, Q., et al.: The detection technology of excavated region in coal mine and case study. J. China Coal Soc. **33**(3), 251–255 (2008)

Wen, L., Mu, Y., Zhang, J., et al.: Optimization of surface detection technology and method of mine goaf. Coal Sci. Technol. **39**(1), 102–106 (2011)

Li, H.: Application of shallow earthquake and transient electromagnetic method in gob exploration. Coal Min. Technol. **18**(1), 17–18, 21 (2013)

Qin, W., Xu, J., Hu, G., et al.: Study on computing method of old goaf methane reserves. J. China Coal Soc. **38**(6), 948–953 (2013)

Gu, D.: Theory framework and technological system of coal mine underground reservoir. J. China Coal Soc. **40**(2), 239–246 (2015)

Xie, H., Hou, Z., Gao, F., et al.: A new technology of pumped-storage power in underground coal mine: principles, present situation and future. J. China Coal Soc. **40**(5), 965–972 (2015)

Landslide Susceptibility Mapping of Tizi-Ouzou Region, Algeria

Amel Kab, Lynda Djerbal(✉), and Ramdane Bahar

LEEGO, University of Science and Technology Houari Boumediene, BP. 32 El Alia, 16111 Algers, Algeria
amoulal684@yahoo.fr, {ldjerbal, rbahar}@usthb.dz

Abstract. Landslides are one of the most catastrophic geo-hazard all over the world. The study of this phenomenon is hampered many difficulties due to the complexity of the instabilities, the loose of data and the important variability of its causes. Geomorphological mapping of landslide susceptibility is a method that relies to the ability of the investigator to actual and potential slope failure according landslide inventory. The landslide susceptibility and hazard assessment approaches need a good knowledge of the previous observed scenarios and include collecting data, construction of spatial database as well as spatial and statistical analyses. Every year, principally in winter, the North of Algeria is affected by instabilities of greater magnitude causing equipment damage and important costs. This research is focused on the study of Tizi-Ouzou region, located at about 90 km east of Algiers, Algeria. Tizi-Ouzou is a region of Kabylia, with an area of 3568 km^2, characterized by a mountainous morphology. This region is characterized by about 84 significant instabilities affecting urban areas. The work aims to make, in the first hand, a spatial and statistical study of instability causes and site sensibility parameters, and in the second hand, the susceptibility assessment and mapping of Tizi-Ouzou region according to the landslide inventory effectuated in this region. The study indicates that the main instability factor is water effect and lets the proposal of Tizi-Ouzou region susceptibility map.

Keywords: Landslide · Sensibility · Susceptibility · Assessment · Mapping

1 Introduction

Landslides causes every year, in the north of Algeria, particularly in winter, severe injuries, property destruction and death. Assessment and management of this phenomenon are very difficult mainly when persons and properties are exposed. The recent development of data acquisition systems and computer skills, including geographic information systems GIS, enabled the development and implementation of numerous tools to hazard assessment and mapping through the world (Carrara et al. 1995; Sabatakakis et al. 2013; Mergili et al. 2014; Passalacqua et al. 2014). These tools led analyzing previous observed instabilities, building a spatial database, extracting landslide related factors, calculating the frequency ratios, make maps.

J. Wasowski et al. (eds.), *Engineering Geology and Geological Engineering for Sustainable Use of the Earth's Resources, Urbanization and Infrastructure Protection from Geohazards*, Sustainable Civil Infrastructures, DOI 10.1007/978-3-319-61648-3_10

Landslides constitute a significant problem to the economy and development of the North of Algeria. Several regions are affected by major instabilities (Machane et al. 2008; Bouhadad et al. 2010; Bourenane et al. 2014; Djerbal et al. 2014; Bahar et al. 2014). The wilaya of Tizi-Ouzou, belonging to Kabylia region, is composed of 67 municipalities and has 1127607 inhabitants spread over an area of 3568 km^2. This region is the most affected by land movements. It is characterized by several active instabilities, such as Ain El Hammam landslide, Azazga landslide and Illilthen mud-slide. This work aims to study the susceptibility of Tizi-Ouzou region to instabilities taking into account the previous instabilities observed. The study concerned 39 municipalities (among the 67 municipalities that constitute this wilaya).

2 Methodology

Landslide susceptibility is the possibility that a landslide occurring in an area, on the basis of local terrain conditions (Brabb 1984). It is the degree to which a terrain can be affected by slope movements (an estimate for where landslides are likely occurs). Susceptibility does not consider the temporal probability of failure, nor the magnitude of the expected landslide. In mathematical language, landslide susceptibility is the probability of spatial occurrence of slope failures, given a set of geo-environmental conditions. This is also called "landslide analysis" by Vandine et al. (2004).

The adopted methodology consists to make a landslide susceptibility analysis according to landslide inventory of Tizi-Ouzou region. The landslide inventory for susceptibility assessment include the evaluation of landslide triggering conditions or reactivation conditions, landslide activity state, movement extends, building at risk, etc. Then spatial and statistical analysis of the sliding triggering causes and the amplitude of the movement will be given according to the landslide inventory effectuated. Then site susceptibility map can be realized for Tizi-Ouzou region using GIS techniques and taking into account the previous instabilities characteristics (Fig. 1).

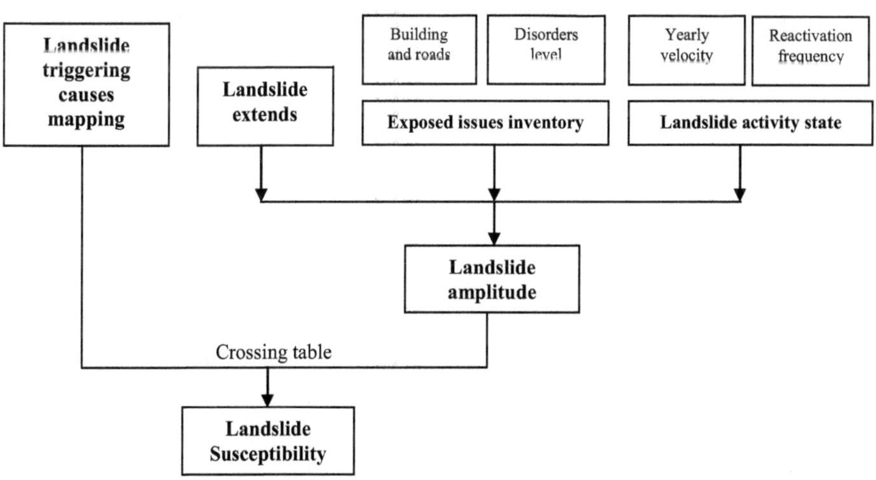

Fig. 1. Landslide sensitivity (susceptibility) assessment

3 Tizi-Ouzou Region Instabilities

The study is to conduct a census of unstable slopes, which endured marked instabilities during the recent years, in the wilaya of Tizi-Ouzou. This census has concerned 39 cities that represent an area estimated to about 53% of the total area of this region. The studied sites have a homogeny distribution over Tizi-Ouzou region territory (Fig. 2). The study led identifying 84 instabilities affecting urban sites and/or road infrastructure. We distinguished among these 84 landslides, more than 34 major movements whose extent exceed 2 ha and reaching in some cases more than 50 ha, as the landslide Tigzirt extending over 136 ha (Guirous et al. 2014) and Azazga landslide extending over an area of 260 ha (Djerbal et al. 2014). This kind of instabilities affects dozens of homes and endangers people. The materials affected by instabilities are: Shale, marl and clay (Table 1).

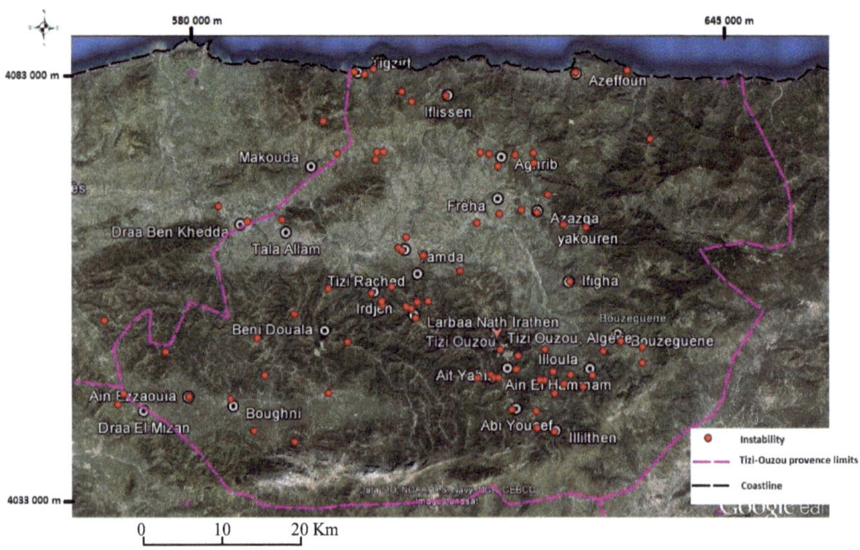

Fig. 2. Landslides spatial distribution in Tizi-Ouzou.

Table 1. The materials affected by instabilities in Tizi-Ouzou region.

Affected soil	Number of variables	Number of cases observed	Percentage (%)
Clay	66	16	32
Marl	66	21	24
Shale	66	29	44

4 Tizi-Ouzou Landslides Causes

A statistical study of the landslides triggering causes was conducted on a sample of 84 instabilities occurring in the wilaya of Tizi-Ouzou during the period ranging between 2000 and 2016. The census study involved 39 of the 67 towns composing this wilaya. The results show that the studied instabilities are significantly influenced by the effect of climate and human activities. Three categories of causes of landslides have been identified:

- **Category 1:** It corresponds to landslides triggered or reactivated by major human activities and climate action (major earthworks in winter, high overload).
- **Category 2:** It corresponds to landslides triggered or reactivated by the effect of climate and minor human activity (deforestation, low earthworks).
- **Category 3:** It consists on movements triggered and reactivated by the only effect of climate action (torrential rain, snow…). These cases correspond to sites located in a precarious state of stability slopes or a reactivation of old instabilities.

The results of this study show that over 65% (category 3) of the land movements are activated by the climate effects, and about 31% (category 1) of landslides activation is due simultaneously to human activities and climate effects (Table 2).

Table 2. Tizi-Ouzou landslides triggering causes analysis.

Instability cause	Number of variables	Number of cases observed	Percentage (%)
Category 1	84	26	30.95
Category 2	84	3	3.57
Category 3	84	55	65.48

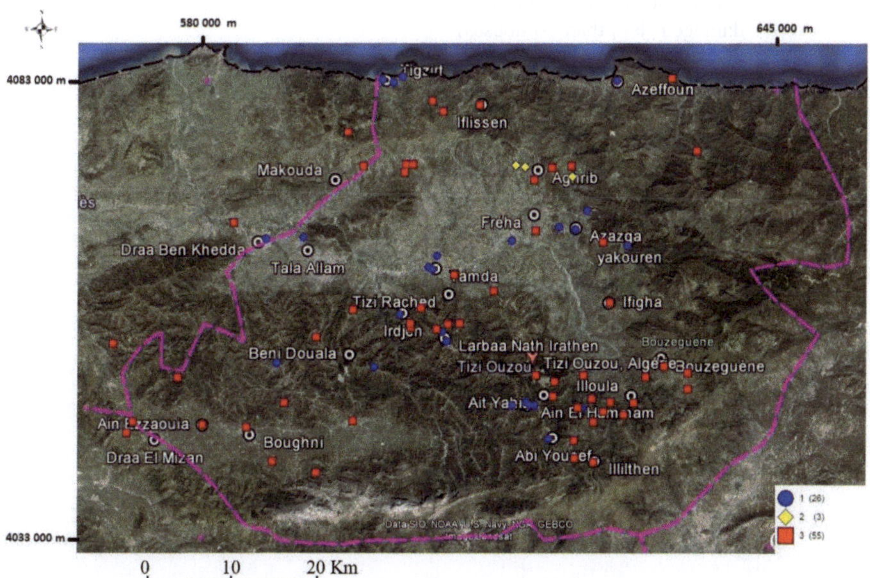

Fig. 3. Spatial distribution of landslides causes.

The spatial distribution of the landslides causes in the wilaya of Tizi-Ouzou (Fig. 3) shows that landslides triggered by the only effect of climate are concentrated in the southern part of the wilaya. While landslides triggered by the joint effect of human activities and climate actions are located in the zone linking Draa Ben Khedda to Azazga, on the coast zone as well as at Ait Yahia and Abi Youcef (in the South).

5 Tizi-Ouzou Landslides Amplitude

The instabilities amplitude was divided into four (04) categories based on the extent and magnitude of disorders induced (Table 3). The results observed led the determination of 34 major landslides, characterized by an extent greater than 2 ha and in some cases reaching several dozen hectares (Categories C3 and C4) (Fig. 4).

Table 3. Landslides amplitudes evaluation.

Landslide amplitude	Parameter description	Number of variables	Number observed cases	Percentage (%)
Category C1	- Instability of low extent (rockslides, collapses) - Houses are not affected - Roads can be partially affected	84	17	20.23
Category C2	- Instability of medium extends (less than 1 ha) - Slow displacements - Some houses are partially damaged (less than 10 houses) - Roads can be damaged or cut for a few times - Medium material disorders - A few persons are exposed to risk	84	33	39.28
Category C3	- Instability of high extends, which can reach about 20 ha - Centimeter to decimeter displacements - Properties and people are exposed to risk - Movement is reactivated every winter - Roads cut traffic - Dozens of houses partially damaged - Persons are exposed to significant risk	84	23	27.38

(*continued*)

Table 3. (*continued*)

Landslide amplitude	Parameter description	Number of variables	Number observed cases	Percentage (%)
Category C4	- Very active instability extending over an area greater than 20 ha - Metric displacements - Several millions of m^3 of material moved - Security of materials and people is highly threatened - Roads totally damaged and cut Traffic	84	11	13.11

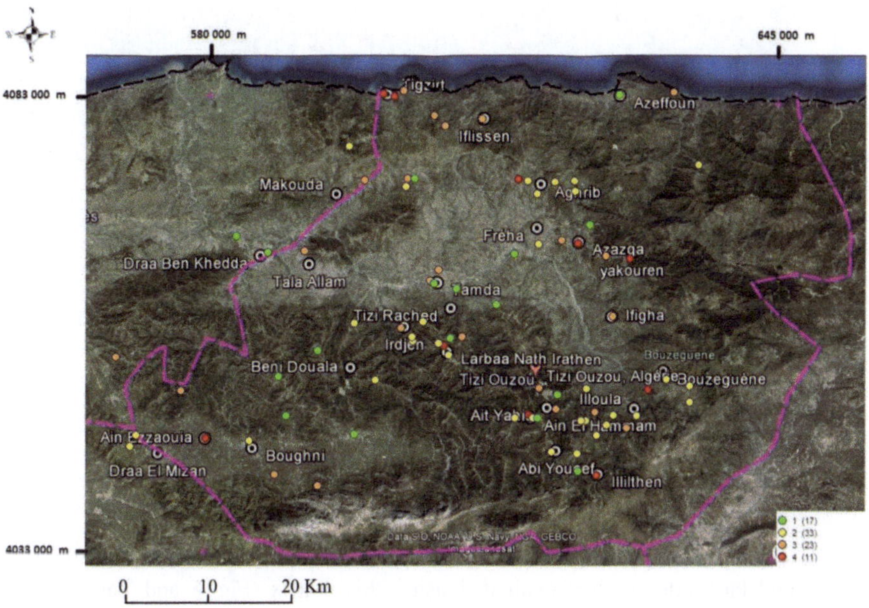

Fig. 4. Spatial distribution of the landslides amplitude.

6 Site Susceptibility Assessment and Mapping

The site sensitivity (susceptibility) assessment is carried out by performing a combination between the amplitude and the causes of instabilities using GIS tools (Table 4). This combination is based on a dual input matrix, which is a technique practiced, using other inputs, to make regulatory risk and hazard mapping in Italy and Switzerland (Guzzetti et al. 1999; Laletlin et al. 2005; Luzi and Pergalani 1999). The amplitude of

Table 4. Combination parameters matrix to assess the site susceptibility to instabilities.

Amplitude	Cause		
	Category 1	Category 2	Category 3
C1	Negligible susceptibility	Low susceptibility	Medium susceptibility
C2	Low susceptibility	Medium susceptibility	High susceptibility
C3	Medium susceptibility	High susceptibility	Very high susceptibility
C4	High susceptibility	Very high susceptibility	Very high susceptibility

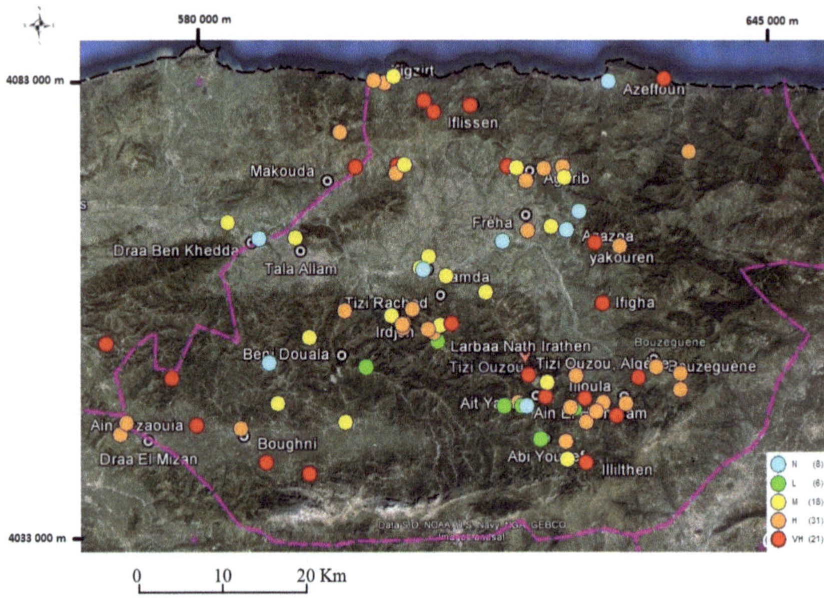

Fig. 5. Spatial propagation of the landslides sensitivity classes.

the studied landslides is then evaluated using this matrix (Fig. 5 and Table 5). The obtained results led to classify 21 movements in Very High sensitivity and 31 in High sensitivity. In order to map the landslide sensitivity propagation (susceptibility), a spatial interpolation of the of the 84 Tizi-Ouzou landslides sensitivities is effectuated using a triangulation with smoothing procedure. The results led to classify an important area of Tizi-Ouzou region in very high-to-high sensitivity to instabilities (Fig. 6).

Table 5. Tizi-Ouzou landslides sensitivity analysis.

Sensitivity	Number of variable	Number of cases observed	Percentage (%)
Negligible	84	8	9.52
Low	84	6	7.14
Medium	84	18	21.43
High	84	31	36.90
Very high	84	21	25.01

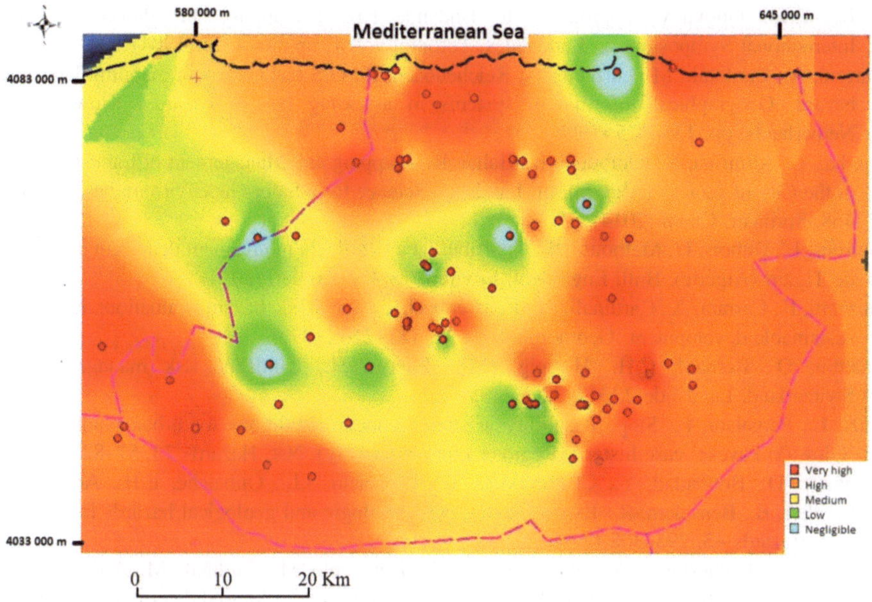

Fig. 6. Map of Tizi-Ouzou region susceptibility to land movements.

7 Conclusions

Tizi-Ouzou is the most affected region by instabilities in Algeria. The census study of unstable sites conducted in this region, considering 39 provinces, led to determine 84 landslides. The statistical analysis of these landslides confirmed the relationship existing between the activity of landslides and climate. In fact, over than 65% of the instabilities are activated by climate effect and 20 major landslides, classed in C3 and C4 amplitude, are triggered or reactivated by the effect of climate. The study of the previous observed landslides led also the assessment and mapping of site susceptibility due to these phenomena using the GIS technique tools and classifying an important area of Tizi-Ouzou in danger zone. Moreover, this work constitutes a first approach to the assessment and mapping of the hazard induced by instabilities in Algeria.

References

Bahar, R., Sadaoui, O., Sadaoui, S.: Landslides induced by intense rainfall and human interventions – case studies in Algeria. Eng. Geol. Soc. Territ. **6**, 1049–1056 (2014)

Bouhadad, Y., Benhammouche, A., Bourenane, H., Ouali, A.A., Chikh, M., Guessoum, N.: The Laalam (Algeria) damaging landslide triggered by a moderate earthquake (Mw = 5.2). Nat. Hazards **54**, 261–272 (2010)

Bourenane, H., Bouhadad, Y., Guettouche, M.S., Braham, M.: GIS-based landslide susceptibility zonation using bivariate statistical and expert approaches in the city of Constantine (Northeast Algeria). Bull. Eng. Geol. Environ. (2014). doi:10.1007/s10064-014-0616-6

Brabb, E.E.: Innovative approaches to landslide hazard mapping. In: Proceedings 4th International Symposium on Landslides, Toronto, vol. 1, pp. 307–324 (1984)

Carrara, A., Cardinali, M., Guzzetti, F., Reichenbach, P.: GIS technology in mapping landslide hazard. Geographical information systems in assessing natural hazards. Kluwer, The Netherlands, pp. 135–175 (1995)

Djerbal, L., Alimrina, N., Melbouci, B., Bahar, R.: Mapping and Management of Landslide Risk in the City of Azazga (Algeria). In: Landslide Science for a Safer Geoenvironment, pp. 463–468. Springer, Cham (2014)

Guirous, L., Dubois, L., Melbouci, B.: Contribution à l'étude du mouvement de terrain de la ville de Tigzirt (Algérie). Bull. Eng. Geol. Environ. **73**(4), 971–986 (2014)

Guzzetti, F., Carrara, A., Cardinali, M., Reichenbach, P.: Landslide hazard evaluation: an aid to a sustainable development. Geomorphology **31**, 181–216 (1999)

Laletlin, O., Bernard, C.H., Haemig, C.H., Raetzo, H.: Landslide risk management in Switzerland. Landslide **2**, 313–320 (2005)

Luzi, L., Pergalani, F.: Slope instability in static dynamic conditions for urban planning: the 'Oltre Po Pavese' case history (Regione Lombardia-Italy). Nat. Hazards **20**, 57–82 (1999)

Machane, D., Bouhadad, Y., Cheikhlounis, G., Chatelain, J.L., Oubaiche, E.H., Abbes, K., Guillier, B., Bensalem, R.: Examples of geomorphologic and geological hazards in Algeria. Nat. Hazards **45**, 295–308 (2008)

Mergili, M., Marchesini, I., Alvioli, M., Rossi, M., Santangelo, M., Cardinali, M., Ardizzone, F., Fiorucci, F., Schneider-Muntan, B., Fellin, W., Guzzetti, F.: GIS-based deterministic analysis of deep-seated slope stability in a complex geological setting. Eng. Geol. Soc. Territ. **2**, 1437–1440 (2014). doi:10.1007/978-3-319-09057-3_254

Passalacqua, R., Bovolenta, R., Federici, B.: An integrated hydrological-geotechnical model in GIS for the analysis and prediction of large-scale landslides triggered by rainfall events. Eng. Geol. Soc. Territ. **2**, 1799–1803 (2014). doi:10.1007/978-3-319-09057-3_318

Sabatakakis, N., Koukis, G., Vassiliades, E., Lainas, S.: Landslide susceptibility zonation in Greece. Nat. Hazards **65**(1), 523–543 (2013)

Vandine, D.F., Moore, G., Wise, M., Vanbuskirk, C., Gerath, R.: Technical terms and methods. In: Wise, et al. (eds.) Landslide Risk Case Studies in Forest Development Planning and Operations. B.C., Ministry of Forests, Forest Science Program, Abstract of Land Management Handbook, vol. 56, pp. 13–26 (2004)

Karst Induced Geo-hazards in Egypt: Case Study Slope Stability Problems Along Some Selected Desert Highways

Ahmed M. Youssef[1,2(✉)], Abdel-Hamid El-Shater[1], Mohamed H. El-Khashab[1], and Bosy A. El-Haddad[1]

[1] Geology Department, Faculty of Science, Sohag University, Sohag, Egypt
bosy_abdelaziz@hotmail.com
[2] Geological Hazards Department, Applied Geology Sector, Saudi Geological Survey, P.O. Box 54141, Jeddah 21514, Kingdom of Saudi Arabia
amyoussef70@yahoo.com

Abstract. Karst environments are characterized by distinctive landforms related to dissolution characteristics. Karst terrains, frequently underlain by cavernous carbonate and/or evaporite rocks, may induce ground instability problems. Impacts and problems associated with karst are rapidly increasing as development expands upon the karst prone areas. This has led to an escalation of karst-related environmental and engineering problems such as landslides developed on rock cuts/slopes weakened by karstification features. In the current work, the effects of karstification and sinkholes on the stability of the rock cuts/slopes along some selected desert highways were evaluated. These highways represent the most used highways in Egypt, connecting most of the Governorates. They represent the backbone of Egyptian transportation and commercial traffic. Finally, the most optimum mitigation/remediation methods were summarized to decrease and minimize the consequences of slope instability.

Keywords: Slope stability · Geohazards · Karstification · Sinkholes · Egypt

1 Introduction

Highways and roads represent an essential part of the economy development in highly populated areas of Egypt, such as the Nile valley area. Many highways connect the Upper Egypt's Governorates with each other and with the capital Cairo, including western desert, eastern desert, and Nile Valley highways. Most of the traffic runs along the western and eastern desert highways. Recently, many development areas have taken place all over the Egyptian territories including new urban areas, reclamation areas, and industrial zones, all of which increase the use of these highways. However, most of these highways move through rugged terrains which are characterized by karst rocks (carbonates) where a variety of different hazards (Parise 2008; Gutierrez 2010) can be presented, and involve the transportation system. Different factors can contribute to landslides (slope stability problems) such as climate, slope conditions, geological

J. Wasowski et al. (eds.), *Engineering Geology and Geological Engineering for Sustainable Use of the Earth's Resources, Urbanization and Infrastructure Protection from Geohazards*, Sustainable Civil Infrastructures, DOI 10.1007/978-3-319-61648-3_11

characteristics, and construction method (Youssef et al. 2009). A landslide can be defined as a rapid disaster which may involve falling, sliding or flowing of soil and rock (Cruden and Varnes 1996). Hungr et al. (1999) indicated that the transportation systems are vulnerable to rockfalls wherever they cut across or skirt along mountains, plateaus, ridges and similar topographic features. In the context of highway rock slopes, potentially landslide zones pose risks to the traveling public, infrastructures, local economies, and the environment (Palma et al. 2012a). Rockfalls may not be considered hazardous unless rocks enter the highways (Chau et al. 2004). Rockfalls remain a serious problem to many transportation agencies, which are responsible for providing and maintaining safe and reliable highways and routes in an economical fashion. Hazards related to landslides can result in traffic accidents, blocked highways, and increased maintenance costs, as well as other hazards to the public (Bateman 2003; Palma et al. 2012a; Youssef et al. 2012). Raju et al. (1999) indicated that landslides cause the death or injury of thousands of people and significant property loss every year in India. However, prediction of landslides (slope stability problems) is very difficult because of the complexity of the factors causing these problems (Yuan and Mohd 1997).

The fundamental role played by karst in predisposing and/or favoring slope stability problems has been barely explored, largely due to the difficulty of its investigation and assessment, but in recent years several works have dealt with the issue (Parise and Gunn 2007; Parise 2008; Gutierrez 2010; De Waele et al. 2011; Gutiérrez et al. 2014). Slope movements in carbonates and evaporites are generally studied and described without paying much attention to the voids, cavities, open joints, dissolution features, and filled sinkholes produced by karst processes and their detrimental influence on the mechanical properties of the rock masses. Dunne (1990) indicated that the presence, distribution, and frequency of voids and/or caves of different sizes may have an essential impact on the slope stability and its hydrology. All types of landslides proposed by Cruden and Varnes (1996) could occur in karstified rock masses.

The environmental fragility of karst settings, together with their endemic hazardous processes, has received an increasing attention from the scientific community in the last decades (Ford and Williams 1989; Williams 1993; Gillieson 1996; Vermeulen and Whitten 1999; Barany-Kevei and Gunn 2000; Waltham et al. 2005; Parise and Gunn 2007). Concurrently, the interest of planners and decision-makers on a safe and sustainable management of karst lands is also growing. Gutiérrez et al. (2014) indicated that a correct understanding of the sinkhole typology constitutes a crucial step for a proper hazard assessment and the design of effective mitigation measures. Many examples of landslides and/or rockfalls occurred in karst rocks (carbonate and evaporate rocks) were documented by different authors (Santo et al. 2007; Prager et al. 2008; Ivy-Ochs et al. 2009; Iovine et al. 2010; Jaboyedoff et al. 2011; Parise and Lollino 2011; Gutiérrez et al. 2008a, 2012; Palma et al. 2012a; Carbonel et al. 2014. Santo et al. (2007) mentioned that there are few reported cases in which a natural cave has been involved in the formation of a slope failure. Pánek et al. (2009) proposed the role played by karstification as a preparatory factor of the development of catastrophic slope failures in mountainous areas. Jaboyedoff et al. (2009) indicated that the development of large and catastrophic translational landslides in carbonate successions is typically favored by the presence of laterally extensive and extremely planar bedding planes such as the Frank rockslide-avalanche, which destroyed the southern end of the town of Frank in southwestern Alberta, Canada. One of the most documented example of landslide in karstic rocks was the 1963 Vajont translational slide in northeastern Italy, which moved into a reservoir, thus causing a huge impulse water wave that overtopped the concrete dam leading to a sudden catastrophic flood destroying some villages and killing over 2000 people (Semenza and Ghirotti 2000; Kilburn and Petley 2003). Another situation in which karst features (differential erosions and dissolution cavities) are the main factors causing slope movements in coastal cliffs, due to deepening of coastal caves and/or notches (Delle Rose and Parise 2004; Youssef et al. 2009; Vallejo 2012).

The main objectives of this research are the assessment of the slope stability problems of the carbonate rocks due to the presence of sinkholes and karst features. Three highway sections were selected, two zones along eastern desert highway and one along the western desert highway (Fig. 1). The term slope stability problems in this work represent any type of landslides that could be registered due to the effect of karstification features, including rockfalls and rock slides of all kinds (free falling, toppling, bouncing, rolling or sliding). Field investigations were carried out in different areas along rock cuts to understand and determine the most influential karst features that have a strong impact on slope stability. The detailed objectives were to discuss: (1) the origin of karst in the study area and their classification; (2) describe the most relevant karstification features that could impact the slope stability; (3) describe the types of slope stability problems related to karstifiation; and (4) develop a remediation/mitigation strategy to minimize any future problems.

2 Study Area

In the recent years, the Egyptian government established many highways all over the country. This will increase the tourist activities, transportation between different areas, goods incomes and outcomes, and will contribute to open new areas for future development. Because of the nature of the terrain and population distribution along Egyptian territories, a significant number of highway projects were constructed in mountainous and/or hilly areas. These hilly areas are characterized by frequent slope stability problems (landslides). The current study deals with the eastern and western desert highways (eastern desert Sohag – Red Sea – Cairo highway, western desert Sohag – Cairo highway; and eastern desert Assuit – Cairo highway) (Fig. 1). The geology of the area is composed mainly of Eocene limestone. Generally, this limestone in the study area is dissected by many structural elements (faults and/or joints) affected by sinkholes. The general appearance of limestone is white to gray in color with weathered reddish surfaces.

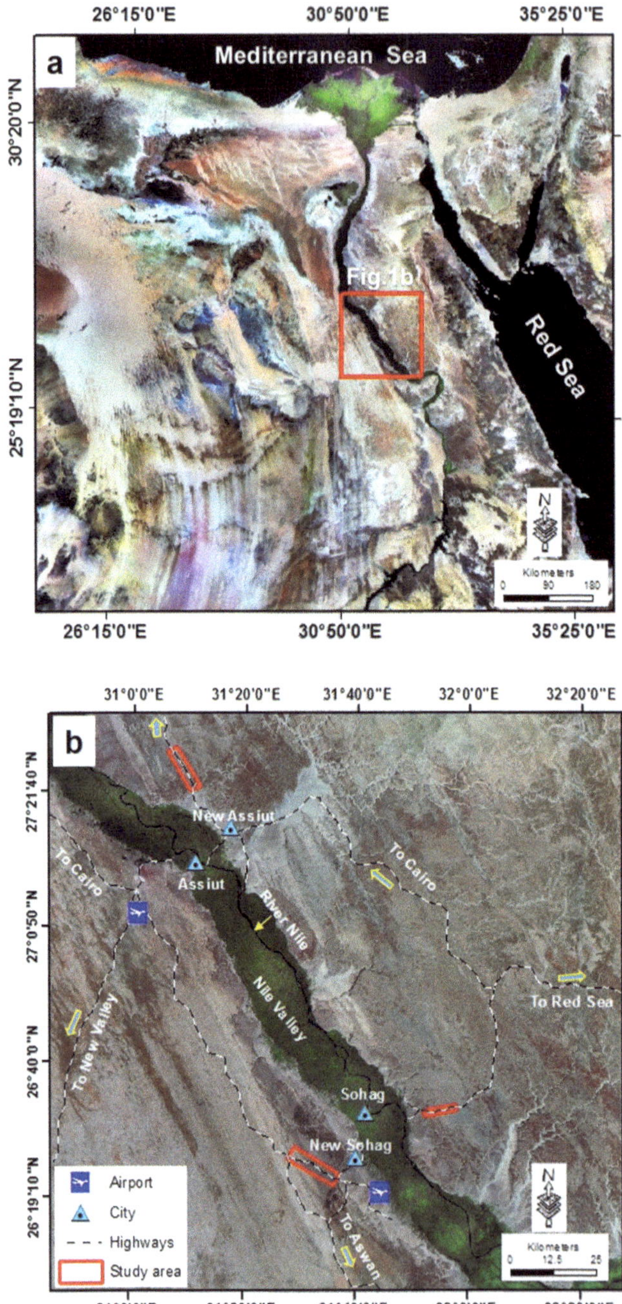

Fig. 1. (a) Study area location in Egypt map. (b) Detailed characteristics of the study area.

3 Data Collection and Methodology

Multiple field visits were carried out in the study area in order to investigate and evaluate the slope stability hazards and their relationship with karstification processes. Observations were conducted to identify potential hazard zones while considering the most influencing parameters. Later these hazard zones were located and documented with GPS coordinates. Field data collection includes (a) rock and material types (the main rock and material types at each site have been collected, identified and investigated); (b) weathering characteristics were identified by observing the degree of change in color and the rock fabric; (c) compressive strength (the strength of the intact rocks and the soil matrix in the filled sinkholes was measured using geological hammer (Burnett 1975)); (d) slop/cut face characteristics were determined (direction and slope); and (e) discontinuities characteristics were identified (fracturing degree, dip and dip direction values). Laboratory work was conducted including; (a) Jar slake test was conducted for shale samples according to Santi (1998) test procedure. The test depends on collecting samples, each sample weight 50 g which is oven-dried at 110 °C for 16 h, and allowed to cool for 20 min then immersed in distilled water for 30 min and 24 h. The results can be described according to Santi (1998) into six categories: (1) degrades to a pile of flakes or mud, (2) breaks rapidly and/or forms many chips, (3) breaks slowly and/or forms few chips, (4) breaks rapidly and/or develops several fractures, (5) breaks slowly and/or develops few fractures, and (6) no change; (b) slake durability test was carried out on marly limestone of the study area. These samples were subjected to 4 cycles of slake durability; (c) determine friction angle using tilt test; and (d) other physical properties were determined for the limestone samples such as unit weight.

4 Results and Discussions

4.1 Origin and Classification of Sinkholes

Sinkholes are depressions with internal drainage, widely regarded as one of the main diagnostic landforms of karst (Ford and Williams 2007). Sinkhole is the most common term in the international literature dealing with engineering and environmental issues (Beck 1984, 1988; Parise and Gunn 2007). Sinkholes can be distinguished in different shapes, such as cylindrical, conical, bowl- or pan shaped, and varying in size from less than a meter up to hundreds of meters across and with a depth from a few to tens of meters (Youssef et al. 2016). Sinkholes are often related to dissolution of carbonate and/or evaporite rocks/deposits (Fookes and Hawkins 1988; White 1988; Ford and Williams 2007; Palmer 2007). There are some crucial differences in sinkhole related to carbonate and evaporite rocks (Gutiérrez et al. 2008a; Gutiérrez and Cooper 2013): (1) Dreybrodt (2004) indicated that evaporites dissolve much more rapid than carbonates. Evaporites, such as gypsum and halite, have significantly lower strength and more ductile than most carbonate rocks. In the current study, the genetic classifications of sinkholes proposed by Gutiérrez et al. (2008b, 2014), and by Gutiérrez and Cooper (2013) were used. In these classifications, two terms are used, the first term referring to material type including cover (unconsolidated deposits or residual soil material), bedrock (karst rocks), and caprock (non-karst rocks). The second term represents the subsidence mechanisms including collapse (brittle deformation of soil or rock material), sagging (ductile bending of sediments and rocks due to the absence of basal support), and suffusion (downward migration of cover deposits through voids). Complex sinkholes could be used in which different material types and several mechanisms could be detected. In the current study, according to the genetic classification, different types of sinkholes were detected in the study area as shown in Fig. 2. These types of sinkholes are the most important from a hazard and engineering

Fig. 2. Sinkholes classification in the study area (a) Cover collapse sinkhole, (b) Cover sagging sinkhole, (c) Bedrock collapse sinkhole, (d) Bedrock sagging sinkhole, (e) Caprock collapse sinkhole, (f) Caprock sagging sinkhole, (g) Cover suffosion sinkhole, (h) Solution sinkhole

perspective including; Cover collapse sinkhole (Fig. 2a), Cover sagging sinkhole (Fig. 2b), Bedrock collapse sinkhole (Fig. 2c), Bedrock sagging sinkhole (Fig. 2d), Caprock collapse sinkhole (Fig. 2e), Caprock sagging sinkhole (Fig. 2f), Cover suffosion sinkhole (Fig. 2g), and Solution sinkhole (Fig. 2h).

4.2 Main Characteristics of Karst Features Along the Desert Highways

In the current study, different processes can increase the slope stability problems including: active weathering processes (differential erosions); breakdown processes of the rocks that lead to increase the joints opening; weathered surfaces with the presence of reddish soil zone; and voids, cavities, and sinkholes (empty or filled). Fill materials in sinkholes include boulders (up to 1.5 m diameter) and fine grained matrix (weak and easily eroded) (Fig. 3). The rock and materials along the three selected highways are characterized by; massive limestone with flint is characterized by strong to very strong rock (50–150 MPa), low degree of fracturing, slightly weathered, and unit weight (2.4–2.5 kg/m^3); bedded limestone is moderately strong to strong rock (25–75 MPa), highly jointed, slightly to moderately weathered, and unit weight (2.35–2.45 kg/m^3); marly limestone is moderately weak rock (5–12.5 MPa), moderately to highly jointed, moderately to highly weathered, and unit weight (2.3–2.4 kg/m^3); shale is weak rock (1.25–5 MPa), highly fractured, slightly to moderately weathered, and unit weight (1.7–1.9 kg/m^3); and block-in-matrix deposits are characterized by limestone blocks embedded in weak cemented deposits (very weak material < 1.25 MPa). Several studies have postulated that differential erosion contributes to rockfalls (Hampton et al. 1693). Erosion of the weak materials (marly limestone and shale) along the rock cuts/slopes of the three selected highways causes overhanging of the massive rocks and eventually collapse. The shale layers have color ranges from green to yellow. The sensitivity of a rock type against erosion is usually described by a durability parameter (Khalily et al. 2013). In the current study, 10 jar slake tests were carried out for the green and yellow color shale layers to determine the durability degree according to Santi (Santi 1998) test procedure. The results indicated that yellow shale degrades from a pile of flakes or mud and green shale breaks to form many chips. In addition to that the slake durability test was carried out on 30 samples of the marly limestone of the study area. The marly limestone were subjected to 4 cycles of slake durability test. Results indicate that the slake durability index ranges from 97.5% (high durability) after the fourth cycle to 99% (very high durability) after the first cycle. The presence of these filled sinkholes during excavation, have an adverse effect on the slope stability. In general, during blasting the explosion gasses will force their way out of the rock mass via the karstic discontinuities rather than by breaking intact rock. There are many links between karst features and slope stability problems (landslides) such as solution enlarged discontinuities in the rock mass. The limestone along the rock cut/slope of the three selected sites is highly fractured, and sometimes sinkholes filled with highly weathered materials. The limestone of the rock cut and slopes along the three selected highways is characterized by the presence of filled sinkholes (Fig. 3b). They have a significant impact on slope stability in the study area. The sinkhole package comprises different materials ranging in size from fine matrix materials to coarse boulders as large as 1–2 m in diameter. They are composed of angular materials cemented by reddish, sandy, mud sediments that are weak and easily eroded (by wind and/or water). The presence of these filled sinkholes during excavation, have an adverse effect on the slope stability. In general, during blasting the explosion gasses will force their way out of the rock mass via the karstic discontinuities rather than by breaking intact rock. There are many links between karst features and slope stability problems (landslides) such as solution enlarged discontinuities in the rock mass. The limestone along the rock cut/slope of the three selected sites is highly fractured, and sometimes sinkholes filled with highly weathered materials.

Fig. 3. (a, c) rock cut and slope along the eastern desert highway (Sohag- Cairo); (b) rock cut along the eastern desert highway (Sohag- Red Sea- Cairo); and (a, d) rock cut and slope along the western desert highway (Sohag- Cairo).

4.3 Landslide Types and Their Mechanisms in Karst Zones

Rockfall mechanisms are controlled by different factors including geological characteristics of the area, climate conditions, and weathering processes (differential erosions). Kay et al. (2006) believed that water, lithology, geological structure, and karstification are important factors in triggering rockslides. Perret et al. (2004) indicated that direction/type of discontinuities and the volume of the overhang materials play an important role in the size, shape, and the detachment place of the blocks. In the study areas, fallen boulders and blocks of different sizes and shapes were recognized. These rockfall events are triggered by different factors including discontinuities, wind action, surface water, human activities, and differential erosion among the different rock units. Based on the detailed field investigations, different slope stability models were explained that represent the slope stability problems along the selected highways.

Rockfall and Raveling Model

Two models were proposed for the rockfall activity in the study area including: (1) The first model (Fig. 4a, b) is related to (a) the karstic effect along tectonic fractures and/or faults can slowly widen the width to create open fractures and empty caves that can be some tens of meters long and more than 1 m wide. These might favor the start of rock falls especially along the upper part of the limestone cliffs in the three selected sites. (b) Rockfalls are also associated with the presence of overhanging, the result of erosion of the shale and marly limestone beds separating massive limestone beddings. The shale and marly limestone layers have low resistance to erosion processes. This will leave a large volume of rocks overhanging. Due to the weight of overhanging blocks, tension cracks will appear at the top of the slope. These tension cracks represent an evidence of a deformation on the rock mass. (2) The second rockfall model is particularly problematic in the filled sinkholes where large blocks embedded in weak materials. The fill materials in these sinkholes consist of large boulders (up to 2 m diameter) embedded in fine weakly matrix. The fine materials (matrix) are weak (1.25–5 MPa) and easily eroded by seepage water and wind action. The rockfalls are associated with the presence of overhanging of the large blocks (boulders) due to the erosion of fine materials (Fig. 4c, d).

Sliding Type Models

Adverse geologic structures including open joints, shear zones, and empty cavities exist along the study area. In order to assess the potential mode of failures and the stability of slope faces in the selected highways along limestone rocks, field investigations were performed to map discontinuity distributions (dip direction and dip angle), to calculate the friction angle along the sliding planes, and to determine the main characteristics of the rock masses (Harrison et al. 2002; Waltham 2002; Parise 2008; De Waele et al. 2011; Parise and Lollino 2011; Gueguen et al. 2012; Palma et al. 2012a, 2012b; Pepe and Parise 2014; Andriani and Parise 2015). Field data of discontinuities were

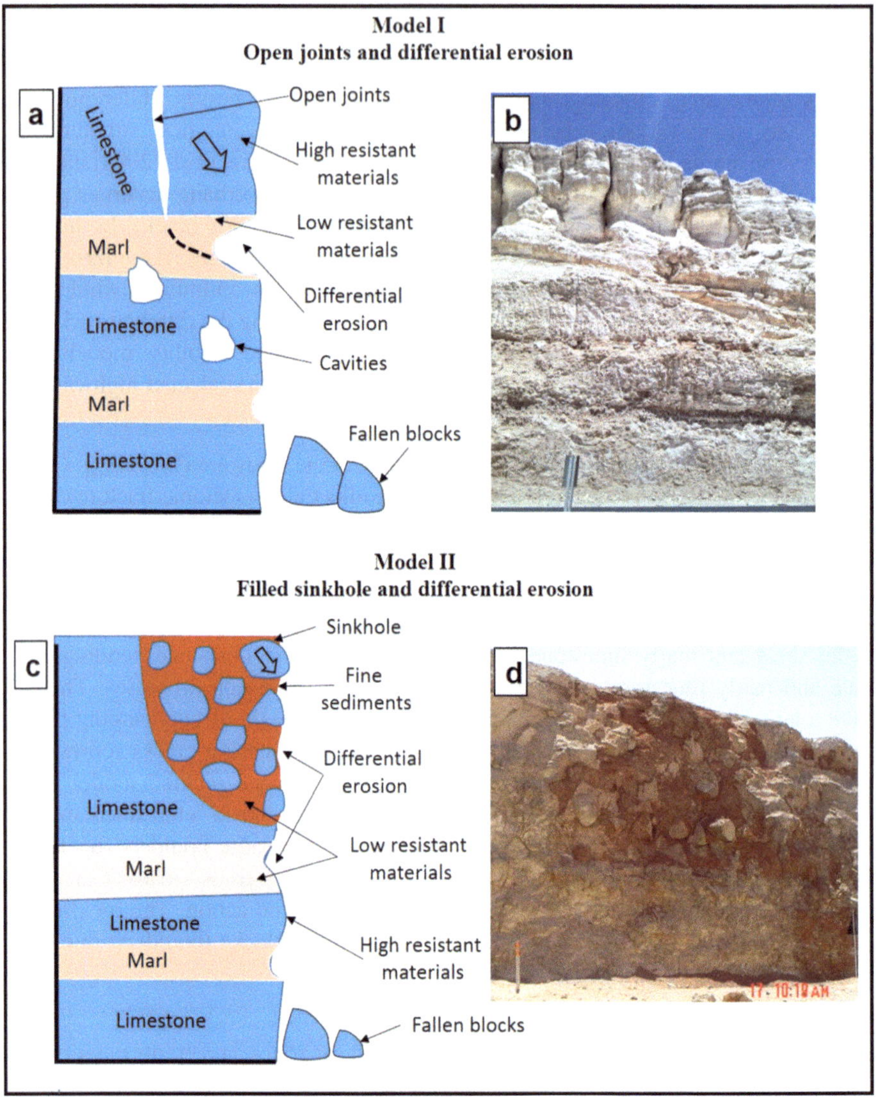

Fig. 4. Models of rockfall and raveling (a, b) due to open joints and differential erosions, (c, d) due to sinkhole and differential erosion.

statistically analyzed for the three selected highways. For the field method, friction angle has been measured using a tilt test, two contiguous blocks are extracted from the exposure, and the upper is laid upon the lower as it was in the rock mass. Both are tilted, and the angle at which sliding occurs was recorded. The results of tilt test of different blocks for the three selected highways in the study area showed that, the value of the tilted angle ranges from 19º–28º. To estimate the friction angle Eq. (1) was used.

The final friction angle for the three selected highways ranges from 22° to 33°. To be conservative a friction angle of 22° has been used in this study.

$$\phi_b = \tan^{-1}(1.155 \tan\alpha) \quad (1)$$

Where alpha (α) is the tilting angle and $Ø_b$ is the friction angle

The calculated friction angle is related to the limestone with weathered surfaces. Consequently, the kinematic method using stereographic projection method was used (Markland 1972), a friction angle of 22° was incorporated into the analysis. The markland test compares the orientation of the rock cut with the orientation of the discontinuities and the friction angle. Results of the test, along the three selected highway zones, indicate that most of the potential failures are of the planar, wedge and circular types (Fig. 5a, b, c, d, e). Planar, wedge and circular failure types are associated chiefly with the presence of joint systems. Weathering effect and karstification features play an important role in these stability mechanisms. Karstic dissolution play fundamental roles in the processes of joint widening, clastic movement, increase of the

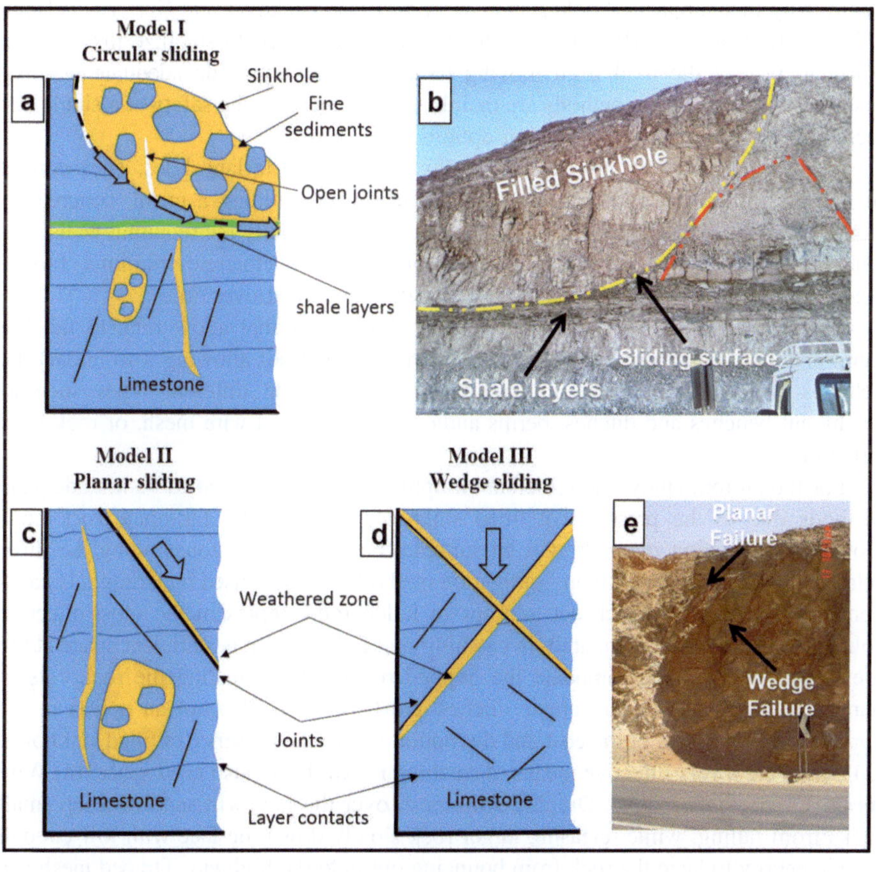

Fig. 5. Models of sliding (a, b) Circular sliding, (c, d, e) planar and wedge sliding.

voids and the reduction of cohesive strength (Fookes and Hawkins 1988; White 1988; Ford and Williams 2007; Palmer 2007). Discontinuities keep widening and extending to the deep of rock mass by constant dissolution and disgregation of underground water (Santo et al. Santo et al. 2007). This leaves cavities supported by remnant points of contact across opened discontinuities leading to decrease in the shear strength, and points of contact may breakdown due to overstressing. The failure potential increases due to the weathering materials along the joint surfaces due to dissolution (clay materials) (Fig. 5e). Another type of failure includes the circular failure along the contact between filled sinkholes and massive rocks (Fig. 5b). The presence of shale layers under the filled sinkhole will accelerate the circular failure to occur. This is related to the rapid disintegration of shale due to water effect. The jar slake tests, of the green and yellow color shale samples, indicate that the shale samples degrades from a pile of flakes or mud (for yellow shale) to form many chips (for green shale).

5 Remediation/Mitigation Options and Recommendations

Many authors suggested different methods that will help in preventing rockfalls or minimizing their impacts (Fookes and Sweeney 1976; Peckover and Kerr 1977; Hoek 1977, 2001; Hoek and Bray 1980). These methods are used to stabilize the rocks in place, to protect the road from moving rocks, and/or to provide adequate warning system to vehicles. Two methods are mainly used to increase the safety at a single site including remediation and mitigation options.

Remediation methods are applying a physical restraint to landslides. They consist of the application of rock bolting and/or shotcrete, or building a metal, concrete or stone retaining wall. These methods are very costly; especially the retaining walls, and tend to be used in very small areas. For the bolted or shotcrete solutions, further deterioration of the rock will make these efforts unpractical over time.

Mitigation Options of landslides consist of methods that allow rock to fail but minimize the damages by containing the fallen rock and not allowing it to impact the vehicles and highways. The mitigation methods include different types such as catchment benches and ditches, berms and/or fences, draped wire mesh, or rock shed structures.

For the current study, the remediation options were considered to be not adequate methods due to the presence of filled sinkholes all over the carbonate rock cuts. However, mitigation methods will be suitable. For the karstic zones along the three selected highways, the following mitigation methods could be used including: (1) catch benches and ditches, which act as traps of fallen rock, represent the most effective solutions. These work adequately in catching rockfalls if there is sufficient space at the toe of the slope to accommodate the trapped rocks. (2) Retreating the rock cuts in narrow roads by blasting to create ditches and benches is also not an option in the current study, because of the cost and disruption due to the presence of filled sinkholes. (3) Berms are a very effective means of catching rockfalls but also need space. (4) Wire mesh is a good alternative. Draping wire mesh over the face will serve to keep small rocks from falling, while releasing larger rock slowly down the face with low enough kinetic energy to keep the rock from bouncing out onto the highway. Draped mesh is a good economical option, but larger rock failure can creates significant damage if the draped mesh is not routinely inspected and maintained. (5) Catch fences or barrier fences are effective way to increase the ditch capacity (volume and width) if the current ditches are inadequate. Fences and barriers such as jersey barriers are ideal solutions for this situation since they occupy small space. (6) Line the ditch with an energy damping material such as gravel or sand. It requires also routine inspection and maintenance (remove of fallen rock).

6 Conclusions

In the current study, three different highway zones were selected to study karst induced slope stability problems. The study reveal that, there is a crucial impact of the karst features on the slope instability phenomena. Many sections of the rock cuts along these selected highways are not stable and may endanger the traffic safety if the rocks are not stabilized. It was found that different karst features contribute the slope stability problems such as differential erosions, open joints, empty cavities, filled sinkholes, and weathering effect along discontinuities. Different landslides types were detected in the study area including rockfalls due to differential erosions and filled sinkholes, planar, wedge, and circular failures. It was found that mitigation methods are most effective to increase the rock stability.

References

Andriani, G.F., Parise, M.: On the applicability of geomechanical models for carbonate rock masses interested by karst processes. Environ. Earth Sci. **74**, 7813–7821 (2015). doi:10.1007/s12665-015-4596-z

Barany-Kevei, I., Gunn, J.: Essays in the Ecology and Conservation of Karst. Acta Geographica Szegedensis, **XXXVI**, (Special Issue) (2000)

Bateman, V.: Development of a database to manage rockfall hazard: the Tennessee rockfall hazard database. In: Proceedings of 82nd Annual Meeting of Transportation Research Board, Washington, D.C. (2003)

Beck, B.F.: Sinkholes: their geology, engineering and environmental impact. In: Proceedings of the First Multidisciplinary Conference on Sinkholes, Orlando, Florida 117 (1984)

Beck, B.F.: Environmental and engineering effects of sinkholes, the processes behind the problems. Environ. Geol. Water Sci. **12**(2), 71–78 (1988)

Burnett, A.D.: Engineering geology and site investigation - part 2: field studies. Ground engineering, pp. 29–32, July 1975

Carbonel, D., et al.: Sinkhole characterization combining trenching, ground penetrating radar (GPR) and electrical resistivity tomography (ERT). Earth Surf. Process Landf. **39**, 214–227 (2014)

Chau, K.Y., et al.: GIS based rockfall hazard map for Hong Kong. Int. J. Rock Mech. Min. Sci. **41**(3), 846–851 (2004). doi:10.1016/j.ijrmms.2004.03.146

Cruden, D.M., Varnes, D.J.: Landslides types and processes. Landslides Investigation and Mitigation. Transportation Research Board, Special report, p. 247 (1996)

Delle Rose, M., Parise, M.: Slope instability along the Adriatic coast of Salento, southern Italy. In: Proceedings IX International Symposium on Landslides, Rio de Janeiro (Brazil), 28 June–2 July 2004, vol. 1, pp. 399–404 (2004)
De Waele, J., et al.: Geomorphology and natural hazards in karst areas: a review. Geomorphology **134**(1–2), 1–8 (2011)
Dreybrodt, W.: Dissolution: carbonate rocks. In: Encyclopedia of Caves and Karst Science, p. 298. Fitzroy Dearborn, New York (2004)
Dunne, T.: Geomorphic significance of subsurface flow. Groundwater geomorphology: the role of subsurface water in earth-surface processes and landforms, (252), 1 (1990)
Fookes, P.G., Hawkins, A.B.: Limestone weathering: its engineering significance and a proposed classification scheme. Q. J. Eng. Geol. **21**, 7–31 (1988)
Fookes, P.G., Sweeney, M.: Stabilization and control of local rock falls and degrading rock slopes. Q. J. Eng. Geol. Hydrogeol. **9**(1), 37–55 (1976)
Ford, D.C., Williams, P.W.: Karst Geomorphology and Hydrology. Unwin Hyman, London (1989)
Ford, D.C., Williams, P.: Karst Hydrogeology and Geomorphology. Wiley, Chichester (2007). 576 p. ISBN: 978-0-470-84996-5
Gillieson. D.: Caves. Blackwell, Oxford. IUCN. 1996. Guidelines for Cave and Karst Protection. The World Conservation Union, Gland, Switzerland (1996)
Gueguen, E., et al.: Geological controls in the development of palaeo-karst systems of High Murge (Apulia). Rend. online Soc. Geol. It. **21**(1), 617–619 (2012)
Gutiérrez, F.: Hazards associated with karst. In: Alcántara, I., Goudie, A. (eds.) Geomorphological Hazards and Disaster Prevention, pp. 161–175. Cambridge University Press, Cambridge (2010)
Gutiérrez, F., et al.: Geological and environmental implications of the evaporite karst in Spain. Environ. Geol. **53**(5), 951–965 (2008a)
Gutiérrez, F., et al.: A genetic classification of sinkholes illustrated from evaporite paleokarst exposures in Spain. Environ. Geol. **53**(5), 993–1006 (2008b)
Gutiérrez, F., et al.: Late Holocene episodic displacement on fault scarps related to interstratal dissolution of evaporites (Teruel Neogene Graben, NE Spain). J. Struct. Geol. **34**, 2–19 (2012)
Gutiérrez, F., Cooper, A.H.: Surface morphology of gypsum karst. In: Frumkin, A. (ed.) Treatise on Geomorphology, vol. 6, pp. 425–437 (2013). doi:10.1016/B978-0-12-374739-6.00114-7
Gutiérrez, F., et al.: A review on natural and human-induced geohazards and impacts in karst. Earth Sci. Rev. **138**, 61–88 (2014). doi:10.1016/j.earscirev.2014.08.002
Hampton, M.A., et al.: Processes that govern the formation and evolution of coastal cliffs. In: Hampton, M.A., Griggs, G.B. (eds.) Formation, Evolution, and Stability of Coastal Cliffs - Status and Trends. U.S. Geological Survey Professional Paper 1693, pp. 7–38 (2004)
Harrison, R.W., et al.: Karstification along an active fault zone in Cyprus. In: Kuniansky, E.L. (eds.) Proceedings of USGS Karst Interest Group, Shepherdstown (West Virginia), Water Resources Investigations Report 02-4174, pp. 45–48 (2002)
Hoek, E.: Rock mechanics laboratory testing in the context of a consulting engineering organization. Int. J. Rock Mech. Min. Sci. Geomech. **14**, 93–101 (1977)
Hoek, E.: Rock Mass Properties for Underground Mines, Underground Mining Methods: Engineering Fundamentals and International Case Studies, Colorado: Society for Mining, Metallurgy, and Exploration (SME) (2001). http://www.rocscience.com/hoek/references/H2001d.pdf
Hoek, E., Bray, J.W.: Rock Slope Engineering Revised 3rd edn. (ISBN 0-419-16010-8), 1980, Institute of Mining and Metallurgy, E&FN Spon Press (1980)

Hungr, O., et al.: Magnitude and frequency of rock falls and rock slides along the main transportation corridors of southwestern British Columbia. Can. Geotech. J. **36**, 224–238 (1999)

Iovine, G.G.R., et al.: Coupling limit equilibrium analyses and real-time monitoring to refine a landslide surveillance system in Calabria (southern Italy). Nat. Hazards Earth Syst. Sci. **10** (11), 2341–2354 (2010)

Ivy-Ochs, S., et al.: Surface exposure dating of the Flims landslide, Graubünden, Switzerland. Geomorphology **103**(1), 104–112 (2009)

Jaboyedoff, M., et al.: Use of terrestrial laser scanning for the characterization of retrogressive landslides in sensitive clay and rotational landslides in river banks. Can. Geotech. J. **46**(12), 1379–1390 (2009)

Jaboyedoff, M., et al.: Slope tectonics: a short introduction, vol. 351, pp. 1–10. Geological Society, London, Special Publications (2011)

Kay, D., et al.: Impacts of longwall mining to rivers and cliffs in the Southern Coalfield. In: Aziz, N. (ed.) Coal 2006: Coal Operators' Conference, University of Wollongong and the Australasian Institute of Mining and Metallurgy, pp. 327–336 (2006)

Khalily, M., et al.: Durability characterization of Abderaz Marly Limestone in the Kopet-Dagh Basin, NE of Iran. Int. J. Emerg. Technol. Adv. Eng. **3**(5), 50–56 (2013)

Kilburn, C.R., Petley, D.N.: Forecasting giant, catastrophic slope collapse: lessons from Vajont, Northern Italy. Geomorphology **54**(1), 21–32 (2003)

Markland, J.T.: A useful technique for estimating the stability of rock slopes when the rigid wedge sliding type of failure is expected. Imperial College Rock Mechanics Research Report 19 (1972)

Palma, B., et al.:. Rock-fall hazard assessment along a road in the Sorrento Peninsula, Campania, southern Italy. Nat. Hazards **61**(1), 187–201 (2012a)

Palma, B., et al.: Analysis of the behaviour of a carbonate rock mass due to tunneling in a karst setting. In: Han, K.C., Park, C., Kim, J.D., Jeon, S., Song, J.J. (eds.) The Present and Future of Rock Engineering. Proceedings of the 7th Asian Rock Mechanics Symposium, Seoul, pp. 772–781 (2012b)

Palmer, A.: Cave Geology, 454 p. Cave Books, Dayton (2007)

Pánek, T., et al.: Time constraints for the evolution of a large slope collapse in karstified mountainous terrain of the southwestern Crimean mountains, Ukraine. Geomorphology **108** (3), 171–181 (2009)

Parise, M.: Rock failures in karst. In: Cheng, Z., Zhang, J., Li, Z., Wu, F., Ho, K. (eds.) Landslides and Engineered Slopes. Proceedings of the 10th International Symposium on Landslides, Xi'an (China), 30 June–4 July 2008, vol. 1, pp. 275–280 (2008)

Parise, M., Pascali, V.: Surface and subsurface environmental degradation in the karst of Apulia (southern Italy). Environ. Geol. **44**, 247–256 (2003)

Parise, M., Gunn, J.: Natural and Anthropogenic Hazards in Karst Areas: Recognition. Analysis and Mitigation. Geological Society, London (2007). sp. publ. 279

Parise, M., Lollino, P.: A preliminary analysis of failure mechanisms in karst and man-made underground caves in Southern Italy. Geomorphology **134**, 132–143 (2011)

Peckover, F.L., Kerr, J.W.G.: Treatment and maintenance of rock slopes on transportation routes. Can. Geotech. J. **14**(4), 487–507 (1977)

Pepe, M., Parise, M.: Structural control on development of karst landscape in the Salento Peninsula (Apulia, SE Italy). Acta Carsologica **43**(1), 101–114 (2014)

Perret, S., et al.: Rockfalls into forests: analysis and simulation of rockfall trajectories - considerations with respect to mountainous forests in Switzerland. Landslides **1**(2), 123–130 (2004)

Prager, C., et al.: Age distribution of fossil landslides in the Tyrol (Austria) and its surrounding areas. Nat. Hazards Earth Syst. Sci. **8**, 377–407 (2008)

Raju, M., et al.: A comprehensive report on Landslide Hazard Zonation of south Mizoram (field season 1997–98). Unpublished report of the Geological Survey of India (GSI), pp. 40–66 (1999)

Santi, P.M.: Improving the jar slake, slake index, and slake durability tests for shales. Environ. Eng. Geosci. **3**, 385–396 (1998)

Santo, A., et al.: Karst processes and slope instability: some investigations in the carbonate Apennine of Campania (southern Italy). Geo. Soc. London Spec. Publ. **279**(1), 59–72 (2007)

Semenza, E., Ghirotti, M.: History of the 1963 Vaiont slide: the importance of geological factors. Bull. Eng. Geol. Env. **59**(2), 87–97 (2000)

Vallejo, L.E.: The influence of notches on the stability of soil and rock slopes. In: Proceedings of the 11th International Symposium on Landslides and Engineered Slopes, Banff, Canada, vol. 1, pp. 415–419 (2012)

Vermeulen, J., Whitten, T.: Biodiversity and Cultural Heritage in the Management of Limestone Resources: Lessons from East Asia. The World Bank, Washington, D.C. (1999)

Waltham, A.C.: The engineering classification of karst with respect to the role and influence of caves. Int. J. Speleol. **31**(1/4), 19–35 (2002)

Waltham, A.C., et al.: Sinkholes and Subsidence: Karst and Cavernous Rocks in Engineering and Construction. Springer, Berlin (2005)

White, W.B.: Geomorphology and Hydrology of Karst Terrains, 464 pp. Oxford University Press, Oxford (1988)

Williams, P.W.: Karst Terrains: Environmental Changes and Human Impact. Catena Supplement, 25. Cetena, Cremlingen-Destedt, Germany (1993)

Youssef, A.M., et al.: Geomorphological hazards analysis along the Egyptian Red Sea coast between Safaga and Quseir. Nat. Hazards Earth Syst. Sci. **9**, 751–766 (2009)

Youssef, A.M., et al.: Coupling of remote sensing data aided with field investigations for geological hazards assessment in Jazan area, Kingdom of Saudi Arabia. Environ. Earth Sci. **65**, 119–130 (2012)

Youssef, A.M., et al.: Natural and human-induced sinkhole hazards in Saudi Arabia: distribution, investigation, causes and impacts. Hydrogeol. J. **24**(3), 625–644 (2016). doi:10.1007/s10040-015-1336-0

Yuan, R.K.S., Mohd, M.I.M.: Integration of remote sensing and GIS techniques for landside applications. In: Proceedings of Asian Conference on Remote Sensing (ACRS) PS397-6 (1997)

Shallow Seismic Refraction Tomography and MASW Survey for Investigating the Fractures Along Qena-Safaga Road, South of Egypt

Ahmed M. Abdel Gowad[1(✉)], Michele Punzo[2], Vincenzo Di Fiore[2], Daniela Tarallo[2], Assem El-Haddad[3], and Abdel Hady Al-Akraby[1]

[1] Department of Geology, Faculty of Science, South Valley University, 83523 Qena, Egypt
ahmed.gharebl@sci.svu.edu.eg

[2] Institute of Coastal Marine Environment, IAMC-CNR, Calata Porta di Massa - Porto di Napoli, 80133 Naples, Italy

[3] Department of Geology, Faculty of Science, Assuit University, 71515 Assuit, Egypt

Abstract. Qena-Safaga road, which is one of the vital transportation lines in south of Egypt, is affected by a set of recent NE-SW to ENE-WSW vertical tensile and en echelon fractures. These fractures are distributed for about 12 km along both sides of the road and constitute a major threat to the infrastructure and environment. Some studies suggested the tectonic origin of these fractures while others suggested that they were formed due to geotechnical problems in the shallow subsurface soil. In order to study the nature and distribution of these fractures in the subsurface, low cost seismic surveys were conducted at the area of KM 22 of Qena-Safaga road. Eleven shallow seismic refraction lines as well as three MASW lines were acquired. The seismic refraction data were analyzed using tomographic methods to produce 2D velocity-depth models. To estimate the near surface seismic properties such as Vp/Vs and Poisson's ratio that are important for the geologic interpretation of the data, MASW data where analyzed to obtain 1D shear wave velocity models. 3D images including 3D volume, fence diagram and depth slices were also produced to study the vertical and lateral variation of the P-wave velocity. In the resulted 2D tomographic models, three seismic units were deduced. The first unit represents the thin weathered surface layer having P-wave velocities of 300–700 m/s. The second unit has P-wave velocities of 700–1600 m/s and S-wave velocities of 281–926 m/s. It may represent the marl unit of Pliocene Durri Formation. Its thickness ranges from 11.8 m to 30 m. The bedrock was deduced to be at depths from 13 to 40 m and is characterized by velocities greater than 1600 m/s. The fractures were traced down to the bedrock along the seismic sections. The variable thickness of the marl unit as well as the variable depths of the bedrock were deduced to be resulting from the effect of subsurface.

J. Wasowski et al. (eds.), *Engineering Geology and Geological Engineering for Sustainable Use of the Earth's Resources, Urbanization and Infrastructure Protection from Geohazards*, Sustainable Civil Infrastructures, DOI 10.1007/978-3-319-61648-3_12

1 Introduction

Since 1970, Qena-Safaga road has been subjected to serious damage due to the fractures occurred in the area between Km 18 to Km 30 of the road at 26° 16′ 59″ & 26° 20′ 9″ north and 32° 48′ 17″ & 32° 55′ 12″ east (Fig. 1). Qena-Safaga road is considered one of the main transportation lines and a vital tourist road in southern Egypt. Its importance increased in the last years due to the development of; Qena El-Gadida city at KM 8 in 2000, the New Qena-Luxor desert road that starts from Km 12,

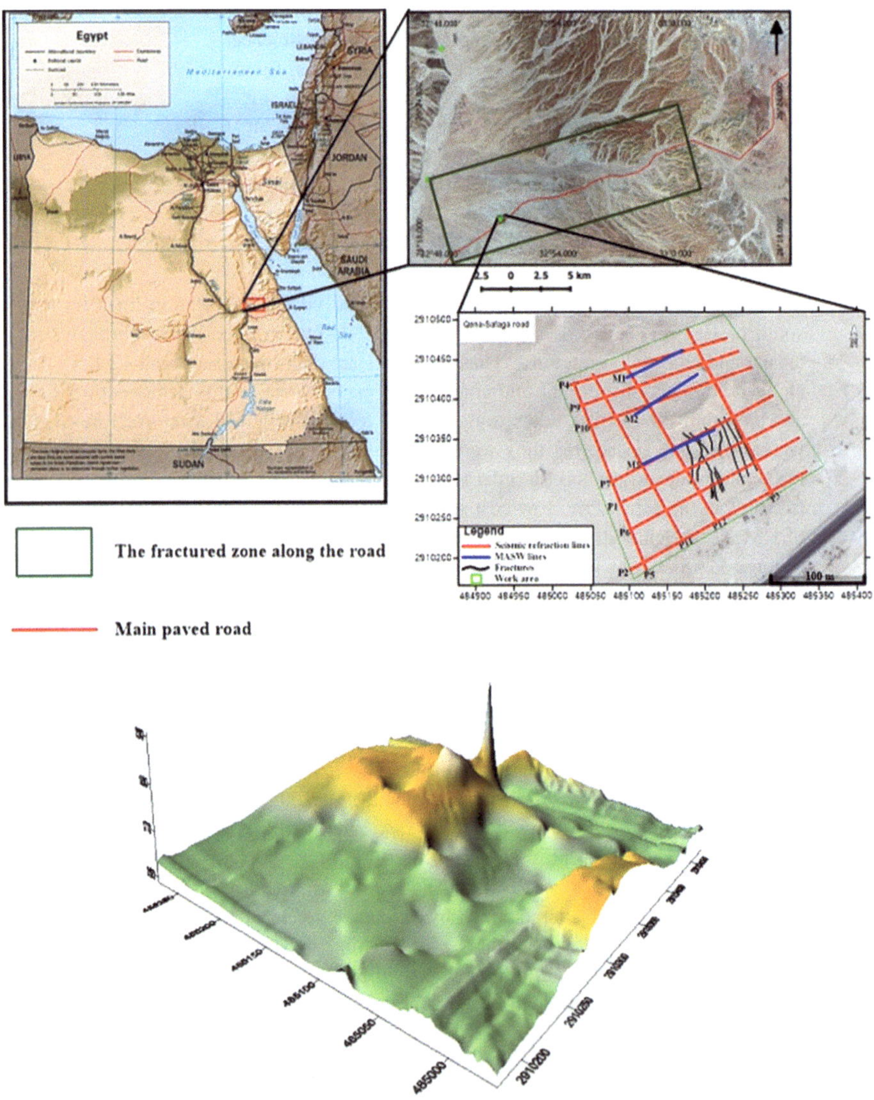

Fig. 1. Location map of the study area and 3D visualization of the elevations

El-Nahda Cement Factory in 2008 at KM 22, the new dessert eastern army road in 2010 at KM 12 and reclaiming land along both sides of the road. The fractures extend along both sides of the road causing severe damages to the local infrastructure. The damage includes cracking and breaking of the asphalt main road, breaking of the potable water pipeline, tilting of power and telephone lines and cracking the nearby buildings (Fig. 2). The fractures are of a progressive nature as new cracks are still forming alongside the road, increasing the threat to the road and nearby structures.

Fig. 2. Some examples of the damages caused by the fractures on the sides of Qena-Safaga Road. In (a), (b) and (c), the fractures damaged the buried water pipes that run parallel to the road. (d) shows cracking of the main asphalt road by the fractures (after Galal 2012)

Many geological and geoengineering studies were performed to explain the origin of these fractures. Earlier studies suggested that the cracks were formed due to the load by heavy trucks on the road or due to solidification and weathering conditions. Several geological studies suggested the tectonic origin of the fractures (Omran et al. 1995a, b; Mesbah et al. 1996; El Kazzaz 1999; Akawy 2002; Moussa and El Arabi 2003; Azab 2009). Such studies indicated that Qena-Safaga road passes a Precambrian brittle shear zone which is composed of a set of strike-slip faults. These faults were reactivated several times and such activity led to the formation of these fractures. In the other hand,

El Akraby (2000) suggested that fractures were formed due to the combined effect of stratigraphic and structural factors. This combined effect can be explained as the sliding of subsurface shale layer along the surfaces of inclined subsurface beds and this is facilitated by the water infiltrated from the water pipe lines. This is accompanied with the effect of the reactivated old tectonic faults. Other recent studies based on geotechnical investigations indicated that the fractures are caused by geotechnical problems such as swelling in the soil and not by tectonic action (Ismaiel et al. 2012; Galal 2012; Makhloof et al. 2013). These studies suggested that the Pliocene sediments around the road were swollen due to water coming from the old damaged water pipes. In addition, drying the saturated Pliocene sedimentary microfacies due to solar effect led to formation of the cracks (large scale mudcracks).

Due to its vital importance, Qena-Safaga road became a target of many governmental projects including widening of the main road, construction of new parallel paved roads, power stations, gas stations, underground water lines, underground gas lines, underground power, telephone and internet cables lines, agriculture reclamation lands and transportation facilities. The fractures that extend for more than 12 km along both sides of the road represent a direct threat to these projects. Thus, investigating these fractures becomes a priority target for many engineering, geotechnical and geophysical studies. In this study, the fractured area around KM 22 is studied using both shallow seismic refraction tomography and multichannel analysis of surface waves (MASW). In this application to ground damage assessment, seismic refraction method along with MASW will detect subsurface conditions and P-wave velocity distribution which is important in evaluating the fractures and the subsurface geologic units. The velocity drop or decrease will give indications regarding the presence of weakness zones (Abiden et al. 2012).

Seismic refraction tomographic method represents one of the most important tools for studying near surface geologic sequences and complex structures. The enormous details produced by 2D and 3D techniques has opened up a huge range of applications in the fields of geology, hydrogeology, environment and engineering. The multichannel analysis of surface wave (MASW) has proven its efficiency for fracture investigation and near surface geotechnical characterization of the ground (Pipan et al. 2005; Tokeshi et al. 2013). The results of MASW can reveal valuable information on the shear wave velocities, Vp/Vs ratio and Poisson's ratio that are important for understanding the properties of the rocks affected by the fractures.

The target of the present study is the analysis of the P-wave velocities to delineate and investigate the subsurface units and structural conditions in the KM 22 area of Qena-Safaga road as well as imaging of the fractures to determine their distribution and depths in the subsurface. This will help in evaluating the area for further projects. In the other hand, the study is also intended to prove the efficiency if the seismic refraction tomography method to image the extensions of the fractures in the subsurface.

2 Geologic Setting

Qena-Safaga district is very complicated from structural point of view with a long tectonic history since the Precambrian to present day. The district is situated within the deltaic mouth of Wadi Qena and made up of Precambrian basement rocks that are found within the fold-and-thrust belt of the Pan African continental margin Orogen (450–650 Ma: El Gaby 1983; Kroner 1984; Shackleton 1986). The basement rocks are unconformably overlain by the Phanerozoic sedimentary rocks that are exposed in two major half grabens dipping to the east and west. The Qena-Safaga road passes a Precambrian NE-SW right lateral brittle shear zone namely Qena-Safaga shear zone (El Gaby 1983; El Gaby et al. 1988), separating the northern from the central Eastern Desert.

According to El Kazzaz (1999), Qena-Safaga shear zone shows dextral sense of shear which initiated during the Neoproterozoic of late Pan African Orogeny (El Gaby et al. 1988). This brittle shear zone is composed of a set of both right-lateral and left-lateral strike slip faults having NW-SE, E-W, NE-SW and N-S trends. The presence of both dextral and sinistral strike slip faults indicates that the reactivation of the faults had taken place at different times with different displacements. In this context, some studies (e.g. Mesbah et al. 1996; El Kazzaz 1999; Akawy 2002) believed that Qena-Safaga shear zone is an active tectonic zone and is now suffering multidirectional extensional stresses. The presence of the opening fractures together with high-magnitude seismicity up to 3.9 with continuously-recorded microseismic events (Basta et al. 1994; Ismail et al. 2000) are evidences supporting this theory (Moussa and El Arabi 2003).

The fractures occurred at the area between KM 18-KM 30 of Qena-Safaga road. They are distributed in sets of NE-SW to ENE-WSW and some have N-S to NW-SE trends. They are mainly tensional fractures with exceptional ones showing an additional strike-slip component – hybrid shear (Akawy 2002). Most of the fractures are opened and filled with siliciclastic materials while the newly-formed fractures are barren with no filling. The fractures are mostly longitudinal tensile with lengths ranging from 1 m to 150 m long (Fig. 3a and b). Widths are ranging from few centimeters up to 1 m width (Fig. 3e and f) and more than 4 m depth. The longer fractures are the oldest (Price and Cosgrove 1990). Variation in length, width and depth indicates the propagation of fractures and reactivation of tectonics (El Kazzaz 1999). Other fractures are randomly oriented en echelon fractures with lengths reach up to 5 m. They are presumably younger and are cross-connected (Fig. 3c and d).

The geologic units outcropped in the area include (Fig. 4):

1. Pliocene sediments represented by the Durri limestone (CONOCO 1987). They are lacustrine deposits cover the Eocene limestone and were formed under arid and semi-arid paleo-environmental conditions. In the Upper Pliocene, the area around Qena was occupied by a fresh water ancient lake interrupting the course of the Nile. This lake was formed when the Mediterranean Sea advanced up to Qena due to a great sinking in the land. As a result, the Nile Valley in Assuit-Luxor stretch flooded

Fig. 3. The different patterns of the fractures along Qena-Safaga road. (a) and (b) show longitudinal tensile fractures. The lengths of these fractures reach up to 150 m. (c) and (d) show the en echelon pattern of the fractures. (e) shows an opened fracture with a width reaches to about 1 m (after Galal 2012). (f) shows a newly formed fracture with width of less than 1 cm

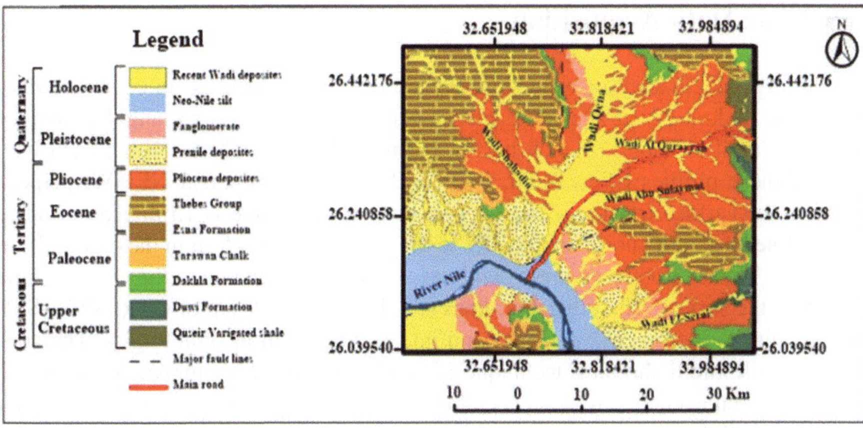

Fig. 4. Geological map of Qena-Safaga district, modified after CONOCO (1987)

forming a narrow and deep estuarine gulf followed by a great movement of uprising forming a series of lakes, or one great lake. The deposits brought to the lakes by lateral streams and by downwash of the surrounded high lands. These sediments were covered by fanglomerates resulting from the erosion of the surrounding Eocene carbonates.

2. The Quaternary sediments:
   a. Prenile sediments (Qena formation). These sediments are divided into two units namely Qena sands at the base and Kom Ombo gravels at the top (Askalany 1988). Qena Sands are the main sediments of Early Pleistocene riverine system (Ahmed 2012) and represent a thick sand unit. Qena sands are unconformably covered by polygenetic gravels (Kom Ombo gravels).
3. Recent to sub-Recent alluvial cover:
   a. Alluvial Fanglomerates formed by the accumulation of series of conglomerates into an alluvial, in rapidly eroding desert environments. They consist of limestone, flint and dolomitic limestone gravels and boulders with thickness of about 10 m.
   b. Wadi Deposits made up of reworked materials from different rock units. They are the most dominant deposits in the wadis occupying low areas with aggressive-bouldery nature. They are composed of silt, sand and gravels of different sizes derived from the nearby Eocene exposures and are intercalated with clay lenses.

The geomorphological features in the study area include hills, pediment (alluvial) and wadis which are mostly controlled by the structural setting of Qena-Safaga district (El Hakim 1978; El Kazzaz 1999; El-Akraby 2000; Akawy 2002; Akawy and Gamal El Din 2006; Azab 2009).

3 Seismic Data Acquisition

In this study, 11 2D shallow refraction seismic lines were acquired to cover the fractured zone at km 22 of Qena-Safaga road (Fig. 1) within a rectangular area of 240 × 240 m. The length of each line is 235 m with 48 geophones along each line. Seven of these profiles (P1, P2, P4, P6, P7, P9, P10) are oriented NE-SW, perpendicular to the surface fractures. The interval between each line is 15 m. The distance between line P10 to the north and line P7 to the south is about 45 m in the east to 75 m in the west. The area between these two lines was not covered due to the elevated hills that hinder the survey. The other four lines (P3, P5, P11, P12) are oriented NNW-SSE perpendicular to the other seven lines and intersecting them at right angles. The continuous profile multi-shooting technique is employed, where a number of 14 in-line shots were taken every 18 m along each line except for line P1 where shots were taken every 20 m. All the lines were recorded in the forward direction, except for the line P4 which recorded in the reverse direction.

The P-wave data were recorded using the 12-channel seismograph (model ES-3000) manufactured by Geometrics. A portable weight dropper of 80 kg designed by El Haddad (2003) was used as an energy source (Fig. 5). The seismic signals were picked up using 14 Hz vertical geophones fixed vertically at 5 m intervals. The first shot for each line is offsetting zero concordant with the first geophone. The shots were stacked from 2 to 20 times. The seismic record length was 0.5 s with a 0.250 ms sample rate. No filters were applied during the acquisition.

Fig. 5. The weight-dropper seismic energy source designed by El-Haddad (2003) that used for the seismic refraction data acquisition

For MASW survey, three lines (SW1, SW2 and SW3) coincident with the refraction lines P4, P9 and P10 respectively, were acquired. The length of each line is 94 m with a number of 48 geophones for each lines. Along each line, two shots offsetting 2 m from each end were taken at the forward and reverse directions. The data were recorded using the 12-channel ES-3000 seismograph with 10 kg sledge hammer as a source. Each shot was stacked 3 times. 10 Hz geophones were used with 2 m interval. The 10 Hz geophones can be also used for MASW survey when investigating for a maximum depth of 30 m (Park et al. 1999). The record length was 1 s and the sample interval was 0.5 ms. No acquisition filters applied.

4 Data Processing and Interpretation

Processing and interpretation of the acquired seismic refraction data was carried out using the commercial software package SeisImager/2D® (Pickwin v.5.1.1.2 2014 and Plotrefa v.3.0.0.6 2014). For all the recorded shots along every line, the traces were normalized and the bad traces were killed. The geometry parameters including the correct shot positions and elevations, geophone positions and elevations, and offset distances were assigned to all the seismic traces in each record.

The seismic records were processed by manually picking the first arrival traveltimes using Pickwin program. Figure 6 shows example of picked data records. Almost 7392 first breaks were picked from more than 154 shots. The picked first arrival times were plotted against the source-to-geophone distances to produce a set of traveltime-distance curves for each line. The traveltime-distance curves were checked and some of them were corrected for precise interpretation of the data. Examples of the traveltime curves are shown in Fig. 7.

The ultimate objective of a refraction seismic survey is to prepare seismic velocity-depth sections specifying the thickness and depths of all layers with the corresponding velocities. Conventional methods for estimating 2D velocity models such as intercept time method (Gurvich 1972), generalized reciprocal method (Palmer 1981) and plus-minus method (Hagedoorn 1959), provide a simplifying rough estimate of the subsurface velocity and structure, but they fail in the presence of strong lateral velocity variations. In the other hand, refraction tomography may give more detailed depth models even in presence of complex velocity structures. Field observations and the previous geological studies show that the area may suffer lateral velocity variation due to the fracturing behavior, lateral facies change and geotechnical characteristics variation in the subsurface (Galal 2012; Ahmed 2012; Ismaiel et al. 2012). Therefore, seismic refraction tomographic inversion was used in the present study for interpreting the acquired seismic refraction data. Depending of the refraction tomography on the initial model can decrease, in some cases, the reliability of the inversion (Bohm et al. 2012). In this prospect, model obtained from another inversion technique could be used as initial model for the tomography in order to avoid this possible drawback.

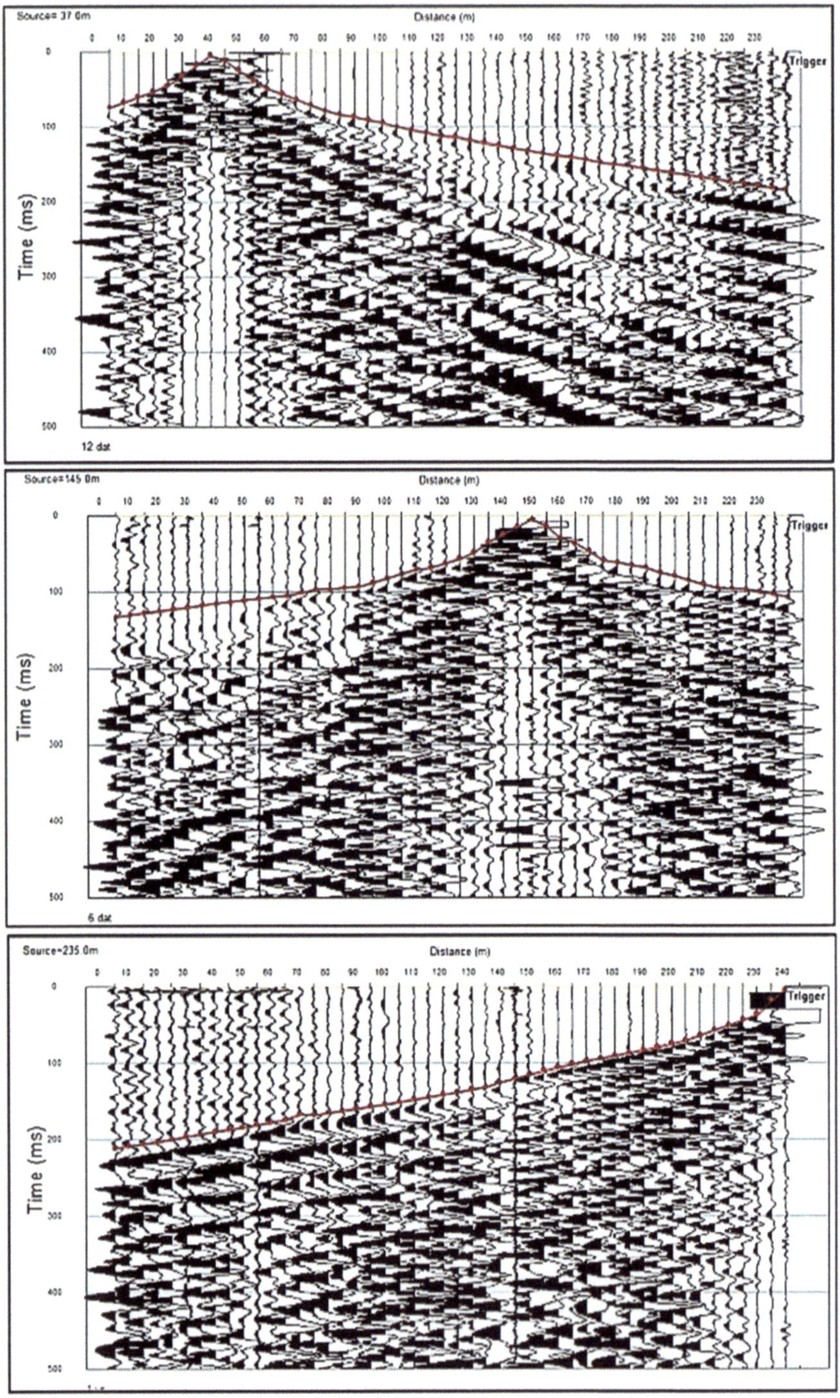

Fig. 6. Example of picked shot records from seismic refraction line P9

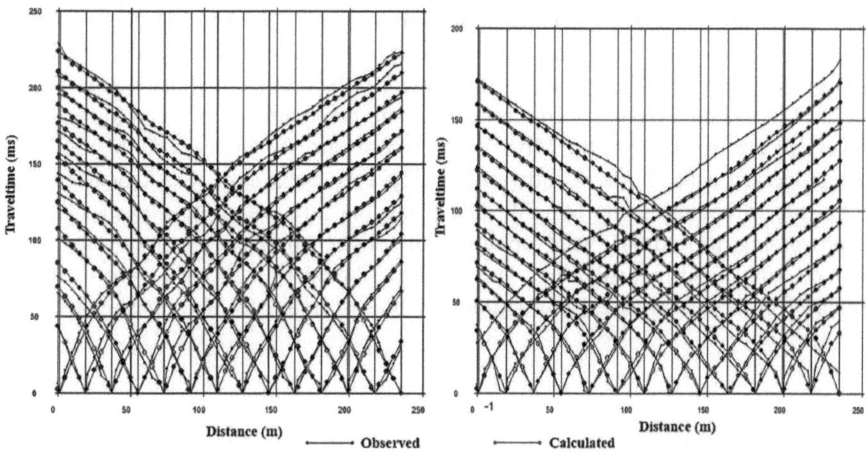

Fig. 7. The observed and calculated traveltime-distance curves of seismic refraction lines P4 and P6

In this study, a first-approximation depth-model was obtained using Godograf software which is designed for traveltime inversion based on the 2D homogenous function method of two coordinates (Piip 1991, 2001). The resulted model is then used as an initial model for tomographic inversion using Plotrefa software. Plotrefa uses nonlinear traveltime tomography (Gebrande and Miller 1985; Rühl 1995) consisting of ray tracing for forward modeling and simultaneous iterative reconstruction technique (SIRT) for inversion. The resulted model is adjusted by iterations to minimize the RMS error between calculated and observed traveltimes in order to produce a velocity model devoid of small scale artifacts.

Fourteen shots in each line were used in the tomographic modelling. The observed and calculated data have a difference of RMS error of less than 4% indicating a best fit

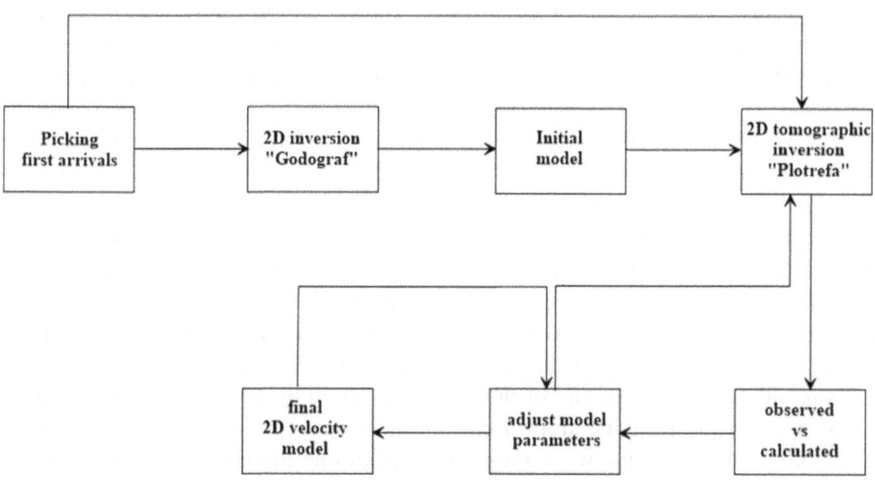

Fig. 8. Processing flow of the tomographic inversion of the seismic refraction data

between the observed and calculated traveltimes (Fig. 7). Eleven 2D depth-velocity models were produced. The models were converted into layered models to better represent the geology. The processing flow used in the method is illustrated in Fig. 8.

The MASW data records were normalized and the geometry assigned to the traces in every record. Example of the MASW shot records are shown in Fig. 9. MASW data were processed using Surfseis software developed by the Kansas Geological Survey, USA. Due to acquiring the MASW data using high-frequency geophones (10 Hz), pre-processing procedures were applied. The higher frequency components were removed with band-pass filtering. The resulted filtered shot records were then preliminarily processed to assess the optimum ranges of frequency and phase velocity.

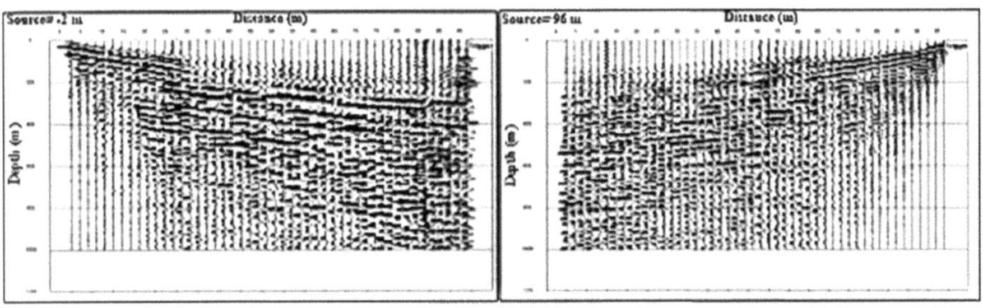

Fig. 9. Two shot records of the MASW line SW2

The overtone image containing the fundamental mode of surface waves and which represents the phase velocity versus frequency, were analyzed for assessing and picking the dispersion curve. This analysis starts with calculating the phase velocities within the specified frequency range. This calculation can be performed several times using different input parameters until an optimum solution is obtained. Generally, the curve with the highest signal-to-noise ratio represents the best solution to be picked. This curve is picked manually, then inverted to generate a 1D shear-wave velocity model (Fig. 10). The inversion uses an iterative process known as the Levenberg-Marquardt least-squares algorithm (Xia et al. 1999) which requires initial earth model (S-wave velocity, P-wave velocity, density and layer thickness) to be specified first. A synthetic dispersion curve is then generated. Due to its influence on the dispersion curve only the shear wave velocity is updated after each iteration until the synthetic dispersion curve closely matches the field curve. To improve the results, the two resulted 1D Vs models from the inversion of each shot were combined together to form a consistent model. These models represent the material directly below the middle of the acquisition line. The final 1D Vs models were superimposed on the coincident 2D Vp models to be used later in the interpretation as a criterion for vertical lithological change (Fig. 11). In these locations, the Vp/Vs ratio which is usually viewed as lithology discriminator (Essien et al. 2014) as well as the Poisson's ratio (σ) were calculated.

Poisson's ratio is a measure of the transverse strain or contraction to longitudinal strain or extension resulting from a change in normal stress under compression or dilatation. A change in the wave propagation from one lithology to another leads to

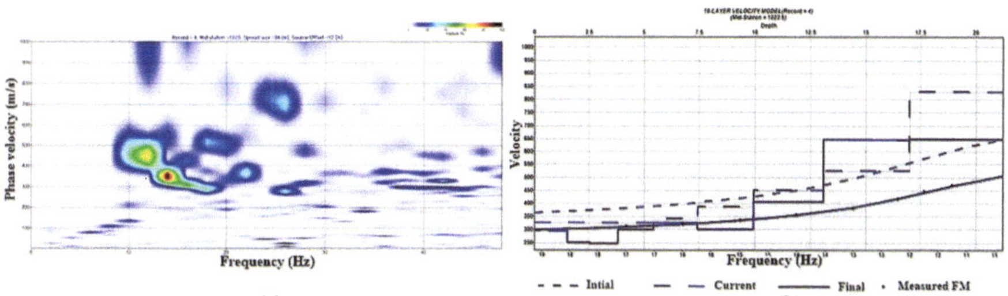

Fig. 10. (a) The overtone image containing the frequency vs. phase velocity (dispersion curve) for MASW line SW3. (b) The resulted 1D Vs model from the inversion of the dispersion curve in MASW line SW3

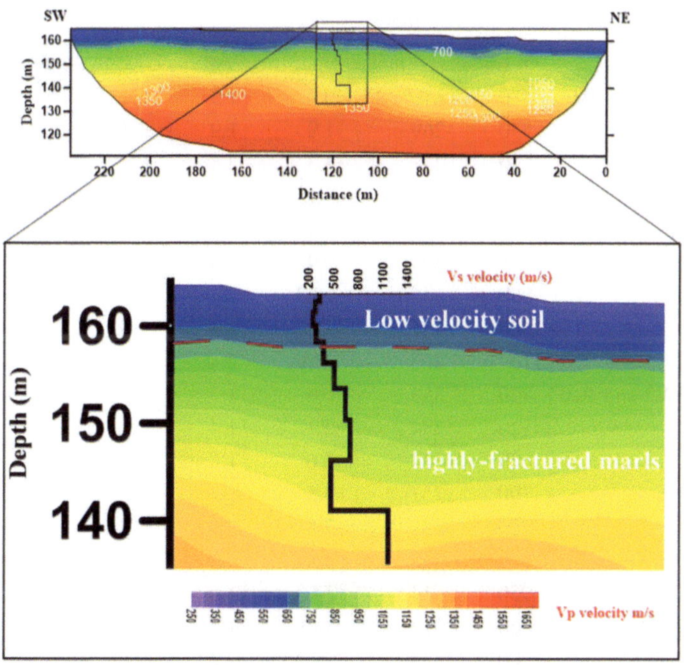

Fig. 11. The 1D Vs model for MASW line SW1 superimposed on the P-wave velocity section for the seismic refraction line P4

variations in σ from 0 (indicating a considerable change in volume) to 0.5 (indicating no volumetric change, corresponding to fluids). Also, σ can have negative values which indicate certain properties (Salem 2000). Poisson's ratio can be calculated from Vp/Vs ratio using the equation:

$$\sigma = \frac{Vp^2 - 2Vs^2}{2(Vp^2 - Vs^2)} \tag{1}$$

Table 1. Typical values of Poison's ratio for some rock types after Sharma et al. (1990) and Davidovici (1985).

Type of material	Sharma et al. (1990)	Davidovici (1985)
Saturated clay	0.40–0.5	0.5
Unsaturated clay	0.10–0.3	-
Medium clay	-	0.40–0.45
Sandy clay	0.20–0.30	0.35–0.4
Silt	0.30–0.35	-
Sand, gravely sand	0.30–0.40	0.25–0.30
Silty sand, clayey sand	-	0.30–0.35
Rock	0.10–0.40	-

Table 2. The measured Vp and Vs velocities in the locations of the 1D S-waves models as well as the calculated Vp/Vs and Poisson's ratios.

Profiles	Vp (m/s)	Vs (m/s)	Depth (m)	Vp/Vs	Poisson's ratio σ
SW1 and P4	470	318	−0.9	1.5	0.1
	530	269	−1.9	1.9	0.3
	610	237	−3.3	2.6	0.4
	668	279	−5.1	2.4	0.4
	720	368	−7.2	1.9	0.3
	775	502	−9.9	1.5	0.1
	880	646	−13.2	1.4	−0.1
	1000	701	−17.4	1.4	0.02
	1100	462	−22.6	2.4	0.4
SW2 and P10	470	353	−0.9	1.3	−0.1
	520	384	−2.1	1.4	−0.1
	620	337	−3.6	1.8	0.3
	700	257	−5.5	2.7	0.4
	780	281	−7.8	2.8	0.4
	850	495	−10.7	1.7	0.2
	950	612	−14.3	1.6	0.1
	1060	615	−18.9	1.7	0.2
	1200	534	−24.6	2.2	0.4
	1370	926	−30.7	1.5	0.1
SW3 and P7	432	298	−0.7	1.4	0.04
	492	292	−1.5	1.7	0.2
	568	249	−2.6	2.3	0.4
	658	245	−3.9	2.7	0.4
	772	298	−5.4	2.6	0.4
	900	323	−7.4	2.8	0.4
	1004	299	−9.9	3.4	0.5
	1131	404	−13.1	2.8	0.4
	1222	644	−17	1.9	0.3
	1250	647	−21.2	1.9	0.3

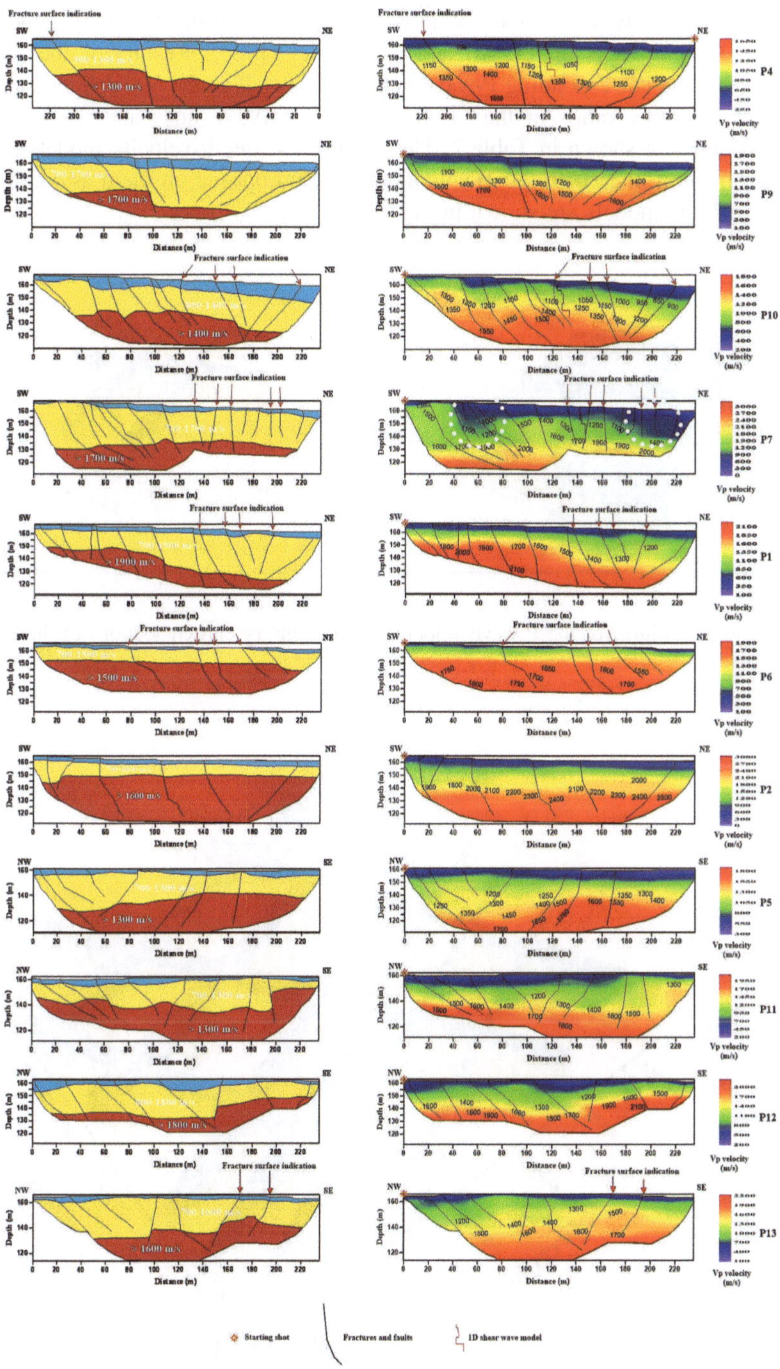

Fig. 12. The generated 2D depth-velocity models resulted from the seismic refraction tomographic inversion and the corresponding layered sections

Table 1 shows typical values of Poisson's ratio for some lithology conditions according to Sharma et al. (1990) and Davidovici (1985). Results of the S-wave velocities and the corresponding P-wave velocities as well as the Vp/Vs ratios and the calculated Poisson's ratios in the locations of the MASW lines and the coincided refraction lines are shown in Table 2. The resulted 2D depth-velocity models for the seismic refraction lines with the 1D Vs models superimposed on them as well as the layered geologic models are shown in Fig. 12.

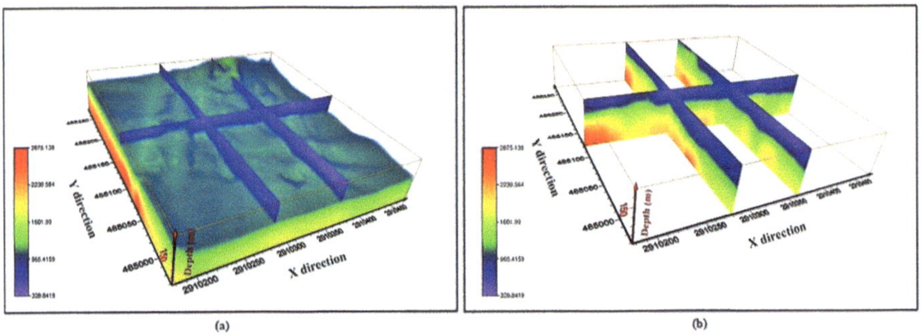

Fig. 13. (a) 3D volume of the P-wave velocities. (b) Arbitrary chosen sections of the P-wave velocities

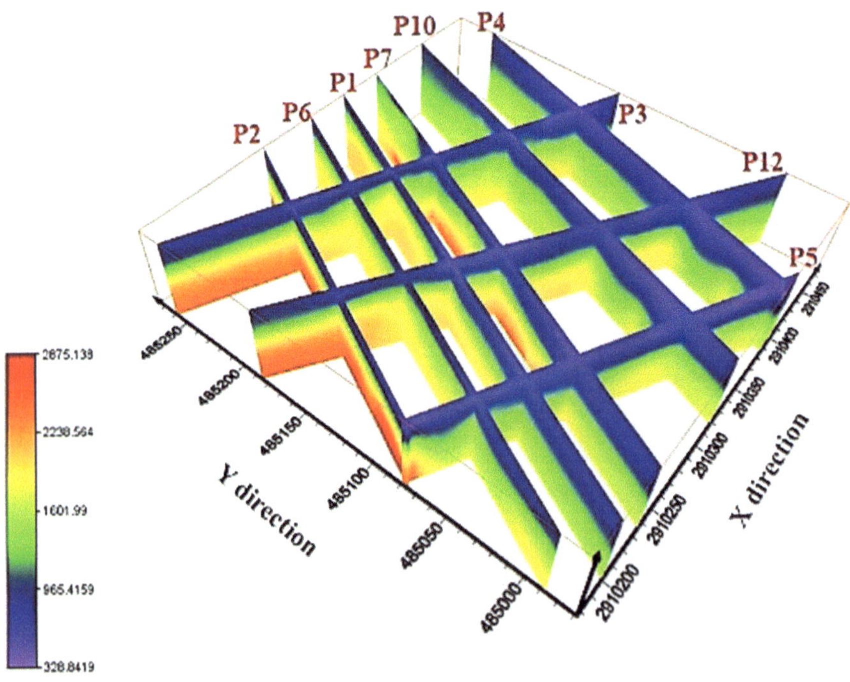

Fig. 14. 3D fence diagram of the seismic refraction lines

The 2D velocity models were then combined via Voxler® to generate 3D images showing the 3D distribution of the P-wave velocity in the subsurface. In this prospect, the seismic data were visualized as follows:

- 3D volume of the P-wave velocity distribution (Fig. 13a).
- Single arbitrary sections (Fig. 13b) allowing any combination of x-, y- or z-cross-sections to be plotted in any orientation. This is important to assess the data consistency from the all measured lines.
- 3D fence diagram (Fig. 14) generated by combining the resulted depth-velocity models. This fence diagram is important to point out the spatial coherence of the P-wave velocities through the survey area.

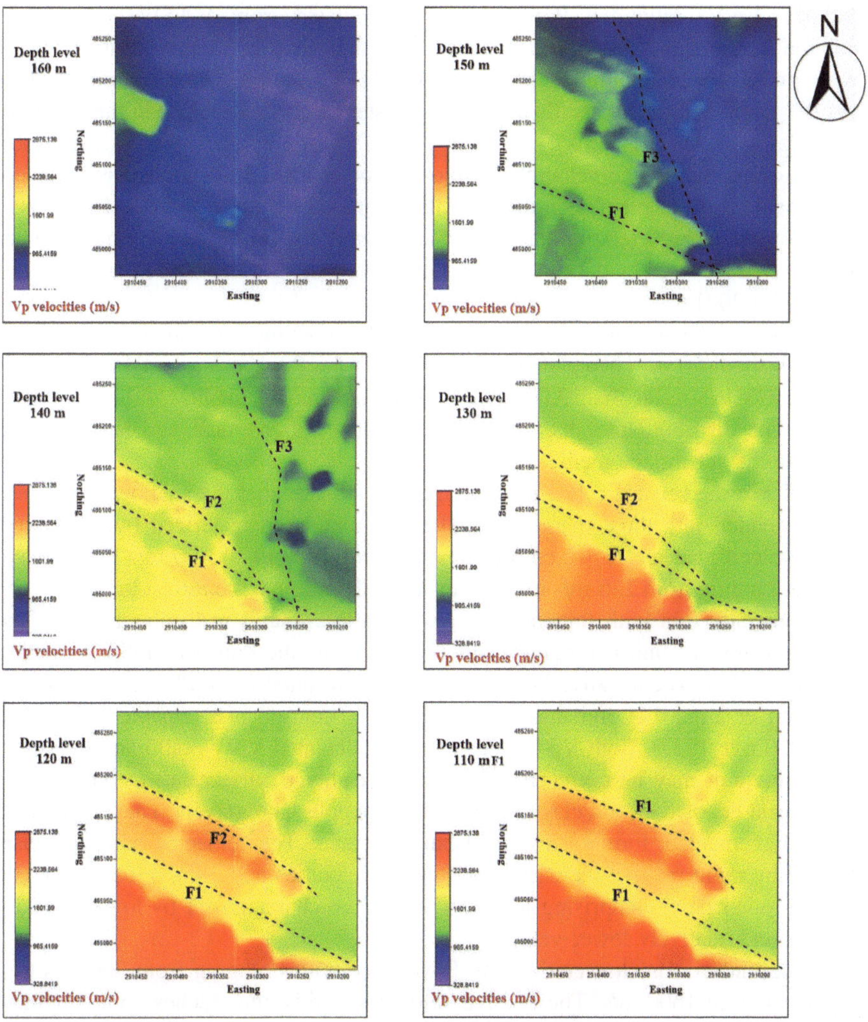

Fig. 15. Depth slices show the P-wave velocity distribution in chosen depth levels

- Depth slices showing the lateral variation of P-wave velocity in the subsurface (Fig. 15). These slices were generated from the 3D volume for depths from 160 m to 110 m with 10 m interval.

5 Results and Discussion

Results of the tomographic inversion and the geologic cross-sections constructed from the estimated velocity models were constrained by; field studies of the outcrops and exposures in the area; previous geological and geophysical studies (El Akraby 2000; Akawy 2002; Akawy and Kamal El Din 2006; Galal 2012; Ahmed 2012); an underway geotechnical study based on core samples (Ahmed pers. Comm.); and shear wave velocities, Vp/Vs ratio and Poisson's ratio calculated from MASW and seismic refraction tomography. As a primary inspection, there is a good conformity between P-wave and S-wave velocities in the locations of 1D Vs models. The determined values of Vp/Vs ratios and the Poisson's ratio are consistent with those found in the literature for similar shallow sediments (Lankston 1990; Almeida et al. 1999; Salem 2000). A visual evaluation of the velocity gradient can be estimated as gradient is inversely proportional to distance between the velocity isolines (El-Haddad et al. 2014). This is also important in characterizing the subsurface seismic zones as each layer has an inner structure, characterized by dominant values of the velocity gradient and its own contour pattern (Piip 2001).

Analysis of the seismic sections shows that the maximum depth of investigation reached is about 50 m. The calculated velocities of the P-waves are ranging from less than 300 m/s to greater than 2,000 m/s which characterizes shallow subsurface succession. The distribution of the velocities indicates both vertical and lateral anisotropism in the subsurface media.

Results of the seismic refraction tomography indicate that the shallow subsurface may be divided into three main units. The first unit is characterized by the lowest P-wave velocities detected which are ranging from 300 m/s to 700 m/s and S-wave velocities ranging from 256 m/s to 352 m/s in the 1D shear wave model locations. The unit is generally thin with thickness ranges from less than 1 m to about 13 m. The Vp/Vs ratio for this unit is ranging from 1.3 to 2.7 and the Poisson's ratio is ranging from – 0.1 to 0.4. The negative Poisson's ratios are resulted for Vp/Vs ratios less than $\sqrt{2}$ (Essien et al. 2014) and are symptomatic of occurrence of ripable anisotropic materials (Love 1927). Both Vp and Vs and the calculated Vp/Vs ratio and Poisson's ratio of this unit may be corresponding to the weathered surface layer composed of loose Recent deposits of silt, clay and sands with pepples and gravels. These sediments are mostly cemented, forming dissected fanglomerates. The cemented parts are characterized by rapid relative increasing of P-wave velocities and high velocity gradient while the other parts comprising loose sediments are characterized by lower velocity gradients and relatively lower velocities.

The second unit is characterized by P-wave velocities ranging from 700 m/s to a mean velocity of 1600 m/s. The P-wave velocities of this unit reaches a maximum of 1900 m/s in many parts of the area. The corresponding shear wave velocities in the locations of MASW lines are ranging from 281 m/s to 926 m/s. This unit occurs at depths of less than 1 m in the southwest to 13 m in the east. It is outcropped on the surface in small zones in the west and southeast. The thickness of this unit is variable and is ranging from 11.8 m in the southeast to more than 35 m in the middle and eastern parts of the study area. The map in Fig. 16 shows the thickness variation of this unit in the study area.

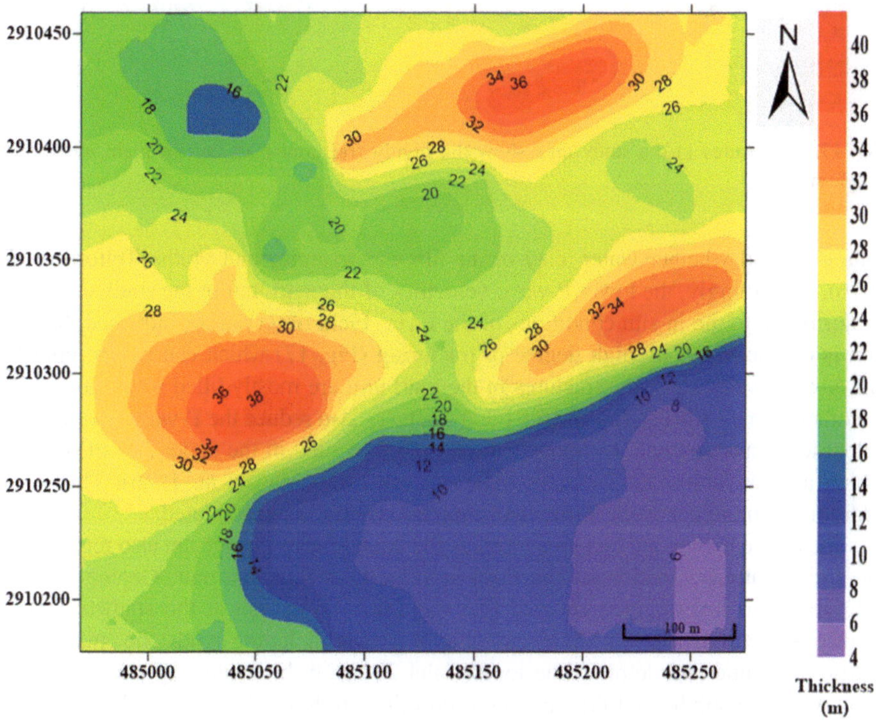

Fig. 16. A contour map showing the variable thickness of the marl unit in the study area

In the MASW lines locations, the Vp/Vs ratio are ranging from 1.02 to 3.7 and the calculated Poisson's ratio is ranging from −0.1 to 0.5. Based on the field observations of the exposures and outcrops in the study area (Fig. 17), the P- and S-wave velocities in addition to the Vp/Vs ratios and Poisson's ratio, this unit may be corresponding to the marl layer represented the Durri Formation of the Pliocene age. The Vp/Vs ratios and their corresponding Poisson's ratio are concordant with those deduced for marls by Dvorkin et al. (2001) and Al-Khersan et al. (2011).

Bourbie et al. (1987) and Pavolovic and Veličković (1998) showed that the typical P-wave velocities for marl are in the range of 2000–3500 m/s and S-wave velocities are ranging from 750 m/s to 1500 m/s. However, the P-wave velocities in the real geologic media is depending on many factors such as rock density, mineral content, lithology, porosity, depth of burial and degree of compaction, degree of fracturing or jointing,

Fig. 17. Exposures in the study area show the highly fractured marl unit overlain by Recent fonglomerates

saturation and whether water, clay or air fills the fractures and voids (Telford et al. 1976). Accordingly, the lower P-wave velocities of the marl unit in the study area may be attributed to the in-situ conditions of the layer. Field observations indicated that the unit is characterized by high degree of fracturing (Fig. 17) which can reduce both P- and S-wave velocities. The fractures in the marl unit are mostly filled with clay, while the others are air-filled. Clay and air-filled fractures also reduce the P-wave and S-wave velocities considerably. Clay intercalations also prevailed in the upper part of the unit which also an important factor in reducing velocities (Han et al. 1986). This is in addition to the upper parts of the marl unit are mostly unconsolidated.

The above mentioned factors as well as the lateral facies change are also reasons for the strong lateral variation of the P-wave velocity within this unit. Geological and petrographical investigation showed that the Durri Formation at the studied area is mainly composed of five sedimentary microfacieses including mudstone, marlstone, marly, dolomitic and detrital limestone (Galal 2012). In his underway study, Ahmed (pers. Comm.) subdivided this formation into three different facies, each has different lithological characteristics. The lateral variation of P-wave velocity can be inspected from the 2D velocity models as well as the estimated Vp depth slices. From the 2D Vp models, it is shown that the P-wave velocities are laterally decreasing from the west to the east. Drop in Vp velocity in the upper part of the layer is recorded in the south-western and northeastern corners of P10 and P7 lines (marked on the Vp models for both lines). This velocity drop may be interpreted as weakness zones due to fracturing.

The third unit deduced from the depth-velocity models is characterized by P-wave velocities that range from a minimum value of 1300 m/s to more than 1900 m/s. The estimated depths of this unit are ranging from 13 m to 40 m. In the depth slices, the higher velocities of this unit are detected at depth level of 140 m, which marked the first occurrence of the unit in the southwest region. The decreasing of depth of this unit towards the southwestern and southern areas may attributed to the uplifting caused by subsurface faults as well discussed later in this section. The P-wave velocities of this unit may be corresponding to the limestone unit deduced from the DC resistivity investigation by El-Akraby (2000). This unit may constitute the bedrock in the study area. This agrees with the underway study by Ahmed (pers. Comm.).

On the generated Vp models, the fractures could be defined as undulations and perturbations in the velocity isolines as well as areas of steep velocity gradients. This may indicate that the fractures are just major cracks with no effective displacement of the subsurface units as revealed by Makhloof et al. (2013) and by the present study. Accordingly, the fractures as defined on the Vp models extend from the surface weathered layer down to the subsurface with different depths. The minimum detected depth of the fractures is about 23 m, while others extend to depths of more than 40 m reaching to the bedrock unit. Fractures with depths less than the wavelength of the recorded seismic waves were not resolved on the velocity sections. The elongated wide cracks could be traced along the all seismic lines that pass them. The three larger fractures in the middle of the study areas were traced from line P7 to P6 as shown in the Vp models in Fig. 18.

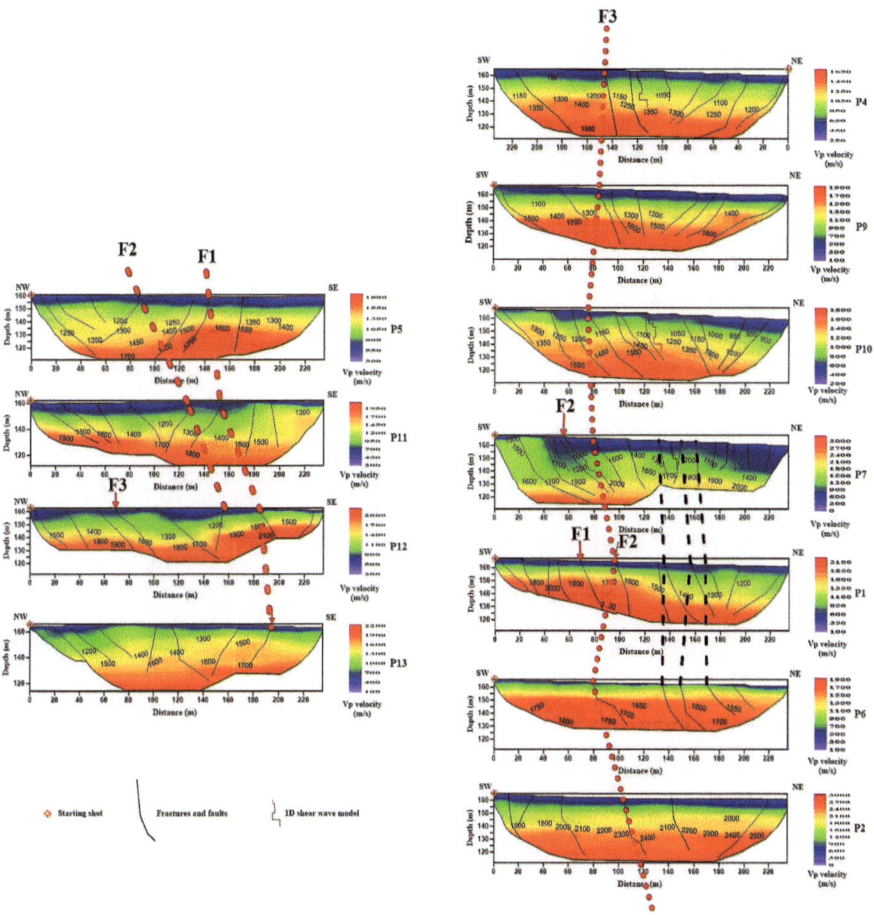

Fig. 18. Tracing fractures and faults along the 2D depth-velocity models

In the other hand, subsurface faults that might cause thinning and variable depths of the subsurface units could be defined as they caused considerable displacements in the velocity isolines. A major fault (F1) trending NW to SE is detected in the south part of the study area. This fault is traced along the Vp models for the seismic refraction lines P5, P11, P12 and P3 as illustrated in Fig. 18. The fault is also defined on the depth slices of the levels 150–110 m. A second fault (F2) is also detected on the NW-SE seismic refraction lines P5, P11 and P12 and is trending NW-SE. It is also defined at the depth slices of levels 140–110. As indicated from the Vp models, the faults F1 and F2 might be responsible for the southward thinning of the marl unit and the variable depths of the bedrock unit.

A third fault (F3) trending N-S is also defined in the perpendicular direction for the two faults F1 and F2. It is detected on the Vp models for the seven NE-SW lines (P4 to P2). This fault only defined on the depth slices of the levels 150 and 140 m. As shown on the Vp models, this fault causes small scale displacement comparable to the faults F1 and F2 and thus might not causes considerable lateral variations of the Vp velocities lower depths. Figure 19 shows the detected faults on the location map of the study area.

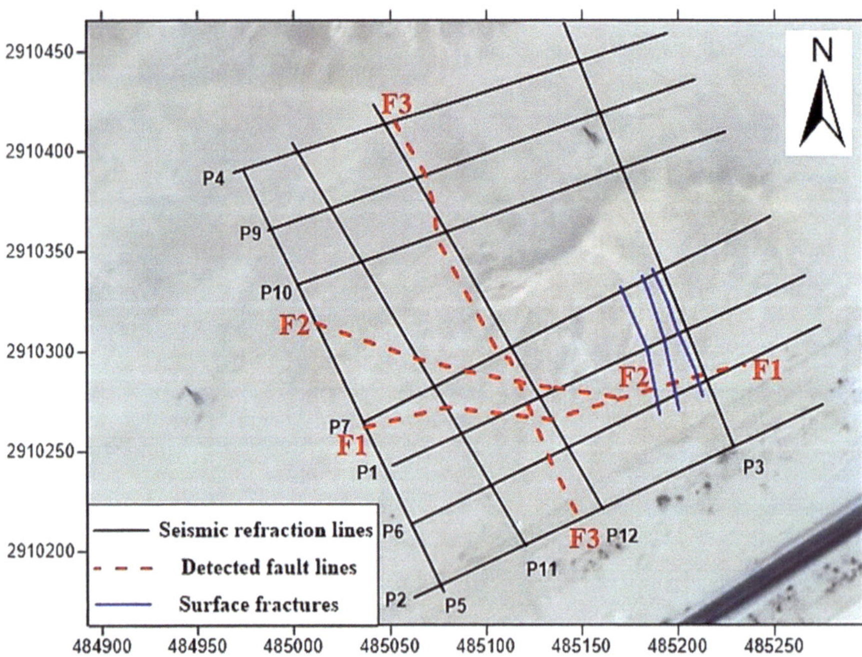

Fig. 19. Location map of the study area illustrating the deduced fault lines

6 Conclusions

Low cost seismic surveys including shallow seismic refraction and MASW were conducted to image and study the fractures in KM 22 area of Qena-Safaga road. The use of 2D seismic refraction tomography as well as 1D shear wave models helps in obtaining useful information on the subsurface conditions in the study area. Generated 3D images including 3D volume, fence diagram and depth slices accompanied with 2D seismic tomography models has made it possible to obtain 2D and 3D distributions of P-wave velocity propagation in the subsurface.

The results indicated that the shallow subsurface of the study area might comprised of three units. The first unit has P-wave velocities of 300–700 m/s and may represent the thin weathered surface layer that composed of loose gravels, sands and silts. The second unit has P-wave velocities of 700–1600 m/s and S-wave velocities of 281–926 m/s. It occurred at depths from less than 1 m to 13 m. The thickness of this unit is variable and is ranging from 11.8 m in the southeast to more than 35 m in the middle and eastern parts of the study area, with considerable thinning southward. This unit may represent the marls of the Pliocene Durri Formation. Strong lateral variation of the P-wave velocities was exhibited within this unit that may reflect lateral facies change. The third unit has P-wave velocities of more than 1600 m/s. It represents the bedrock that might comprised of limestone unit. This unit occurred at variable depths ranging from 13 to 40 m with depth decreases southward. The surface fractures were traced on the velocity-depth models where they extended from the surface to depths from 23 m to 40 m affecting the bedrock in some parts.

Akawy (2002) and Azab (2009) deduced that the fractures along Qena-Safaga road might be formed due to recent tectonic activity in the Red Sea region that led to reactivation of older tectonic trends causing fracturing of the shallow layers along Qena-Safaga road. On the other hand, recent geotechnical studies by Ismaiel et al. (2012), Galal (2012) and Makhloof et al. (2013) indicated that the marls of Durri Formation are the responsible for forming the fractures by swelling action. In the present study, three faults were detected that affected the area causing southward thinning of the marl unit and decreasing depth of the bedrock in the southern part. Two faults are trending NW-SE perpendicular to the major surface fractures. These major fractures are the longest (lengths of more than 90 m) and the widest (widths about 0.5 m) among the other fractures in the area. Likewise, they extend for about 40 m in the subsurface penetrating the upper surface of the bedrock unit as deduced from the 2D depth-velocity models. Accordingly, they are probably the oldest formed fractures in the survey region. On the local scale of the study area, it is deduced that the occurrence of the fractures might be connected with subsurface faults. The NW-SE fault trends were confirmed by the detailed structural study of Akawy (2002) and Akawy and Kamal El Din (2006). The two detected NW-SE trending faults caused relatively considerable displacement in the marl and the bedrock units while the other N-S trending fault causes small-scale displacement. High-magnitude seismicity of 3.9 in the region between Km 20 to Km 30 along Qena-Safaga road was recorded by Basta et al. (1994) as well as recent and ongoing microseismic events of magnitude between 1 and 2 were also recorded (Ismail et al. 2000; Badawy 2005). In addition, the continuous opening and expanding of the older formed fractures as well as forming of new fractures is still ongoing as indicated from the field observations of the present study and previous geological studies. All these mentioned conditions indicate the recent and ongoing activity of the local structural lines in Qena-Safaga district.

Furthermore, Ahmed (2012) indicated that the marls of the Durri Formation in Qena-Safaga district have low swelling ability. Therefore, the authors support the forming of the fractures due to recent activity of subsurface faults. However, this is cannot be fully resolved. Another study using deep seismic reflection and refraction methods accompanied with deep core samples is highly recommended to investigate the deeper subsurface along Qena-Safaga road.

Accordingly, due to the deduced subsurface faults and the depths of the fractures in the subsurface as well as the present ongoing opening and forming of the fractures along the road, urban projects are not recommended in the study area.

Using the seismic refraction tomography to image the shallow fractures and faults could be efficient when surface occurrences of the fractures and faults are present in the study area. Careful mapping of these features on the depth-velocity models could be possible with precise field observations.

Acknowledgments. The authors would like to thank Dr. Ahmed M. Ibrahim Ahmed, Assistant lecturer in Department of Geology, Faculty of Science, South Valley University, Qena, Egypt for sharing valuable information and results of his ongoing geological and geotechnical study on the fractured zone of Qena-Safaga road. Also, we would like to thank Dr. J. Wasowski and an anonymous reviewer for their valuable comments and suggestions.

References

Abidin, M.H.Z., Saad, R., Ahmad, F., Wijeyesekera, D.C., Baharuddin, M.F.T.: Seismic refraction investigation on near surface landslides at the Kundasang area in Sabah, Malaysia. In: International Conference on Advances Science and Contemporary Engineering (ICASCE 2012) (2012). Procedia Engineering, vol. 50, pp. 516–531

Ahmed, A.M.I.: Geotechnical and sedimentological studies of the quaternary sediments at Qena Region, Egypt. M.Sc. Thesis (2012). 125 pages

Akawy, A.: Structural geomorphology and neotectonics of the Qina-Safaga district, Egypt. N. Jb. Geol. Palaont Abh. **226**, 95–130 (2002)

Akawy, A., Kamal El Din, G.M.: Middle eocene - recent tectonics in the Qina area, Upper Egypt. Neues Jahrbuch für Geologie und Paläontologie, Abhandlungen **240**, 19–51 (2006). doi:10.1127/njgpa/240/2006/19

AL-Khersan, E.H., Khorshid, S.Z., Faraj H.H.: Determination some of physical and geotechnical properties of the calcareous rocks in Kufa Quarry using ultrasonic velocities. Iraqi Natl. J. Earth Sci. **11**(2), 11–36 (2011)

Almeida, F., Moura, R., Teves Costa, P., Oliveira, C.S.: Caracterização Dinâmica das Areias de Faro-Quarteira Através de Ensaios Sísmicos. In: 4th Meeting on Seismology and Seismic Engineering, Faro, Portugal, Expanded Abstracts, pp. 233–243 (1999)

Askalany, M.M.S.: Geological studies on the Neogene and Quaternary sediments of the Nile Basin, Upper Egypt. Ph.D. Thesis, Fac. Sci., Assuit Univ. (1988). 254 pages

Azab, M.A.: Assessment and management of natural hazards and disasters along Qena-Safaga road, central eastern desert, Egypt. J. Remote Sens. Space Sci. **12**, 55–70 (2009)

Badawy, A.: Seismicity of Egypt. Seismol. Res. Lett. **76**(2), 149–160 (2005). doi:10.1785/gssrl.76.2.149

Basta, N.Z., Ismail, A.M., Salem, E.M., Eissa, G.K., Mohamed, A.Z., El Hakim, B.A.: Seismic activity at Qena-Safaga region. Unpublished internal report, The Egyptian Geological Survey and Mining Authority, Abbasyia, Cairo (1994)

Böhm, G., Francese, R., Giorgi, M.: Integrated refraction seismics and tomographic study of a gravitational collapse phenomenon. Bollettino di Geofisica Teorica ed Applicata **53**(4), 539–550 (2012). doi:10.4430/bgta0081

Bourbie, T., Coussy, O., Zinszner, B.: Acoustics of Porous Media. Gulf Publishing, Houston (1987). 334 pages

CONOCO: Geological map of Egypt, Sheet of Asuit, scale 1:500 000 (1987)

Davidovici, V.: Génie parasismique. École Nationale des Ponts et Chaussées, Paris, 1105 (1985)

Dvorkin, J., Walls, J.D., Mavko, G.: Rock physics of marls. SEG Technical Program Expanded Abstracts, pp. 1784–1787 (2001)

El-Akraby, A.M.: A contribution to the delineation of the geologic setting of the fracture area (km 19–31) along Qena-Safaga road, using geoelectrical studies. Ain Shams Sci. Bull. **38**, 95–111 (2000)

El-Gaby, S.: Architecture of the Egyptian basement complex. In: Proceedings of 5th International Conference of Basement Tectonics, Cairo, vol. 5, pp. 1–8 (1983)

El-Gaby, S., List, F.K., Tehrani, R.: Geology, evolution and metallogenesis of the Pan-African belt in Egypt. In: El-Gaby, S., Greiling, R.O. (eds.) The Pan-African belt of Northeast Africa and Adjacent Areas, pp. 175–184. Friedr Viewgsohn, Braunschweig/Wiesbaden (1988)

El-Haddad, A.E.: Development of portable weight dropper apparatus and field applications in Dakhla Oasis area, Western Desert, Egypt. Bull. Fac. Sci. Assiut Univ. **2**(1-F), 103–119 (2003)

El-Haddad, A.E., El-Akraby, A.M., Ismail, I.S., Abdel Gowad, A.M.: Investigation of groundwater potentiality in the eastern part of Qena, central Eastern Desert, Egypt, using 2D homogeneous function method. Arab J. Geosci. **7**, 3011–3021 (2014). doi:10.1007/s12517-013-0934-1

El Hakim, B.A.: Study of aeromagnetic survey for Qena-Safaga area (A.R.E.). M.Sc. Thesis, Department of Physics, Faculty of Science, Alexandria University, Alexandria, Egypt (1978)

EL-Kazazz, Y.A.: Active faulting along Qena-Safaga road, Eastern Desert, Egypt. In: The First International Conference on the Geology of Africa, vol. 2, pp. 385–404. Assiut Univ., Egypt (1999)

Essien, U.E., Akankpo, A.O., Igboekwe, M.U.: Poisson's ratio of surface soils and shallow sediments determined from seismic compressional and shear wave velocities. Int. J. Geosci. **5**, 1540–1546 (2014). doi:10.4236/ijg.2014.512125

Galal, A.A.: Surveying and geotechnical studies for the evaluation of deformation along Qena-Safaga road. M.Sc. Thesis, Minia University (2012). 182 pages

Gebrande, H., Miller, H.: Refraktionsseismik. In: Angewandte Geowissenschaften II, Methoden der angewandten geophysik und mathematische verfahren in den geowissenschaften, pp. 226–260 (1985)

Gurvich, I.: Seismic Prospecting. Mir Publishers, Moscow (1972)

Hagedoorn, G.J.: The plus–minus method of interpreting seismic refraction sections. Geophys. Prospect. **7**, 158–182 (1959). doi:10.1111/j.1365-2478.1959.tb01460.x

Han, D.H., Nur, A., Morgan, D.: Effects of porosity and clay content on wave velocities in sandstones. Geophysics **51**, 2093–2107 (1986). doi:10.1190/1.1442062

Ismail, A.M., Basta, N.Z., El-Hakim, B.: Seismic activity at Qena-Safaga region. Ann. Geol. Surv. Egypt **V**(XXIII), 707–712 (2000)

Ismaiel, H.A.H., Makhloof, A.A., Mahmoud, A.A., Galal, A.A.: Geotechnical behavior of pliocene sedimentary rocks exposed along Qena-Safaga Road at Qena region, Egypt. Int. J. Min. Eng. Miner. Process. **1**(2), 84–93 (2012). doi:10.5923/j.mining.20120102.10

Kroner, A.: Late Precambrian plate tectonics and orogeny: need to redefine the term Pan-African. In: Klerkx, J., Michot, J. (eds.) African Geology, Musee Royal de L'Afrique Central, Tervuren, pp. 23–28 (1984)

Lankston, R.W.: High-resolution refraction seismic data acquisition and interpretation. In: Ward, S.H. (ed.) Geotechnical and Environmental Geophysics, vol. 1, pp. 45–73. Society of Exploration Geophysicists (1990). doi:10.1190/1.9781560802785.ch3

Love, A.E.H.: A Treatise on the Mathematical Theory of Elasticity. Cambridge University Press, Cambridge (1927). 104 pages

Makhloof, A.A., Mahmoud, A.A., Ismaiel, H.A.H., Galal, A.A.: Evaluation and interpretation of cracks along Qena-Safaga road using surveying techniques. Minia J. Eng. Technol. (MJET) **32**(2), 194–203 (2013)

Mesbah, M.A., Mousa, M.M., Odah, H.: Fault tracing using detailed land magnetic survey at Qena. Egypt. Egyptian Geophys. Soc. **14**, 141–158 (1996)

Moussa, M.M., El Arabi, A.M.: Soil radon survey for tracing active fault: a case study along Qena-Safaga road, Eastern Desert, Egypt. Radiat. Meas. **37**, 211–216 (2003). doi:10.1016/S1350-4487(03)2900039-8

Omran, A.A., Philobbos, E.R., Riad, S., Othman, A.B.: Subsurface structures and their surface reflection in Wadi Qena and the surrounding area. Egypt J. Geol. **39**(1), 313–338 (1995a)

Omran, A.A., Philobbos, E.R., Riad, S., Othman, A.B.: Sedimentary basins and crustal thickness in Wadi Qena and the surrounding areas as interpreted from Bouguer anomalies. Egypt J. Geol. **39**(1), 377–398 (1995b)

Palmer, D.: An introduction to the generalized reciprocal method of seismic refraction interpretation. Geophysics **46**, 1508–1518 (1981). doi:10.1190/1.1441157

Park, C.B., Miller, R.D., Miura, H.: Optimum field parameters of an MASW survey. SEG-J, Tokyo, pp. 22–23 (Expanded Abstract) (1999)

Pavlović, V.D., Veličković, Z.S.: Measurement of the seismic wave propagation velocity in the real medium. Phys. Chem. Technol. **1**(5), 63–73 (1998)

Piip, V.B.: Local reconstruction of seismic refraction sections on the basis of homogeneous functions. Izvestiya Acad. Sci. USSR Phys. Solid Earth **27**, 844–850 (1991)

Piip, V.B.: 2D inversion of refraction traveltime curves using homogeneous functions. Geophys. Prospect. **49**, 461–482 (2001). doi:10.1046/j.1365-2478.2001.00270.x

Pipan, M., Forte, E., Dal Moro, G., Gabrielli, P.: Integrated seismic and GPR characterization of fractured rocks. In: Workshop on Hydrogeophysics – A Tool for Sustainable Use of Groundwater 2005, Palermo, Italy (2005)

Price, N.J., Cosgrove, J.W.: Analysis of Geological Structures. Cambridge University Press, Cambridge (1990). 502 pages

Rühl, T.: Determination of shallow refractor properties by 3D CMP refraction seismic techniques. First Break **13**, 69–77 (1995). doi:10.3997/1365-2397.1995005

Salem, H.S.: Poisson's ratio and the porosity of surface soils and shallow sediments determined from seismic compressional and shear wave velocities. Geotechnique **4**, 461–463 (2000). doi:10.1680/geot.2000.50.4.461

Shackleton, R.M.: Precambrian collision tectonics in Africa. In: Coward, M.P., Ries, A.C. (eds.) Collision Tectonics, pp. 329–349. Geological Society, Special Publication, London (1986). doi:10.1144/GSL.SP.1986.019.01.19

Sharma, H.D., Dukes, M.T., Olsen, D.M.: Field measurement of dynamic moduli and Poisson's ratios of refuse and underlying soils at a landfill site. In: Landva, A., Knowles, G.D. (eds.) Geotechnics of Waste Landfills: Theory and Practice, pp. 57–70. American Society for Testing and Materials, Philadelphia (1990). doi:10.1520/STP25299S

Telford, W.M., Geldart L.P., Sheriff, R.E., Keys, D.A.: Applied Geophysics. Cambridge University Press, Cambridge (1976)

Tokeshi, K., Harutoonian, P., Leo, C.J., Liyanapathirana, S.: Use of surface waves for geotechnical engineering applications in Western Sydney. Adv. Geosci. **35**, 37–44 (2013). doi:10.5194/adgeo-35-37-2013

Xia, J., Miller, R.D., Park, C.B.: Estimation of near-surface shear-wave velocity by inversion of Rayleigh waves. Geophysics **64**, 691–700 (1999). doi:10.1190/1.1444578

Shallow Seismic Investigation of the Yangtze River Fault at Zhenjiang, Jiangsu Province, China

Guang-ya Wang[1], Greg You[2(✉)], Ming-zhu Chen[1], Guo-xing Zhou[3], and E. Jian[1]

[1] Key Laboratory of Earth Fissures Geological Disaster, MLR (Geological Survey of Jiangsu Province), Nanjing 210049, China
[2] Mining Engineering, Federation University Australia, Ballarat, VIC 3353, Australia
g.you@federation.edu.au
[3] Jiangsu Coal Geophysical Prospecting and Surveying, Nanjing 210046, China

Abstract. Debates on the Yangtze River Fault at Zhenjiang, Jiangsu Province, China have lasted for a long time. In order to clarify the stratigraphy, characteristics and activity of the fault, a large-scale shallow seismic investigation was conducted on the site. It involved 10 onsite field experiments, 5715 m long shallow seismic exploration and five geological drill holes, and consequent data processing and interpretation of the shallow seismic time profiles in combination with one geological profile and a satellite image. It was found that the loose Quaternary strata present good continuity in the area. Four pre-Cenozoic faults present in the bedrock (including the Yangtze River Fault), and two Quaternary faults cut the Quaternary sediments. Therefore, the existence of the Yangtze River Fault is verified, and its activity and characteristics are analyzed in this study.

1 Introduction

Shallow seismic exploration is widely applied for investigations of subsurface structures (Hanebuth et al. 2002; Chourak et al. 2005; Iacono et al. 2009), such as fault (Yang et al. 2004; Peng and Ben-Zion 2006; Zhigulev and Patrikeev 2007), bedrock profile and overlying stratigraphy (Ismail et al. 2012; Rucker et al. 2012), and shallow marine explorations for resources (Klein et al. 2005; Liu et al. 2013), such as gas (Bertin and Chaumillon 2005; Andreassen et al. 2007). In this technique, seismic waves generated by an artificial source are received by arrays of sensors. Different techniques exist, differing for both the kind of waves analyzed, e.g. P-wave, S-wave or surface wave, and the wave propagation exploited, e.g. by reflection or refraction (Al-Saigh and Al-Dabbagh 2010; Cifci et al. 2011; Ismail et al. 2012). As a result, different combinations can be seen in literature (Hanebuth et al. 2002; Zhigulev and Patrikeev 2007; Liu et al. 2013). The principle of wave propagation is well understood. When the source wave propagates in media and it refracts and reflects at interfaces. The reflective wave can be received on ground surface and processed subsequently, resulting in a seismic profile that can be correlated to the subsurface structures and strata

J. Wasowski et al. (eds.), *Engineering Geology and Geological Engineering for Sustainable Use of the Earth's Resources, Urbanization and Infrastructure Protection from Geohazards*, Sustainable Civil Infrastructures, DOI 10.1007/978-3-319-61648-3_13

(Zhigulev and Patrikeev 2007; Cifci et al. 2011; Ismail et al. 2012). This is an effective alternative for investigating shallow geological structures, so its application in geology and engineering has been increasing.

In China, there has been a rapid advance in deep geophysical exploration since 1980s. The deep seismic exploration has created a very important database to study the deep geological structure and its relationship with the shallow one in the earthquake zones. But the target strata in those explorations were mostly deeply-buried, well-layered strata and structures (Wang et al. 1994; Liu et al. 1999). In last 20 years, high resolution seismic technique was used for shallow geological structures, such as faults, caverns, tunnels and pillars in underground coal mine (Xie et al. 2000; Yue and Chai 2008; Chai 2009). Yin et al. (2000) applied a basket of techniques, including high-resolution shallow seismic, for the exploration of deep structures of the Wushan loose deposit. Liu et al. (2012) achieved high Signal to Noise Ratio (SNR) p-wave, s-wave and converted-wave profiles through the application of seismic source of p-waves or s-waves recorded by three-component receivers. In particular, the interpretation on geological stratigraphy and fault identification inferred from the analysis of the horizontal component of transverse and converted waves was in agreement with the geological profile acquired from the cross-fault drilling.

Given the successes of the early applications in China, there is a demand for extending its applications in geological survey, e.g. to study the distribution of Quaternary sediments and their profiles. The purpose of this research work was to investigate the bedrock profile, its overlying sediments and the characteristics of the Yangtze River Fault at Zhenjiang, Jiangsu Province, China and to preliminarily evaluate the activity of the Fault by means of shallow seismic (longitudinal wave) exploration in combination with geologic drilling and satellite image.

2 Geological Background

The research area is located on the south bank of Yangtze River in Zhenjiang city, Jiangsu province, China, as shown in Fig. 1. It is on the top edge of the Yangtze River Delta, a vast alluvial plain covered by Quaternary deposits, which are stratified and composed of clay, sandy clay, and silt and fine sand, medium sand, and gravelly coarse sand. The groundwater table is shallow, about 2.0 to 3.0 m below ground surface. The physical properties of the clay and the sand layers are significantly different, so efficient wave reflection develops on the interfaces of strata.

The area is geologically located on the Ning-Zhen uplift. Lithology mainly consists of Mesozoic Cretaceous igneous rocks and Triassic sandstone and carbonate rocks. Since the physical properties of deep rocks are relatively similar, especially under the shielding effect of igneous rocks, consistent high SNR wave reflections can be hardly obtained from their interfaces.

The exploration line crossed the urban area from southwest to northeast, passing through a main road, a demolition zone, a scenic area outside Jinshan Park, a sluice of rivers and a beach of Yangtze River. The terrain is complex and the composition of shallow sediments is diversified. The local economy is well developed, with high density of buildings and transportation network, heavy traffic and plenty of surface water bodies. As a result, the site working conditions were complex, which caused some difficulties to the fieldwork, such as the stability of wave source and the interference from passing vehicles.

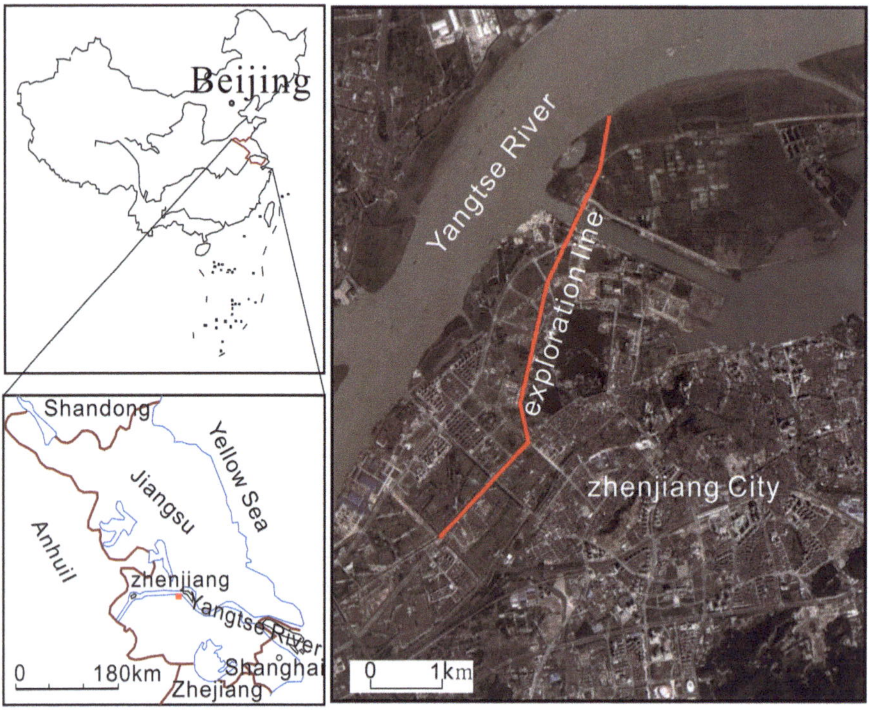

Fig. 1. Location map of the research area

3 Shallow Seismic Exploration

The initial stage of reflection survey involves field trials in the area to determine the most suitable combination of source, offset recording range, array geometry and detector spacing (the horizontal distance between the centers of adjacent geophone arrays, often referred to as the group interval) to produce good seismic data in the prevailing conditions.

To determine appropriate parameters for generating and receiving seismic waves, 10 field experiments were carried out prior to the exploration. The experiment line was located on the north of and paralleled to the exploration line. The recording parameters are detailed in Table 1.

The purpose of the field trials was to determine optimum shot number and scanning length. From a set of comparative field experiments carried out with the receiving frequency of 20 to 120 Hz, the optimum number of shots was found to be three and the optimum scanning length was 12 s (Fig. 2). These parameters were then used for the seismic exploration.

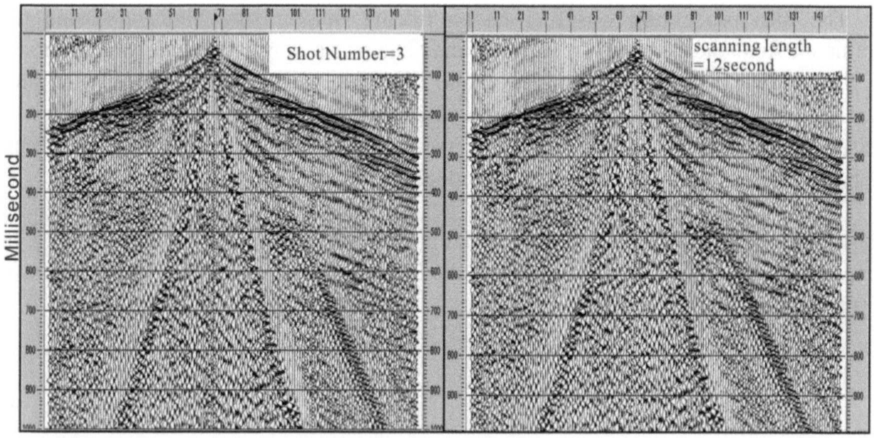

Fig. 2. Seismic travel time and scanning length of single shot in field experiments

The exploration line is 5.715 km long, passing some major infrastructures, such as a main road and a sluice in the city (see Fig. 1). Considering the site condition and interference from passing vehicles, the wave generation and acquisition system was chosen according to what detailed in Table 1.

Table 1. Parameters in field experiments and seismic exploration

	Field experiments	Seismic exploration
Source wave generator	MERTZ vibrator	MERTZ vibrator
Seismometer	French 408UL telemetry	French 408UL telemetry
Sampling interval	0.25 ms	0.25 ms
Record length	1.0 s	1.0 s
Geophone	60 Hz	3 PS-60A
Stack	3	12
Recording channels	120	144
Geophone spacing	5 m	5 m
Shot interval		30 m

4 Exploration Data Processing

The data acquired from the seismic exploration were sent to the computing center of Anhui Province Exploration and Design Institute, where a Promax processing system was used to process the data. Data interpretation was carried on an Utrl60 workstation at Jiangsu Coal Geophysical Prospecting and Surveying, using the IESX software in the Geoframe4.0.4 automatic interpretation system.

The data processing was focused on fidelity, maximizing SNR, assessing the dominant frequency and the frequency band. The processing procedure consists firstly of the module parameters examination and trial-processing, and secondly batch processing. The main processing module and parameters included spatial attributes of seismic data, refraction static correction, recovery of amplitude, filtering, deconvolution, velocity analysis and automatic residual static correction.

Two-dimensional seismic profile was generated after the data processing. Figure 3 shows the superb static correction, and Fig. 4 shows examples of the recovery of amplitude consistent with ground, where the CDP spacing was 2.5 m and the time was 1.0 s. The profile is 5105 m long.

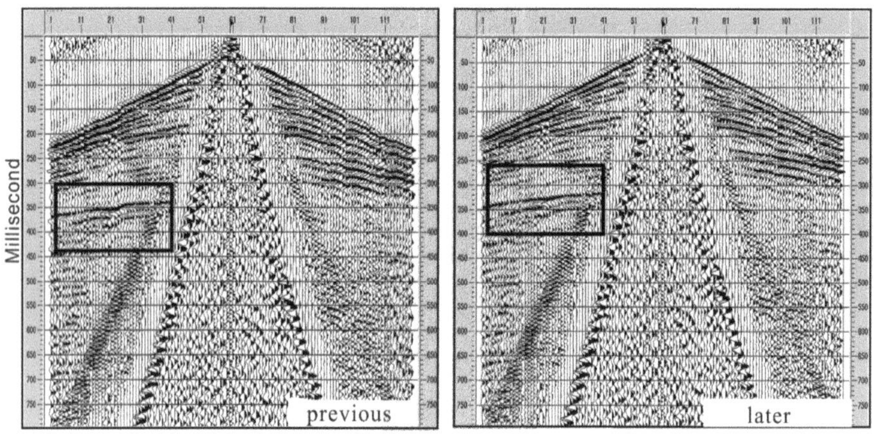

Fig. 3. Comparison of single shot prior and after static calibration of refraction

5 Data Interpreation

Seismic data interpretation is a process of conversion of the seismic data into geologic data. It is based on the comparison between the seismic profile and the known geologic one, to realize the interpretation of the seismic data.

5.1 Seismic Strata Interpretation

Seismic data interpretation started from recognizing the phase axis of the waves reflected by the target stratum. It was divided into three steps, namely, determination of seismic strata, contrasted interpretation of wave groups, and the time-depth conversion of the target wave group.

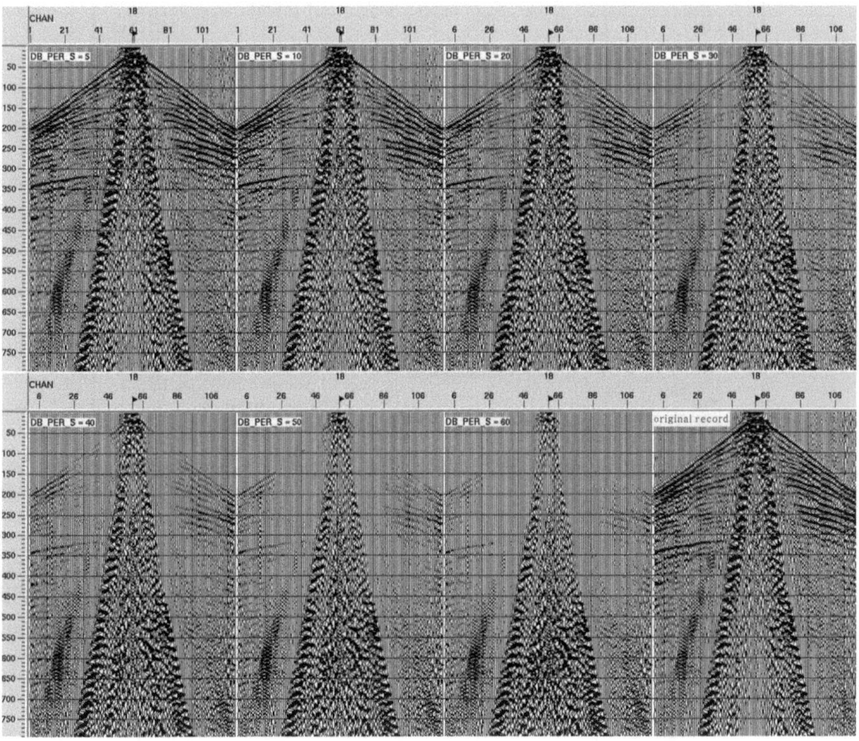

Fig. 4. Comparison of true amplitude recovery parameters

5.1.1 Strata Calibration

Calibration of seismic strata is the first step for an adequate interpretation. The calibration of the phase axis of effective wave was achieved by comparing the seismic profile with the data of the target strata from the nearby drill hole. From the bottom up, TQ wave was formed near the bottom of the Quaternary, TX wave at the bottom of a formation within the pre-Cenozoic (Fig. 5). The TQ wave, which was corresponding to the Quaternary bottom interface (bedrock surface), has high SNR in the northern part of the survey line, and low SNR to the south; the t0 time was 75–130 ms. The pre-Cenozoic strata present a high quality reflection wave TX, which shows dual phases, strong energy and good continuity, forming a successive correlation mostly along the exploration line. Its t0 time was 175–210 ms.

When the characteristics, such as wave propagation velocity, density and impedance, differ substantially between two strata, their interface would be efficient to reflect waves. In the area, the interfaces between clay/silty clay and sand generally formed nice reflection waves, whose phase generally corresponded to the clay. In contrast, interfaces between different sand layers did not form a continuous reflector, or

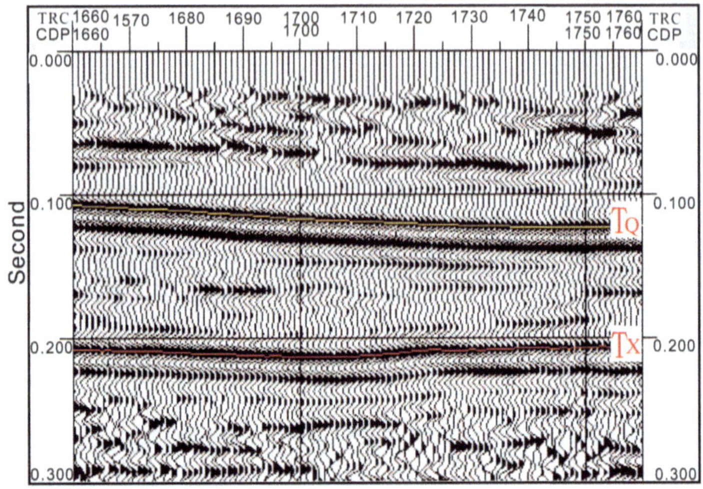

Fig. 5. Schematic diagram of strata calibration

even did not generate any reflection. On the other hand, the spacing between the interfaces has to be greater than one wavelength to generate distinctive reflection waves from each interface. Consequently, the reflection wave from interfaces of multiple thin inter-bedding layers usually appeared as a composite one.

5.1.2 Fault Interpretation

A fault is in essence a broken point on a time profile. The breakpoint on a seismic time profile was recognized through deformations of the phase axis of the reflection wave, such as dislocation, termination, twist, shape change, bifurcation, phase transformation, merger and diffracted waves. In this investigation, the main deformations recognized for fault identification were dislocated reflection wave, phase transition, termination and bifurcation.

5.1.3 Time-Depth Conversion and Velocity

The time-depth conversion was calibrated according to the borehole data and the principle of propagation of seismic waves. The time profile was converted into the depth profile by directly employing the root-mean-squared (RMS) velocity in data processing, resulting in a more visualized display of the geological features and depth of the target layers on the profile.

The frequency for stack velocity analysis was up to once per 100 m in seismic data processing. As the pre-Cenozoic strata tend to be horizontal, the stacking velocity is consistent with the RMS velocity.

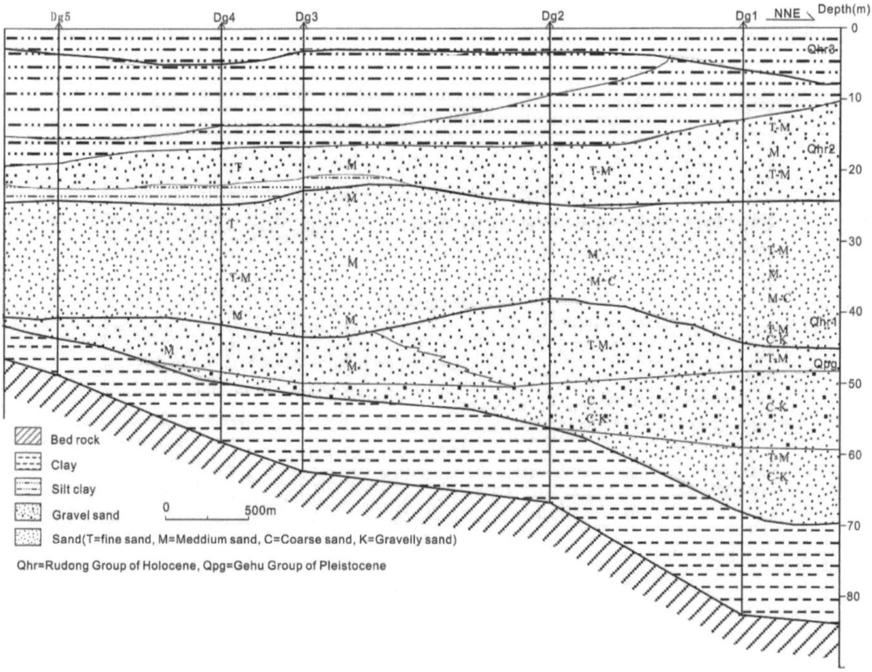

Fig. 6. Quaternary strata profile along the seismic exploration line

5.2 Seismic Time Profile Interpretation and Results

The interface at the bottom of Quaternary becomes gradually deeper from the southwest to the northeast along the exploration line (Fig. 6). The shallowest part is in the southwest, where the minimum value of TQ wave at t0 time is 75 ms and the depth is 45 m; the deepest is located in the northeast end, where the maximum value of TQ wave at t0 time is 130 ms and the depth is 95 m.

In the pre-Cenozoic strata, there is an interface denoted as "X" (Fig. 7), which was dislocated by a fault and whose depth corresponds to 140–175 m, lower in the northeast and higher in the southwest. From the supplementary drill holes DG01, DG02, DG03, DG04 and DG05 along the exploration line (Fig. 6), the depth of loose deposit varies significantly, and the surface of the bedrock, consisting of igneous rocks, presents some undulations.

5.2.1 Faults in pre-Cenozoic

There are four normal faults in the pre-Cenozoic strata named as F_{NZ}, FD1, FD4 and FD5. They were fractured, weathered and eroded during the pre-Cenozoic era. The reflection wave received from the faults was not broken or twisted within the Quaternary strata, suggesting that they were formed prior to Quaternary and not rejuvenated since then, as shown in Figs. 7, 8, 9 and 10. Table 2 summarizes the orientation and displacement of the faults.

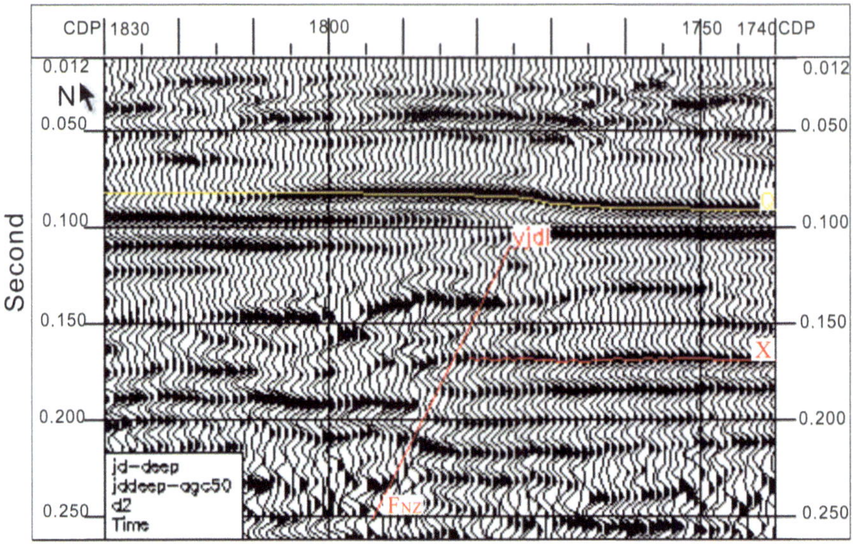

Fig. 7. F_{NZ} Fault on seismic time profile

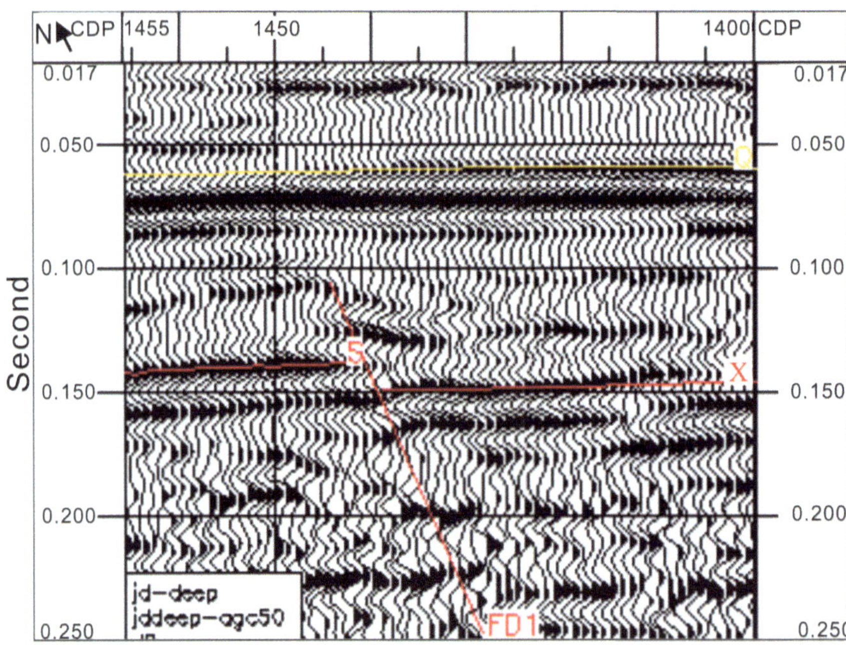

Fig. 8. FD1 Fault on seismic time profile

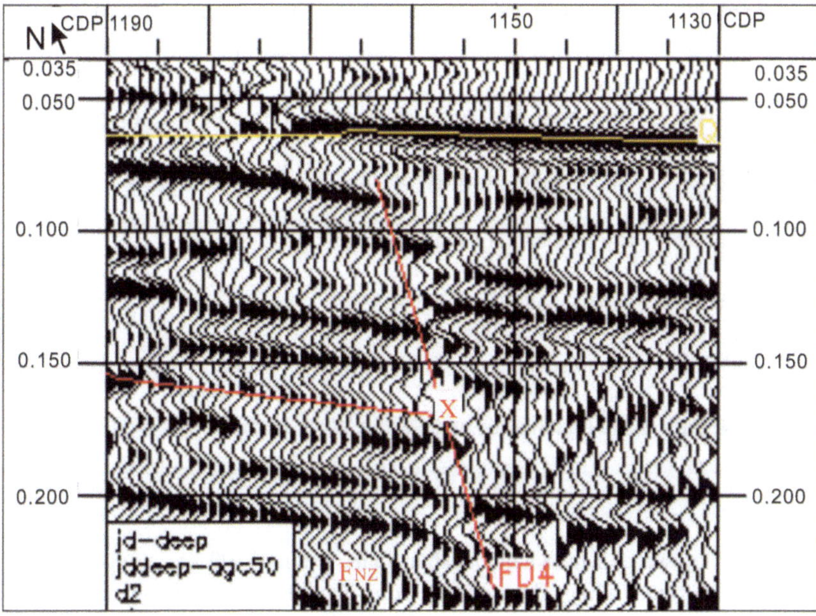

Fig. 9. FD4 Fault on seismic time profile

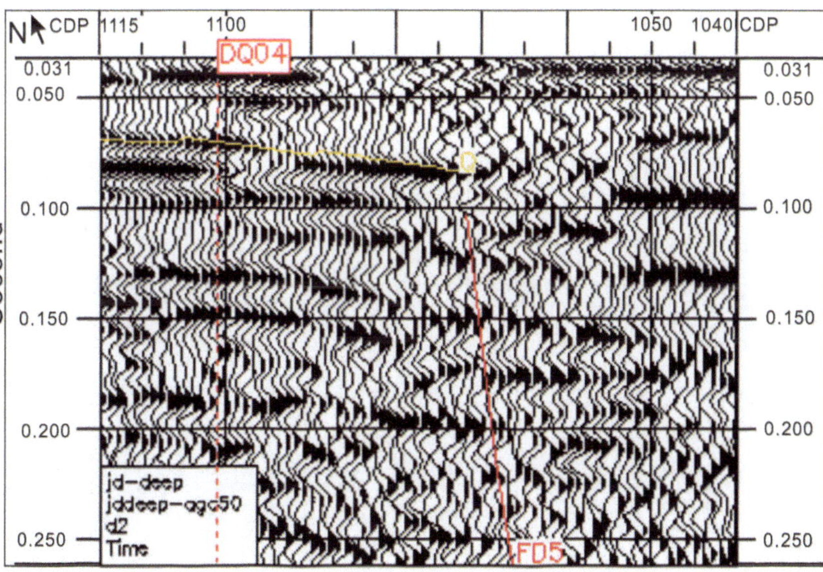

Fig. 10. FD5 Fault on seismic time profile

Table 2. Orientation and displacement of faults in the research area

Fault name	Location	Orientation	Displacement	Era
F_{NZ}	1783CDP (Fig. 7)	Strike near EW, dip to north, dip angle of 65 to 70°	Throw 12 to 15 m	pre-Cenozoic
F_{D1}	1440CDP (Fig. 8)	Dip to south, dip angle about 45°	Throw about 15 m	
F_{D4}	1156CDP (Fig. 9)	Dip to south, dip angle about 55°		
F_{D5}	1070CDP (Fig. 10)	Dip to south, dip angle of approximately 70°		
F_{D2}	1393CDP	Dip to south dip angle about 55° and	Throw are about 10 m	Quaternary
F_{D3}	1254CDP (Fig. 12)	Dip to north, dip angle about 37°	Throw 5 m	

F_{NZ}, also known as Mufushan-Jiaoshan fault, extends 100 km along the Yangtze River in the northern Ning-Zhen uplift, from Nanjing to Zhenjiang (Fig. 11). The direction of extension is broadly east-west. The fault affected the landscape, resulting in a sharp descent of the northern half of Mufushan-Qixiashan anticline (the hanging wall of the fault). This led to the formation of the low-lying Yizheng fault depression basin in the north of Yangtze River, and the formation of Ning-Zhen fault Block Mountains (footwall) in the south of Yangtze River. It demonstrated significant differential ascending and descending movements along the fault.

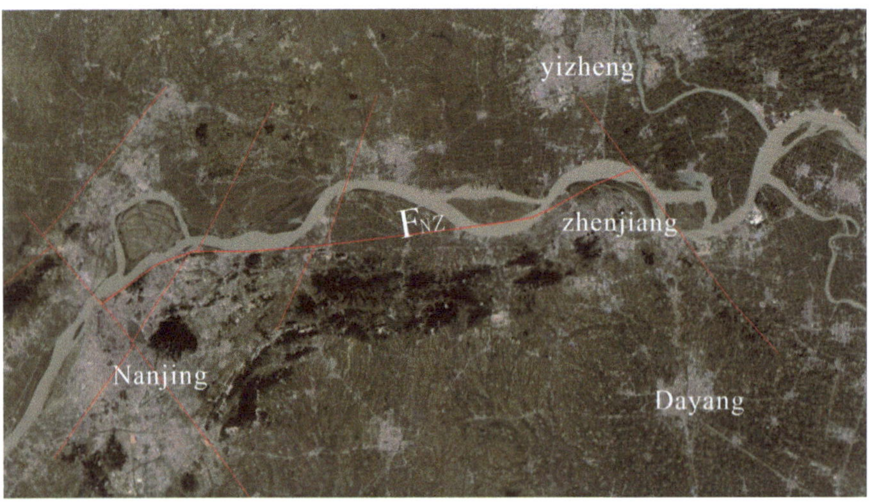

Fig. 11. FNZ Fault on satellite image

Some sections of the fault evidence a linear trace on satellite images (marked red in Fig. 11). Some steep cliffs are observed along the fault zone. There are some significant reflections in the geophysical field, including an abnormal gravity gradient zone and a substantial anomaly in the aeromagnetic map. The magnetic field is smooth in the north of the fault, ranging from 50 nT to 100 nT, but it is disordered in the south with a huge negative anomaly zone (the minimum is −500nT). On the other hand, the rock intruded along the fault zone causes positive anomalies (the maximum is 500 nT–600 nT). The dip angle of the interface between high and low resistants in the electrical sounding profile is very steep.

The artificial shallow seismic exploration revealed that the depth of the bedrock surface is shallow near Qixia District with large undulation, and that the fault zone consists of multiple faults filled by igneous rock. The fracture was developed within the lower middle Jurassic Xiangshan formation, the upper Jurassic Longwangshan group, and the upper Cretaceous Pukou formation. The width of the fault zone is up to hundreds of meters.

5.2.2 Faults in Quaternary

There are two normal faults in Quaternary, namely FD2 and FD3. FD3 faulting is clearly revealed by the dislocation of the reflector at the bottom of Quaternary strata (Fig. 12). Table 2 includes the orientation and displacement of these two faults.

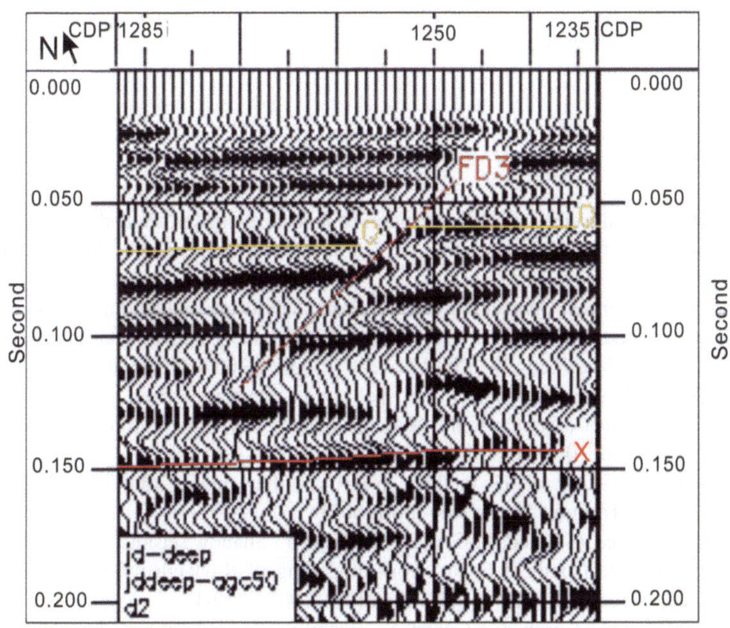

Fig. 12. FD3 Fault on seismic time profile

6 Conclusions

This is a large-scale shallow seismic investigation conducted on the Yangtze River Fault at Zhenjiang, Jiangsu Province, China. It included 10 field experiments to determine the adequate parameters, 5715 m long shallow seismic exploration and five geological drill holes, and consequent data processing and interpretation of the shallow seismic time profiles in combination with the geological profile along the exploration line and satellite images in the region.

During the interpretation, a preliminary analysis was carried out to calibrate the Quaternary seismic stratigraphy with the support of geological drill holes. It aimed to interpret the seismic facies, or ancient stratification, and breakpoints on profiles (faults), using the characteristics of seismic reflection waves.

Based on the seismic time profile, interpretations were made to reveal the stratification and faults along the exploration line. It was found that the loose Quaternary sedimentary strata present good continuity in the area, which provides a reliable database for creating geological cross sections in this area. There are six faults in the area, four of which are pre-Cenozoic faults in the bedrock (including the Yangtze River Fault), and two of which are Quaternary faults that cut the Quaternary strata. Therefore, the study confirmed the existence of the Yangtze River Fault, and furthermore analyzed its activity and characteristics, which may provide technical reference for engineering constructions and geological stability evaluation in the area.

Acknowledgments. The authors gratefully acknowledge the supports of China Geological Survey Project (1212010914005).

References

Al-Saigh, N.H., Al-Dabbagh, T.H.: Identification of landslide slip-surface and its shear strength: a new application for shallow seismic refraction method. J. Geol. Soc. India **76**, 175–180 (2010)

Andreassen, K., Nilssen, E., Ødegaard, C.: Analysis of shallow gas and fluid migration within the plio-pleistocene sedimentary succession of the SW Barents Sea continental margin using 3D seismic data. Geo-Mar. Lett. **27**, 155–171 (2007). doi:10.1007/s00367-4007-0071-5

Bertin, X., Chaumillon, E.: New insights in shallow gas generation from very high resolution seismic and bathymetric surveys in the Marennes-Oléron bay, France. Mar. Geophys. Res. **26**, 225–233 (2005). doi:10.1007/s11001-005-3720-y

Chai, Z.: Application of 2-D seismic methods in Jixi Yongfeng area [二维地震方法在鸡西永丰区的应用]. Coal Technol. **28**(5), 149–150 (2009)

Chourak, M., Corchete, V., Badal, J., Gomez, F., Seron, J.: Shallow seismic velocity structure of the Betic Cordillera (southern Spain) from modelling of Rayleigh wave dispersion. Surv. Geophys. **26**, 481–504 (2005). doi:10.1007/s10712-005-7260-4

Cifci, G., Pamukcu, O., Coruh, C., Copur, S., Sozbilir, H.: Shallow and deep structure of a supradetachment basin based on geological, conventional deep seismic reflection sections and gravity data in the buyuk menderes graben, western anatolia. Surv. Geophys. **32**, 271–290 (2011). doi:10.1007/s10712-010-9109-8

Hanebuth, T., Stattegger, K., Saito, Y.: The stratigraphic architecture of the central Sunda shelf (SE Asia) recorded by shallow-seismic surveying. Geo-Mar. Lett. **22**, 86–94 (2002)

Iacono, D., Zollo, A., Vassallo, M., Vanorio, T., Judenherc, S.: Seismic images and rock properties of the very shallow structure of Campi Flegrei Caldera (southern Italy). Bull. Volcanol. **71**, 275–284 (2009). doi:10.1007/s00445-008-0222-1

Ismail, A., Smith, E., Phillips, A., Stumpf, A.: Pitfalls in interpretation of shallow seismic data. Appl. Geophys. **9**(1), 87–94 (2012)

Klein, G., Bohlen, T., Theilen, F., Kugler, S., Forbriger, T.: Acquisition and inversion of dispersive seismic waves in shallow marine environments. Mar. Geophys. Res. **26**, 287–315 (2005). doi:10.1007/s11001-005-3725-6

Liu, B., Sun, Z., Zhao, C.: Shallow seismic reflection profiling with high resolution in Yanqing-huailai region [延庆怀来地区高分辨率浅地震反射剖面]. Seismolog. Geol. **21**(4), 425–430 (1999)

Liu, B.J., Zhao, C.B., Feng, S.Y.: Application of the three-component shallow seismic reflection method to probing buried active faults [应用三分量浅层地震反射方法探测隐伏活动断裂]. Chin. J. Geophys. **55**(8), 2676–2686 (2012)

Liu, H., Hu, Y., Yin, Y., Wang, L., Tong, S., Ma, H.: Shallow water body data processing based on the seismic oceanography. J. Ocean Univ. China (Oceanic Coast. Sea Res.) **12**(3), 319–326 (2013)

Peng, Z., Ben-Zion, Y.: Temporal changes of shallow seismic velocity around the karadere-düzce branch of the north anatolian fault and strong ground motion. Pure Appl. Geophys. **163**, 567–600 (2006). doi:10.1007/s00024-005-0034-6

Rucker, M., Meyers, R., Lommler, J.: Assessing shallow rock conditions over a brine cavern using seismic methods. Carbonates Evaporites **27**, 199–205 (2012)

Wang, C., Zhang, X., Wu, Q., Zhu, Z.: Seismic evidence of detachment in north china basin [华北盆地滑脱构造的地震学证据]. Chin. J. Geophys. **37**, 613–620 (1994)

Xie, Z., Zhang, G., Fu, X.: Progress of 2-D seismic exploration in onshore, Beipu region [北堡陆地二维地震进展]. Oil Gas Geol. **21**(4), 324–326 (2000)

Yang, X., Li, D., Zhao, C., Liu, B., Sun, Z., Zhao, J.: The movement age of hidden fault and analysis on width of its effect zone from shallow seismic sounding and drilling data. Acta Seismol. Sin. **17**, 94–102 (2004)

Yin, Y., Zhang, J., Chen, B.: Formation mechanism of large-scale loose sediment at the relocation sites of Wushan County on the three-gorges [三峡库区巫山移民新城址松散堆积体成因机制研究]. Chin. J. Eng. Geol. **8**(3), 265–271 (2000)

Yue, Y., Chai, Z.: Application of 2-D seismic method in looking for coal in Dayanzhani river area of Inner Mongolia [二维地震方法在内蒙古大雁扎尼河区找煤中的应用]. Coal Technol. **27**(3), 129–130 (2008)

Zhigulev, V.V., Patrikeev, V.N.: Shallow seismic refraction analysis: Application to studying the active north sakhalin fault. Russ. J. Pac. Geol. **1**(1), 15–22 (2007). [Original Russian Text © V. V. Zhigulev, V.N. Patrikeev, 2007, published in Tikhookeanskaya Geologiya, 2007, Vol. 1, No. 1, pp. 20–28.]

Normal Fault Movement Propagation in Overlying Seabed Deposits

Lama Thebian(✉), Salah Sadek, Shadi Najjar, and Mounir Mabsout

Department of Civil and Environmental Engineering,
American University of Beirut, P.O. Box 11-0236 Riad El-Solh,
1107-2020 Beirut, Lebanon
{ltt00, salah, sn06, mounir}@aub.edu.lb

Abstract. This paper addresses the response of seabed sands subjected to underlying normal fault movement. This problem is relevant to the design of overlying offshore structures and subsea oil/gas pipelines connecting offshore platforms to the shoreline. The propagation of the faulting offset in seabed sediments is explored using 2D finite element modeling. Abaqus© is used as a numerical platform in modeling this complex problem, while accounting for nonlinear soil behavior with strain softening. Different dip angles and vertical fault displacements of up to 10% of the soil layer thickness were considered. The results include the effect of the relative density of the seabed sands on the extent and magnitude of ground surface deformations. The required bedrock displacement/offset for the rupture to reach the surface and the length and location of the distorted zone are also reported. The results show that the cases of loose sands and larger soil layer thicknesses result in larger distorted zones and that larger bedrock displacements are required for the fault base rupture to propagate to the surface. At low dip angles, graben formation is observed at small bedrock displacements. Based on the parametric analyses and results presented in this paper, observations related to the potential magnitudes and extents of surface deformations for various conditions of seabed densities and thicknesses are provided. These would be of importance in determining likely effects of distortion/loading on pipelines and offshore structures crossing the fault zone.

1 Introduction

Historically, earthquakes that were triggered by dip slip fault systems (normal or reverse) have caused significant damage to structures, human lives, and the environment. Specific examples of cases of normal faults include the Manhattan earthquake (1867), Atalanti earthquake (1894), L'Aquila earthquake (2009), Hamadori earthquake (2011), Norcia earthquake (2016). Damage of onshore oil and gas facilities by normal faults have led to large economical losses due to service shut down, business interruption, property losses, replacement costs. One example is the damage of the largest refinery in Korfez, Turkey during the 1999 Kocaeli earthquake. Despite of the wide knowledge that has been gathered thorough the years in the field of offshore

J. Wasowski et al. (eds.), *Engineering Geology and Geological Engineering for Sustainable Use of the Earth's Resources, Urbanization and Infrastructure Protection from Geohazards*, Sustainable Civil Infrastructures, DOI 10.1007/978-3-319-61648-3_14

geotechnics and pipeline design, the experience with seismic design of offshore pipelines is limited since most of these facilities have been constructed in zones of low seismicity as is the case in North Sea, Western Australia, and the Gulf of Mexico. As a result, investigations related to offshore fault propagation have not attracted as much attention as the onshore context and the mechanism of fault propagation in the offshore environment remains less understood than the onshore cases.

The risk of offshore pipelines crossing faults is present in several locations in the world. One specific example is the case of the Mediterranean sea where reserves of oil and gas have been discovered in the last decade. Another relevant case where gas pipelines cross active seismic zones is the Japan Sakhalin Gas Pipeline case that crosses along the Pacific Ocean and intersects an active normal fault offshore Chiba prefecture to the east of Tokyo. Other active normal faults that could cause risk to subsea pipelines are located in the Euboean Gulf in Greece and in the Caspian sea.

Normal faults are classified as low-angle normal faults or lags for cases with dip angles less than 45° or else they are named high-angle normal faults. Although studies have shown evidence for active low-angle normal faults like the cases of Italy and Tibet, limited research was done to investigate the propagation of low-angle normal faults in subsea sediments in relation to pipeline design and stability.

The possibility that offshore pipelines could cross active normal fault zones necessitates special design considerations to ensure the safety and stability of these systems. Several research studies targeted this problem using experimental programs. Sanford (1959), Belousov (1961), Emmons (1969), Horsfield (1977), Cole and Lade (1984), Lade et al. (1984), Bray et al. (1993), Johansson and Konagai (2004) conducted 1-g small scale experiments, while Bransby et al. (2008a, b) used centrifuge experiments to model dip slip fault propagation. Other researches modeled the problem using numerical methods. Scott and Achoustra (1974) simulated vertical fault propagation through 800 m deep alluvium cohesionless dry soil using numerical analysis. Roth et al. (1982), Scott (1987), White et al. (1994), Nakai et al. (1995), Loukidis (1999), Loukidis et al. (2009) used the finite difference method for the same purpose. Bray et al. (1994b), Taniyama and Watanabe (2000), Lin et al. (2006), Anastasopoulos et al. (2007) conducted experiments coupled with finite element models.

Unlike other published studies that are generally restricted in their scope to one relative density (Johansson and Konagai 2007), limited fault displacements (Loukidis et al. 2009), onshore applications (Anastasopoulos et al. 2007), and high-angle faults, the analyses in this study targeted fault propagation in offshore sand deposits with different relative densities (dense, medium dense, and loose sand), different dip angles (30°, 45°, 60°) for normal faulting, and a wide range of fault displacements (1% to 10% of the overlying sand layer thickness). The offshore fault propagation problem is realistically investigated through a numerical finite element analysis that accurately depicted the geometry, boundary conditions, and soil properties, while ensuring that the proper effective stress state was considered through the incorporation of 8 node plain strain quadrilateral, biquadratic displacement, bilinear pore elements (CPE8P).

2 Numerical Modeling

The response of offshore sands that overly normal faults is examined numerically using the general purpose finite element software Abaqus v.6.14. The adopted model simulates the behavior of the saturated sand using effective stress analysis by considering 8 node plain strain quadrilateral, biquadratic displacement, bilinear pore elements (CPE8P) assuming drained conditions.

2.1 Model Geometry, Mesh, and Boundary Conditions

A 2D plain strain model of dimensions 20 m × 140 m is depicted to simulate the fault propagation mechanism in the offshore sand sediments. The model is shown in Fig. 1 and illustrates the overlying soil layer of thickness H = 20 m which is susceptible to the fault displacement and the corresponding finite element mesh. 0.5 m × 0.5 m quadratic elements are employed in the region at the vicinity of the fault where the failure plane and maximum deformations are, and the mesh becomes coarser as we move away from the fault.

The following 2 steps have been adapted in the simulation: (1) application of the self-weight and water load, and (2) application of fault displacement by subjecting the external nodes of the moving part to displacement controlled loading in the direction of a pre-determined dip-angle. To cater for discontinuity and convergence problems, the fault displacement was applied within a width of 1 m. At any displacement increment, a linearly varying fault movement is forced along this 1-m width, by fixing the right end of this width and forcing the left end to move in accordance with the applied fault displacement.

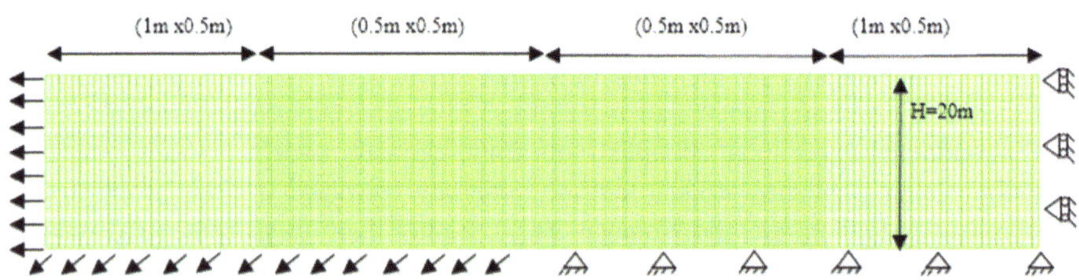

Fig. 1. The 2D fault propagation FE model

2.2 Soil Constitutive Model

Elastoplastic material behavior that is based on a modified Mohr-Coulomb model that is capable of modeling strain softening is considered to model the sand behavior. The strain softening response was modeled by reducing the peak friction and dilation angles as a function of the deviatoric plastic strain such that:

$$\Phi_{mob} = \begin{cases} \Phi_p - \frac{\Phi_p - \Phi_{cr}}{\gamma^f_{dev}} \gamma^p_{dev} & for \quad 0 \le \gamma^p_{dev} \le \gamma^f_{dev} \\ \Phi_{res} & for \quad \gamma^p_{dev} \ge \gamma^f_{dev} \end{cases} \tag{1}$$

$$\Psi_{mob} = \begin{cases} \Psi_p \left(1 - \frac{\gamma^p_{dev}}{\gamma^f_{dev}}\right) & for \quad 0 \le \gamma^p_{dev} \le \gamma^f_{dev} \\ \Psi_p & for \quad \gamma^p_{dev} \ge \gamma^f_{dev} \end{cases} \tag{2}$$

2.3 Properties

Sands with three relative densities (20%, 50%, and 80%) were considered in the analysis to represent loose, medium dense, and dense sands, respectively. The sand was assumed to be a round-shaped, uniform, medium sand with a coefficient of uniformity $C_u = 3$, with a typical specific gravity of $G_s = 2.67$. The maximum and minimum void ratios were estimated based on the EPRI design manual (Kulhawy and Mayne 1990). Knowing the relative density and maximum and minimum void ratios, the void ratio of each sand type was calculated.

The strength properties of the sands were assumed to vary with depth and are presented in Table 1. The Young's modulus (E), peak friction angle (Φ_{peak}), peak dilation angle (Ψ_{peak}), and Poisson's ratio (υ) are a function of the confining pressure and therefore of depth. Typical values of E_0 of 15600 kPa, 22000 kPa, 33000 kPa were assumed and the stiffness was increased with depth according to the following equation for triaxial tests from Schanz and Vermeer (1998) and a typical value of 0.5 was assumed for the exponent *m*. σ_v represents the vertical stress and σ_a stands for the atmospheric pressure.

Table 1. Saturated sand properties

Density	Dense	Medium dense	Loose
RD	80%	50%	20%
K_0	0.35	0.42	0.5
E (KPa)	$11209\sqrt{z}$	$7300\sqrt{z}$	$5063\sqrt{z}$
C (KPa)	2	2	2
Φ_p (°)	$51 - 2.4\ln(6.85z) \le 42$	$42 - 1.5\ln(6.9z)$	30
Φ_{res} (°)	30	30	30
ν	$0.49 - 0.036\ln(6.85z)$	$0.355 - 0.0225\ln(6.9z)$	0.18
Ψ_p (°)	$26.25 - 3\ln(6.85z)$	$15 - 1.88\ln(6.9z)$	0
Ψ_{res} (°)	0.1	0.1	0.1
γ^p_f	0.1	0.1	–

$$E = E_0\left(\frac{\sigma'_v}{\sigma_a}\right)^m \tag{3}$$

The critical friction angle was assumed equal to 30°. The peak friction angle (ϕ_p) was calculated from Eq. (4) from Bolton (1986).

$$\phi_{peak} - \phi_{cr} = 3\left(D_r\left[10 - \ln\left(100P_f/P_a\right)\right] - 1\right) \tag{4}$$

It is worth noting that the upper bound of the peak friction angle was taken as 42°. The dilation angle (Ψ) was estimated from Bolton's (1986) equation as $\Psi peak = (\phi_{peak} - \phi_{cr})/0.8$. Poisson's ratio was estimated based on EPRI (Kulhawy and Mayne 1990) as $\upsilon_d = 0.1 + 0.3\phi_{rel}$ where $\phi_{rel} = (\phi_{peak} - 25)/(45 - 25)$.

3 Numerical Results

In the first section, a model with a height of 20 m and length of 140 m was adopted as a reference to perform finite element analyses for dense, medium dense, and loose sand for a normal fault having a dip angle of 45°. The reference model is compared with cases of dip angles of 30° and 60° to assess the effect of the fault dip angle on the propagation mechanism. In the second section, the height of the overlying soil was varied from H = 20 m to values of H = 40 m and 80 m and the influence of each parameter was examined. Vertical bedrock displacements (U_2) up to 10% of the soil layer thickness were enforced. The results included an analysis of the vertical ground surface deformation profile (y), the horizontal distance (S) from the location of the shear band to the center of the fault (O) (see Fig. 4), the critical inclined bedrock displacement required for the shear band to reach the surface (d_{crit}), and the length (L) and center (C) of the distorted zone. In the presentation of the results, most of the parameters defined above were normalized with respect to the soil layer thickness.

3.1 Effect of Relative Density

Ground Surface Deformation Profiles

Ground surface deformation profiles that depict the variation of the vertical normalized ground surface deformations (y/H) for dense, medium dense, and loose sand for normal faults with dip angles of 45° are presented in Fig. 2 for vertical bedrock displacements ranging from 2% to 10% of the soil layer thickness (H). Similarly, the normalized ground surface deformation profiles for dense sand for normal faults with different dip angles (30°, 45°, and 60°) are presented in Fig. 3. These surface deformation profiles provide essential input for mitigating the fault rupture geohazard which requires accounting for fault-induced ground surface deformation (Kelson et al. 2001; Chang et al. 2015).

Results on Fig. 2 indicate that the inclination of the ground deformation profile in the shearing zone is greater for the case of dense sand than medium dense sand and loose sand, respectively. The higher inclination angles for dense sands could be attributed to the higher soil stiffness and higher dilatancy. For dense sands, an investigation of the concentration of plastic shear strains in the fault-propagation zones (Fig. 4) indicate narrower and concentrated shearing zones. A general analysis of the distribution and concentration of plastic shear strains at a vertical bedrock displacement of 10% of the soil layer thickness indicates that the lower the relative density, the wider the extent of the zone with relatively large plastic strains. Similar results were obtained by Loukidis et al. (2009), Anastasopoulos et al. (2007).

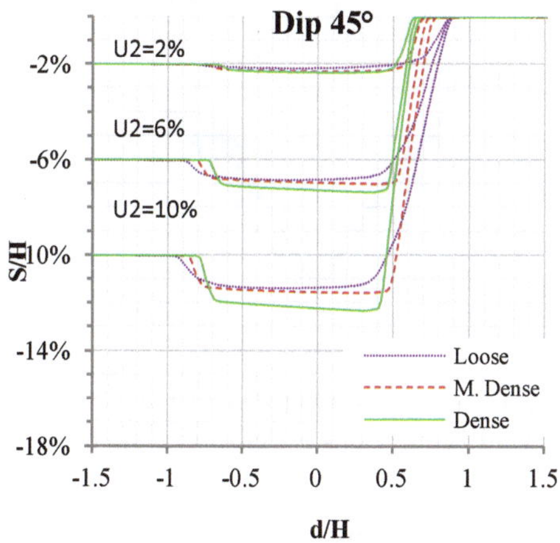

Fig. 2. Ground surface deformation profiles for a seabed soil layer of thickness 20 m subjected to normal fault of dip angle 45° for cases of dense, medium dense, and loose sand.

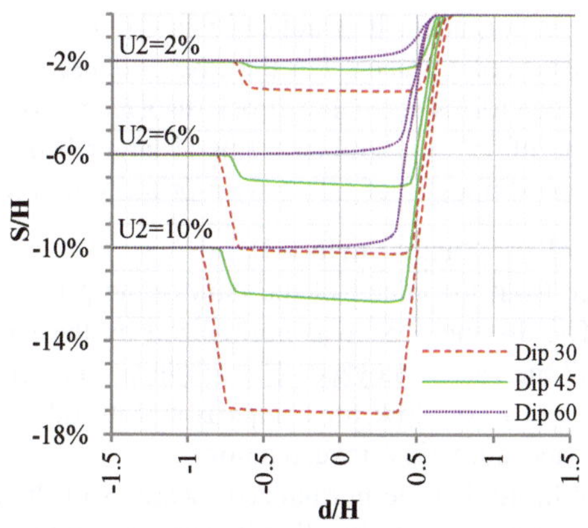

Fig. 3. Ground surface deformation profiles for a seabed dense sand layer of thickness 20 m subjected to normal fault of dip angles 30°, 45°, and 60°.

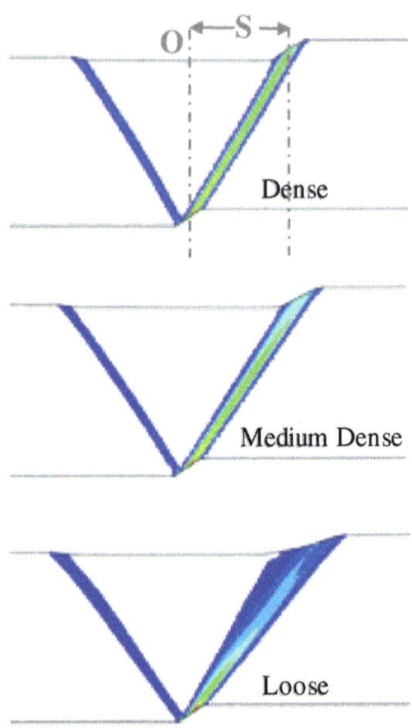

S: horizontal distance between the location the primary shear band reaches the ground surface and the center of the fault
O: Center of the fault on the surface

Fig. 4. Plastic shear strains for a seabed subjected to normal fault of dip angles 45° at a vertical bedrock displacement U2 = 10%H for cases of dense, medium dense, and loose sand.

By comparing the curves on Figs. 2 and 3, it is noticed that ground surface deformations are more sensitive to the dip angle than the relative density. As the fault displacement increases, a shear band forms between the moving block and the fixed block. For the cases with dip angles of 30° and 45°, once the shear band reaches the top surface, a secondary shear band in the opposite direction of the first shear band is formed and propagates to the top ground surface at low bedrock displacements inducing the formation of a graben. It is noticed that the shallower the dip angle, the more likely is the formation of a graben. For the case of dense sand with the shallower dip angle of 30°, the graben forms at small bedrock displacements in the order of 0.12% H. For the steeper angle of 45°, the graben forms at larger bedrock displacements of 0.65%H. For the steepest dip angle of 60°, only a primary failure plane forms.

Extent and Location of the Shear Surface/Band

The horizontal distance from the center of the fault to the primary shear surface/band was defined as "S". The distance "S" was normalized by the soil thickness "H" and plotted on Fig. 5 for all the cases analyzed. The results are presented for a normalized fault displacement of 2.75% to allow for comparison with other published data from previous experimental and numerical investigations.

Results on Fig. 5 indicate that the normalized locations of the primary shear bands for the case of normal faults seem to be slightly sensitive to the sand relative density and to the dip angle. For the shallowest dip angle of 30°, *S* decreases from a value of 0.75% to 0.60% as the sand relative density is changed from loose to dense. As the dip

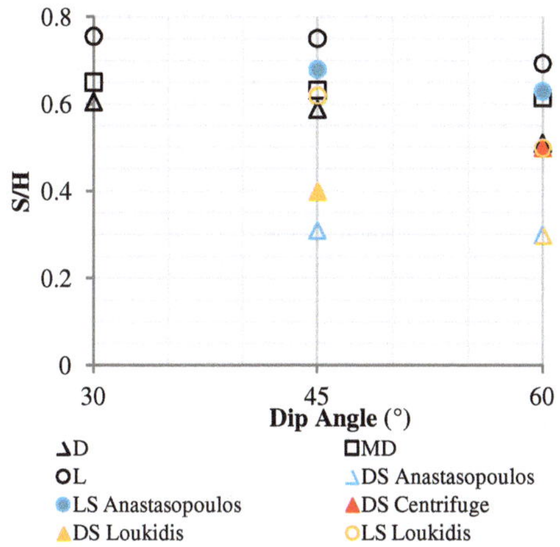

Fig. 5. Normalized primary shear band locations from the center of the fault for dense (D), medium dense (MD), and loose sand (L) at a vertical bedrock displacement $U_2 = 2.75\%H$ versus the fault dip angle compared with results from the literature.

angle of the fault increases, the normalized distance to the shear band is found to slightly decrease, irrespective of the sand relative density. The decrease in *S* for steeper fault angles is expected given that smaller values of *S* are indicative of steeper shear bands which go hand in hand with steeper dip angles.

A comparison between the results presented in Fig. 5 and similar published results indicate that the normalized shear band locations that were obtained for the case of loose sands from published numerical studies are in the range of values obtained in our study. The same applies to the centrifuge tests that were conducted with dense sands. For the case of published numerical studies involving dense sands, results on Fig. 5 indicate differences with the results of our study, with the published results showing normalized shear band locations that are consistently smaller than our results. The difference in the results between our model and other published models is expected to be associated with how the soil parameters were defined in the finite element analysis. In the studies conducted by Anastasopoulos et al. (2007), Loukidis et al. (2009) the dilation angle and friction angle for dense sand were assumed to be constant with depth with assumed values of 15 to 18° for the dilation angle and 45° for the friction angle. In our study, all soil properties (modulus of elasticity, dilation angle, and friction angle) were assumed to vary with depth to provide a realistic model of soil behavior for granular materials that are dependent on confining pressure. A sensitivity study done by the authors demonstrated that the stiffness, dilation angle and the friction angle of the sand are critical factors that govern the fault propagation mechanics and shear band formation.

Bedrock Displacement Required for Rupture to Reach the Surface

Upon fault rupture and depending on the magnitude of the fault displacement, the shear band could propagate to the surface resulting in risk of potential damage to overlying structures (Chang et al. 2015). In this paper, d_{cr} is denoted as the inclined bedrock displacement required for the rupture to propagate to the surface. Values of d_{cr} were calculated for all the cases analyzed and the normalized critical distance d_{cr}/H was plotted on Fig. 6 for dense, medium dense, and loose sand cases under different fault dip angles considering an overlying soil thickness of 20 m.

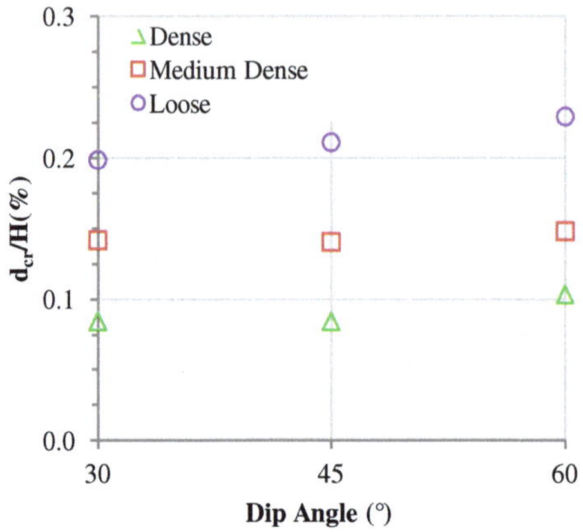

Fig. 6. Normalized inclined bedrock displacement required for the rupture to reach the surface versus the normal fault dip angle for cases of dense, medium dense, and loose sand of a seabed of thickness H = 20 m

Results on Fig. 6 indicate that the required bedrock displacement for the rupture to reach the surface is relatively small (<0.3%H), with minor variations for different fault dip angles (difference in the order of 0.03%H) and differs by a maximum range of 0.13% between dense and loose sand. The higher bedrock displacements that are required for the shear bands to reach the surface in loose sands could be attributed to the fault propagation mechanism that is characterized by the formation of wider multiple shear bands and to compressive volume changes that would accompany the shearing mechanism necessitating larger fault movements at the source to push the failure mechanism to the surface.

Length and Center of Distorted Zone

A realistic definition of the distorted zone that is affected by the fault propagation mechanism is essential to determine the lateral extent of zones that would ensure minimal interference with the fault propagation mechanism. The distorted zone was defined by Loukidis et al. (2009) as the length of the zone within which the ground surface inclination (slope) exceeds 0.2%. In this zone, it could be argued that the soil would have suffered from distortion, softening, strength loss, etc., resulting in reductions in bearing capacity and stiffness. The numerical results were used to define the length (L) and center (C) of the distorted zone which were then normalized by the height of soil H. The normalized values of L and C are plotted as a function of the fault dip angle for all the analyzed cases at a vertical bedrock displacement of 10% of the soil layer thickness (Fig. 7).

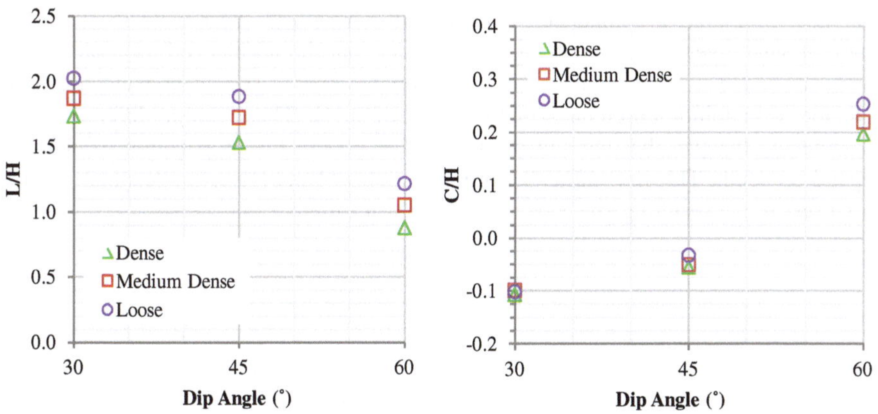

Fig. 7. Normalized length (L/H) and center of the distorted zone (C/H) versus the normal fault dip angle for normal faulting cases of dense, medium dense, and loose sand at a vertical bedrock displacement $U_2 = 10\%H$

Results on Fig. 7 indicate that the normalized length of the distorted zone for normal faults is more affected by the dip angle of the fault than it is affected by the density of the soil. For a normal fault with a dip angle of 30°, L/H ranges from about 1.7 to about 2.0 for the dense and loose sand cases, respectively at a vertical bedrock displacement of 10%H. These numbers decrease for steeper dip angles, reaching values as low as 0.9 to 1.3 for dense and loose sand cases with a steep fault dip angle of 60°. The larger values of L/H for lower dip angles are related to the formation of graben structures at certain bedrock displacements. The formation of these structures at larger bedrock displacements magnify the length of the distorted zone by a factor of about 2.0 compared to the cases where these structures do not form. The secondary shear bands form in the opposite direction to the primary shear band and propagate to the surface forming a graben and shifting the location of the center of the distorted zone towards the center O. For the steeper dip angles that did not exhibit the formation of a graben, the center of the distorted zone is located on the side of the footwall.

3.2 Effect of Thickness of Overlying Sand Layer

Since the majority of the results of the previous sections were presented in a normalized form (normalized with respect to the height of the overlying sand layer), it is important to investigate whether the results of the fault propagation analysis remain valid for different soil overburden heights. For this purpose, the finite element analyses that were conducted in the previous sections with a soil height of 20 m were repeated for cases involving soil thicknesses H of 40 m and 80 m.

An investigation of the sensitivity of the normalized ground surface deformation profiles to the height of the overlying sand layer (not presented for length limitations) leads to the conclusion that the normalized ground surface deformation profiles are relatively insensitive to the thickness of the sand layer for the different dip angles analyzed. Similarly, a comparison between the normalized shear band locations for different sand layer thicknesses (Fig. 8) indicates that the normalized locations of the shear bands are not sensitive to the soil layer thickness. The shear band locations coincide with minor differences in the order of 0.1H. These minor differences are attributed to the fact that the soil properties that govern the propagation mechanism depend on depth. This may affect the process of shear band formation leading to small variations in the resulting shear bands in layers of different thicknesses.

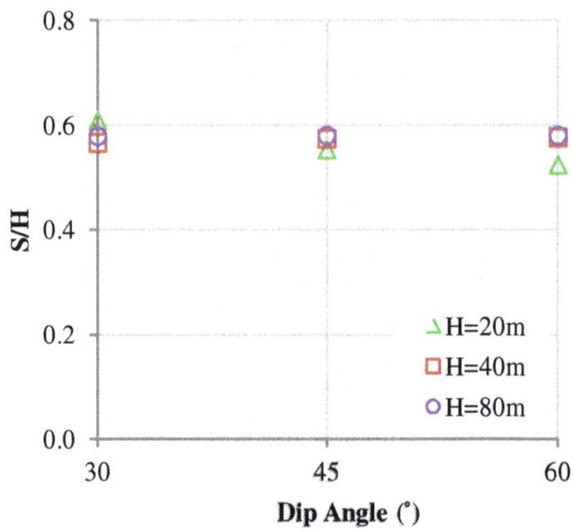

Fig. 8. Normalized shear band location for different dense sand layer thicknesses measured at a vertical bedrock displacement U2 = 10%H

Figure 9 shows the normalized critical inclined bedrock displacement for all sand types and sands layer thicknesses. The critical bedrock displacements for the normal fault cases are relatively small and are slightly sensitive to the thickness of the overlying sands. The range of the difference between H = 20 m and H = 80 m is ≈ 0.1%H for dense sand. The difference between the critical bedrock displacements is expected

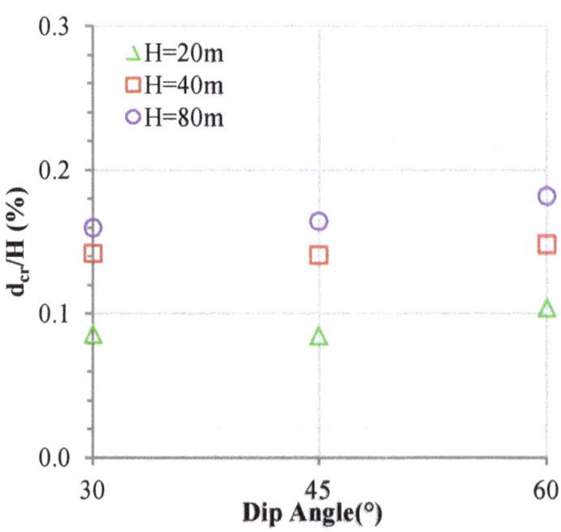

Fig. 9. Normalized inclined bedrock displacement required for the rupture to reach the surface

to be larger for loose sand which requires additional bedrock movements for the fault to reach the surface. Normalizing the required displacement by the height of the overburden soil does not lead to a constant d_{cr}/H in the different cases analyzed.

Finally, the variation of the normalized length and center of the distorted zone with fault dip angle for the cases involving different soil thicknesses is presented in Fig. 10. In general, results indicate that the normalized lengths of the distorted zones are more-or less similar and independent of the soil height for cases involving normal faults. This is attributed to the fact that the ground surface deformation profiles normalize for these cases, irrespective of the soil height. The normalized length of the distorted zone for H = 80 m is found to be slightly larger than for H = 40 m and H = 20 m, respectively. This is attributed to the fact that for a height of 80 m, the shear bands are slightly farther for larger soil thicknesses and that is accompanied by larger distorted zones. It is worth noting that the normalized locations of the center of the distorted zone were also relatively insensitive to the soil thickness.

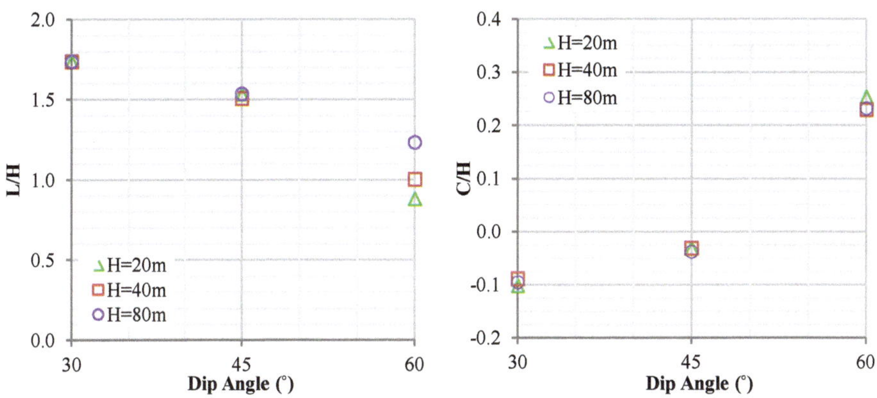

Fig. 10. Normalized length (L/H) and center of the distorted zone (C/H) versus the normal fault dip angle for dense sand at a bedrock displacement U_2 = 10%H.

4 Conclusions

In this paper, a comprehensive investigation of the mechanism of fault propagation in offshore sand deposits that overly dip-slip normal faults is conducted. The investigation was conducted using the finite element method with a model consisting of a sand layer that is modeled by the Elastic-Plastic Modified Mohr Coulomb constitutive model to capture the adequate behavior of the sand. The cases considered included different cases of the sand relative density, the faulting dip angle, and the sand layer thickness. Based on the analysis of the results, the following can be concluded:

- The normalized ground surface deformation profiles for normal faults are more sensitive to the dip angle than the relative density.
- Grabens are more likely to form with normal faults having shallow dip angles. The width of the graben is larger for cases of low dip angles.

- The shear band locations are farther from the center of the fault for loose sands, larger soil layer thicknesses, and lower fault dip angles.
- The shear band reaches the top surface at small values of bedrock displacement. The normalized bedrock displacement required for the rupture to reach the surface is slightly larger for lower relative densities, shallower dip angles, and larger soil layer thicknesses.
- The length of the distorted zone is more sensitive to the dip angle than to relative density and soil layer thickness. Cases that include graben formation include distorted zones with length that are twice the lengths for cases where only one failure plane is formed. In the former, the center of the distorted zone is shifted towards the center of the fault. The normalized length of the distorted zone is slightly larger for cases of lower relative densities and larger soil layer thicknesses.
- The results of the fault propagation study for normal faults could be presented in normalized format since the normalized results are relatively insensitive to the height of the overlying soil layer.

Acknowledgments. The authors would like to acknowledge the support of the Lebanese National Council for Scientific Research, the Masri Institute, and the University Research Board at the American University of Beirut for supporting the research of the geotechnical group in the Department of Civil and Environmental Engineering.

References

Anastasopoulos, I., et al.: Fault rupture propagation through sand: finite-element analysis and validation through centrifuge experiments. J. Geotech. Geoenviron. Eng. **133**(8), 943–958 (2007). doi:10.1061/(ASCE)1090-0241(2007)133:8(943)

Belousov, V.V.: Experimental geology. Sci. Am. **204**(2), 96–107 (1961)

Bolton, M.D.: The strength and dilatancy of sands. Gotechnique **36**(1), 65–78 (1986)

Bransby, M.F., et al.: Centrifuge modelling of normal fault-footing interaction. Bull. Earthq. Eng. **6**(4), 585–605 (2008a). Springer, doi:10.1007/s10518-008-9079-0

Bransby, M.F., et al.: Centrifuge modelling of reverse fault-foundation interaction. Bull. Earthq. Eng. **6**(4), 607–628 (2008b). Springer, doi:10.1007/s10518-008-9080-7

Bray, J.D., et al.: lg small-scale modelling of saturated cohesive soils. Geotech. Test. J. **16**(1), 46–53 (1993). doi:10.1520/GTJ10266J

Bray, J.D., et al.: Analysis of earthquake fault rupture propagation through cohesive soil. J. Geotech. Engrg. **120**(3), 562–580 (1994b). ASCE, doi:10.1061/(ASCE)0733–9410(1994) 120:3(562)

Chang, Y.Y., Lee, C.J., Huang, W.C., Hung, W.Y., Huang, W.J., Lin, M.L., Chen, Y.H.: Evolution of the surface deformation profile and subsurface distortion zone during reverse faulting through overburden sand. J. Eng. Geol. **184**, 52–70 (2015). doi:10.1016/j.enggeo. 2014.10.023

Cole, D.A., Lade, P.V.: Influence zones in alluvium over dip-slip faults. J. Geotech. Engrg **110**(5), 599–615 (1984). ASCE, doi:10.1061/(ASCE)0733–9410(1984)110:5(599)

Emmons, R.C.: Strike-slip rupture patterns in sand models. Tectonophysics **7**(1), 71–87 (1969). doi:10.1016/0040-1951(69)90065-1

Horsfield, W.T.: An experimental approach to basement-controlled faulting. Geol. Mijnbouw **56** (4), 363–370 (1977). Haarlem, The Netherlands

Johansson, J., Konagai, K.: Fault surface rupture experiments: a comparison of dry and saturated soils. In: Proceeding of 27th Symposium on Earthquake Engineering, JSCE, Paper No. 271 (2004)

Johansson, J., Konagai, K.: Fault induced permanent ground deformations: experimental verification of wet and dry soil, numerical findings' relation to field observations of tunnel damage and implications for design. J. Soil Dyn. Earthq. Eng **27**(10), 938–956 (2007). doi:10.1016/j.soildyn.2007.01.007

Kelson, K.I., et al.: Representative styles of deformation along the Chelungpu fault from the 1999 Chi-Chi (Taiwan) earthquake: geomorphic characteristics and responses of man-made structures. Bull. Seismol. Soc. Am. **91**(5), 930–952 (2001). doi:10.1785/01200000741

Kulhawy, F.H., Mayne, P.W.: Manual on estimating soil properties for foundation design. Report EL-6800, p. 306. Electric Power Research Institute, Palo Alto (1990)

Lade, P.V., et al.: Multiple failure surfaces over dip-slip faults. J. Geotech. Engrg. **110**(5), 616–627 (1984). doi:10.1061/(ASCE)0733-9410(1984)110:5(616)

Lin, M.L., et al.: Deformation of overburden soil induced by thrust fault slip. Eng. Geol. **88**(1), 70–89 (2006). doi:10.106/j.enggeo.2006-08.004

Loukidis, D.: Active fault propagation through soil. Thesis, School of Civil Engineering, National Technical University, Athens, Greece (1999)

Loukidis, D., et al.: Analysis of fault rupture propagation through uniform soil cover. J. Soil Dyn. Earthq. Engrg. **29**(11), 1389–1404 (2009). doi:10.1016/j.soildyn.2009.04.003

Nakai, T., et al.: Numerical calculations of soil response over a displacing basement. Soils Found. **35**(2), 25–35 (1995). doi:10.3208/sandf1972.35.2_25

Roth, W.H., et al.: Numerical modeling of fault propagation in soils. In: Proceedings of 4th International Conference Numerical Methods and Applications in Geomechanics, Edmonton, Canada, pp. 487–494 (1982)

Sanford, A.R.: Analytical and experimental study of simple geologic structures. Geol. Soc. Am. Bull. **70**(Jan), 19–52 (1959). doi:10.1130/0016-7606(1959)70[19:AAESOS]2.0.CO;2

Schanz, T., Vermeer, P.A.: On the stiffness of sands. Géotechnique **48**, 383–387 (1998)

Scott, R.F., Achoustra, J.J.: Nuclear power plant siting on deep alluvium. J. Geotech. Engrg. Div. **100**(GT4), 449–459.a66 (1974). ASCE

Scott, R.F.: Failure. Geotechnique **37**(4), 423–466 (1987). London, England

Taniyama, H., Watanabe, H.: Deformation of sandy deposits by fault movement. In: Proceedings of 12th WCEE, vol. 6, pp. 325–336. Mc Graw Hill, Tokyo (2000). CD, code

White, R.J., et al.: Effect of particle size on localization development in model tests on sand. In: Leung, C.F., Lee, F.H., Tan, T.S. (eds.) Proceedings of Institute Centrifuge Conference, Balkema, Rotterdam, The Netherlands, pp. 817–822 (1994)

Hydrodynamic Performance of Coastal Geotube Embankment with and Without Gabion Boxes

S. Sherlin Prem Nishold, R. Sundravadivelu(✉), and Nilanjan Saha

Department of Ocean Engineering,
Indian Institute of Technology Madras, Chennai, India
rsun@iitm.ac.in

Abstract. Erosion of shoreline is an important coastal phenomenon which takes place due to the movement of sand mass by tides, wave action, and wave induced currents. The often used materials, like artificial armour units or rubble mounds, for coastal protection, are very expensive and time-consuming to install, apart from not being readily available. Therefore Geo-synthetics, like geo textile-tube embankments are, innovative solutions for protection from coastal erosion when compared to other conventional materials and methods. Geo-tube with gabion is an excellent solution for scour protection applications as well as soil with low bearing capacity. It has noted from previous studies, high reflection coefficients of coastal structures can lead to poor performance during rough weather for conventional embankment. This results in increased possibility of scour and failure of coastal sea walls and quays. One way of dealing with this problem is to deploy embankments with high energy dissipation characteristics, using unique geometries of geo-tube embankments. Hence, detailed studies are carried out for a geo-tube embankment with ten geo-tubes in a four layer configuration. These studies involve a scaled model (1:10) with and without gabion boxes to understand the wave energy dissipation for such configurations better. The scaled model in the mentioned configuration of geo-tubes was of woven geotextile fabric. The gabion box is made up of eco-friendly polypropylene tar-coated rope which consists of small rubble stones that increase the porosity when compared to the conventional monolithic rubble mound. Then scaled model was examined for design water level for different wave heights and different wave periods and the results were analysed for various hydrodynamic coefficients. Based on these results, a prototype of this configuration is intended to construct for field trials on the coast of Pentha (Odisha, India).

1 Introduction

Pentha (20°32.5′N 86°47.5′E) is a small coastal village in Kendrapara District of Odisha State at about a distance of 8.6 km from Rajnagar Town. About 50,000 fishermen live here. Agriculture is a secondary occupation for these people along with fishing. The coast near the Pentha is a village subjected to severe erosion for the past 25 years. The saline embankment to protect the village eroded. Therefore, a retarded embankment was built. The retarded embankment is likely to erode if not protected.

J. Wasowski et al. (eds.), *Engineering Geology and Geological Engineering for Sustainable Use of the Earth's Resources, Urbanization and Infrastructure Protection from Geohazards*, Sustainable Civil Infrastructures, DOI 10.1007/978-3-319-61648-3_15

The Government of India has received a credit from the International Development Association for the Integrated Coastal Zone Management Project (ICZMP) and Government of Odisha intends to utilise a portion of the funds for the construction of suitable Geo tube embankment on the seaside of the retarded embankment. The proposed geo-tube embankment lies between the two points 20°32′21.57″N – 86°47.15.12″E and 20°32′39.08″N – 86°47′30.58″E for 505 m length.

1.1 Causes for Erosion

The high water level about the site is about 3.6 m above MSL. Therefore vast quantities of tidal waters enter into the rivers for more than 20 km distance from their mouths, since the gradient between the tidal entrance and upland river mouth is very gentle near the coast. The major reason for erosion is, it lies between two rivers is due to the circulation of currents in between the two river clusters which discharge water into the sea. The bathymetry is perfectly parallel to the shoreline, and the beach slope is gentle, results in the formation of regular waves at equal intervals, since the slope is gentle the wave breaks in the longshore bar and due to higher wave celerity it plunges over foreshore up to the berm. This results in the movement of sediments from onshore and transported back to foreshore during backwash and cross-shore transport (littoral drift).

Further the site is continuously affected by cyclones, storm surges, associated with a low-pressure weather system. The storm surge causes the water to pile up higher than the average sea level and tends to increase the wave height is the predominant reason for the erosion of beach berms and sand dunes. Since storm surge waves are of high intensity and breaks after the longshore bar, the gradient in transport rate in the direction of the net transport has increased.

2 Literature Review

2.1 Geotextile Tube

Geotextile tubes are made up of synthetic fibres which are sustainable, permeable textile fibres those can contain, filter, and reinforce soil. The integrity of the geotextile structure depends on the type of infill material and type of geosynthetics used. The permeability of the fill, as well as geotextile, has a significant influence on the tube structure and the apparent opening size (AOS) at the rate filter cakes forms. These are some of the factors that affect the water outflow. In consequence, the strength of the soil infill in geotextile tubes with high moisture content will not be sufficient to support geo tube stacking (Shin and Oh 2007). (Leshchinsky et al. 1996) Developed a computer program for calculating the geometry and stresses on the encapsulating geotextile of the tubes based on design parameters such as pumping pressure, allowable circumferential tensile force, and unit weight of the fill material along with the tube height. Numerous studies on the stability of stacked geotextile tubes under wave actions can be found in the works of (Van Steeg et al. n.d.). (2011), (Das Neves et al. 2009), (Bezuijen et al. 2004), (Pilarczyk 2008), (Cho 2009). However, there are only a few studies on the stability of stacked geotextile tubes subjected to scouring. Kim et al. (2013) recently performed finite-element analyses (FEAs) on ground modification techniques for improved stability of geotextile tube–reinforced reclamation embankments subject to scouring. Significant scouring commonly takes place at the sides of the geotextile tubes facing the shore.

2.2 Gabion Box

Considering previous literature, we have found two major factors affect the geotechnical stability of the structure viz. scouring and formation of the filter cake. This paper presents the stability of Geo tube embankment with infill material as coarse river sand, it allows free drainage and to prevent the scouring action. For this, a scour apron had designed to a depth of −3 m to protect stacked Geo tube, based on Hudson's formula (Hudson 1959). Flexible tar coated polypropylene gabion in the form of gabion box were used to encapsulate the stacked geo tube. These gabions will dissipate the wave energy because of the porous nature and protect embankment from scour.

2.3 Need for Geo-tube Embankment

The shoreline at Pentha was 500 m away from the saline embankment built in 1975. Since this original embankment eroded, a retarded embankment has been constructed 60 m behind that. The erosion is about 10 to 15 m per annum, and on 21st Nov 2009 the shoreline was at 50 m from the retarded embankment and on 23rd Oct 2011, the shoreline was at 33 m from retarded Embankment. About 300 m stretch of the retarded embankment was subjected to storm waves from 2009. It had confirmed from the soil investigations that there is a possibility for development of pore water pressure and differential settlement. For a conventional gravity structure, these are the reasons for a possible failure due to differential settlement. The geotextile tube is made of woven geotextile sheets those are flexible and perforated, which allows water to exit. Thus the development of pore water pressure will be avoided. The strength of the sheet entraps the solid particles inside the container. Even though there may be a differential settlement, the geo-tube will adjust with soil bed profile because of the flexibility and porous nature of geo-tube. Thus, to protect the retarded embankment, a standalone geo-tube embankment was designed with 30 m base width has aligned about 5 m to 10 m away from the retarded embankment for a length of 700 m in 2011. However, due to the subsequent erosion of coast, the base width of the integrated geo-tube embankment is altered to 24 m. The detailed topography is shown in Fig. 1.

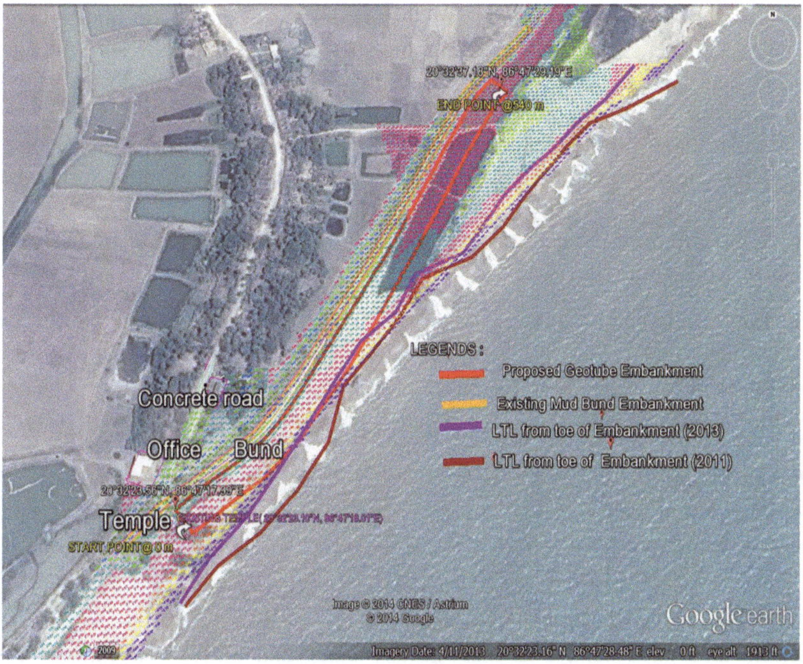

Fig. 1. Topographical view of proposed geo-tube embankment

2.4 Factors Influencing Stability of Geotube

Hydrodynamic characteristics such as inertia and drag due to the waves, induce lateral and vertical loads which have to be balanced by the soil bed profile. However, depends upon the bearing capacity as well as other geotechnical characteristics of the ground behaviour. The following mechanisms have been reported to be of importance for geo tube embankments.

2.4.1 Hydrodynamic Failure Mechanism

Among the various failure mechanisms reported, sand loss and sand migration due to aggressive waves, currents through the geotextile pores have been determined to be a cause for the failure of Geocontainers. This mechanism is closely related to the footprint of the geo-tube on the soil. To ascertain this, (Restall et al. 2002) have conducted an extensive study on pullout failure, failures due to material friction, wave force, and pressure. As a result, it found that filling ratio of the geo-tube should be within 80% of the volume available within the geo-tube cross section. It helps in maintenance of an elliptic cross section with a wider footprint over the soil base.

Another major factor prevents sediment loss is the fineness. Loss of fineness of the contained sediments through the geotextile pores during wave action is reported to be from a minimum of 8% to a maximum 27% of the original size (Lawson 2008).

To prevent the sand loss, the grade of the particle size should be higher than the geotextile aperture size. The Smaller grain size of the sand particles will lead to increased sand loss. Other reasons for sand loss include the damage to the geotextile due to vandalism, bursting, puncturing, etc.

The other major problem encountered by a coastal structure is scouring (Kim et al. 2013). Scour can, however, be prevented by various ground modification techniques as well as decreasing the velocity of the impinging wave and current.

Also, the failure of the geo-tube due to sand loss and beach deformation manifests itself in the following effects, viz.

- Flapping of Geotextile tube
- Caterpillar Effect
- Sand bye pass
- Sliding and Reshaping
- Group Failure
- Pullout failure
- Scour of Foundation
- Non-structured vortices.

2.4.2 Geotechnical Failure Mechanism

Geotextile tubes will behave as mass-gravity units, and thus a conventional design approach following standard procedure of assessing the possible modes of failure or deformation can be employed to arrive at a safe design solution (Lawson 2008), Engineering and Physical behavior of soil will vary with respect to location, environment and with the influence of load acting upon it (Lawson 2008). The behaviour and physical change of soil will vary with the degree of interaction with the structure. Usually, the properties modified during soil strength-deformation are a shear strength, stiffness, and permeability (Nishold et al. 2014). Coastal structures were constantly exposed to wind, waves, and currents in different environment such as cyclone and storm surge, in which these characteristics will have a significant influence on the change in soil properties and the stability of structures. The features and natural variations in the properties of soil due to the external forces (induced by wave inertia and drag along with tidal fluctuation under various environments) and deformations can make a structure susceptible to a failure mode. This change in soil property manifests itself in the following forms viz.

- Ultraviolet Instability
- Sliding stability
- Stability against overturning
- Bearing stability
- Foundation Settlement
- Global stability.

2.5 Design of Geo-tube Embankment

The geo-tube embankment had proposed for a length of 505 m with a base width of 30 m as shown in Fig. 2 for a design water depth of 5 m, with a scour apron for depth of −3 m below MSL and a toe mound of 3 m above MSL. The toe mound and scour apron will act as a protection to the structure from scouring action of waves. The width of toe mound should be maximum of twice the significant wave height or 0.4 times design water depth. The height of the mound should be 50% of the width. Four layers of sand filled geo-tubes are aligned parallel to the shore and gabions boxes were stacked up on it. The sand filled geo-tubes will act as a protective barrier against tidal waves and stacked gabion boxes will absorb the wave energy.

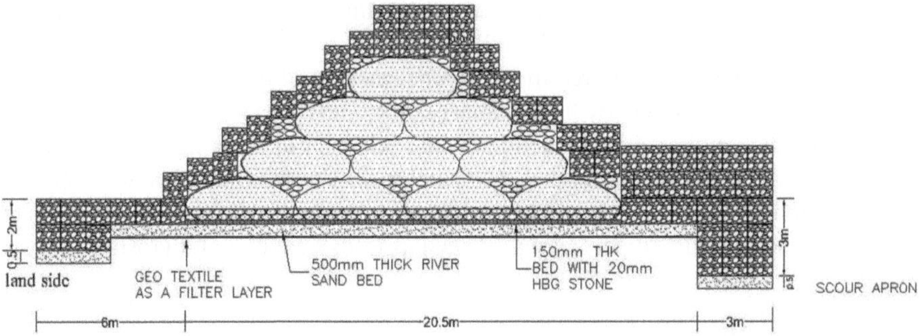

Fig. 2. Cross-sectional view of proposed geo-tube embankment

2.6 Applications of Geotextile Tube & Gabion Boxes

Geo-tubes have been gaining popularity in usage due to the continuing land expansion activities by reclamation. It is made up of polypropylene woven geotextile are strong, non-corrosive and eco-friendly to the marine environment (Pilarczyk 1996). In harbours, coastal dykes equipped with geosynthetics for reinforcement purposes. Geo-tubes offer an economical alternative to the conventional use of rock or precast concrete block units for shoreline protection of marine embankments and dykes (Yan and Chu 2010).

Currently, rock or precast concrete block units used in the construction of Shoreline Protection for Marine embankment and dykes have increasingly become expensive to construct. So geo-tube embankment, made of permeable and soil-tight geotextile material, is alternatively the most economical solution for coastal protection, Geotubes are hydraulically filled with dredged marine soil or sand (Dassanayake and Oumeraci 2012). Essentially a geo-tube is a single construction unit block containing soil. Currently, it has used in coastal engineering projects such as shoreline protections and breakwaters. These sand-filled geo-tubes will act as a monolithic core in the embankment and will serve as an impervious medium.

However, these geo-tube elements continuously subjected to various static and dynamic forces such as wave load in addition to the lateral and overburden pressures that it supports the weight of embankment fill and superimposed surface loads. A combination of these applied forces and loading may contribute to potential problems. Therefore, to counterbalance these forces gabion boxes will be used to dissipate the large kinetic wave forces. Gabion boxes are made up of flexible polypropylene tar coated ropes filled with stones of nominal diameter ranging from 200 mm to 250 mm.

2.7 Major Objectives of Geo-tube with Gabion Embankment

Gabion boxes filled with a smaller range of stones are more porous those are capable of dissipating large kinematic wave forces. Stacking of gabion boxes with each other in various interlocking patterns is equivalent to installing the armour units for the conventional constructions. All the Gabion Boxes should be tie manually to the adjoining Boxes on all sides using PP tarred rope after the stacking of gabion box in position, so as to protect the gabion boxes from damages from movement in case of large wave forces. Even though there may be a differential settlement of the soil, the geo-tube will adjust with soil bed profile because of the flexibility and porous nature of geo-tube. The geo-tube embankments also protect the inland area from erosion and storm water inundation and provide proper coastal protection from severe in-situ erosion.

3 Experimental Investigations

To understand the hydrodynamic characteristics of geo tube embankment, a series of experiments have been performed for two different model,

1. Geo-tube Section
2. Geo-tube with Gabion Section

3.1 Test Facility

The experiments conducted in a wave flume at the Department of Ocean Engineering, Indian Institute of Technology Madras, India. The flume is 72.5 m long, 2 m wide and 2 m deep. A hydraulic piston wavemaker is installed at one end of the flume and has been used to generate waves with predefined characteristics for these set of experiments. A personal computer, connected to the servo actuator was used to input the time history of the signal to the wave maker as well as for the data acquisition of the signals from wave gauges through an amplifier. An artificial beach consisting of a combination of a parabolic perforated steel sheet and a rubble mound is provided at the other end of the flume to absorb the generated waves efficiently.

3.2 Details of Prototype and Scaled Model

The prototype geo-tube embankment parameters had been scaled to model using Froude scaling. Using a scaling ratio of 1:10 and model dimensions had arrived, and details had furnished in Table 1.

Table 1. Details of prototype and scaled model

Type of structure	Prototype	Model (1:10)
Geo-tube circumference	9 m	0.9 m
Geo-tube diameter	3 m	0.3 m
Gabion Box dimension	2 m × 1m × 1 m	0.2 m × 0.1 m × 0.1 m
Slope	Sea Side-1:1	Sea Side-1:1
	Lee Side-1:2	Lee Side-1:2
Water level at sea side	5 m	0.5 m
Maximum Height of Water Depth (D_{max}) = High Tide Level + Strom Surge	4.8 m	0.48 m

3.3 Model Setup and Test Condition

The positions of the wave gauges and the erected model in the wave flume are shown in Figs. 3, and 4 show the top view and side view, respectively, of the proposed model in a wave flume. The length of an individual gabion boxes is installed across the width of the wave flume as shown in Fig. 4. Moreover, the 2 m width of the flume is split along the middle of the flume with a 2 mm thick galvanised iron sheet for a distance of 18 m. It separates the wave flume into two parallel channels for the models to study. The first channel of the flume has a Geotube section type installed while the other has Geotube with Gabion section. The clearance between the model and either side of the flume wall as 2 cm. This configuration studied for various hydrodynamic coefficients and dissipation parameters under regular waves. Further details are shown in Figs. 5 and 6. The performance of the structure for the design water depth of 0.5 m tested for a different range of wave period ranging from 1.5 s to 4.7 s under regular wave condition of varying wave heights.

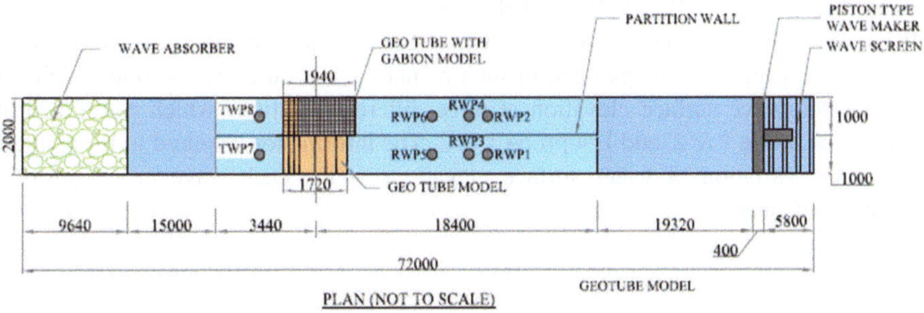

Fig. 3. Plan view of wave flume with models arrangement

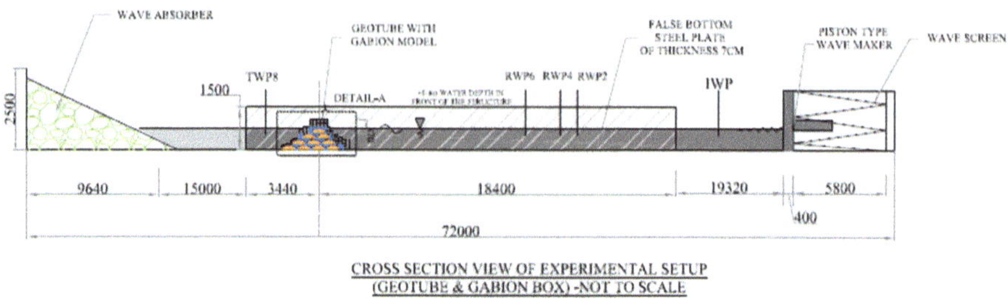

Fig. 4. Typical cross section of wave flume with models

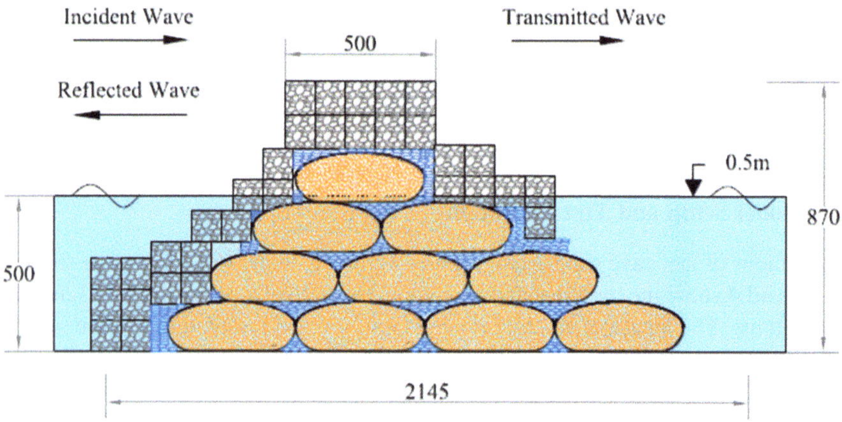

Fig. 5. Typical cross section of geo tube with Gabion section (GGTS)

3.4 Data Acquisition

The scaled model was set-up on wave on the wave flume as Geotube section and Geotube with gabion section approximately 32 m from the wave paddle in a way such that the reflection from the structure will not affect the measurement.

Physical modelling and flume setup had been made based on the guidelines provide by (O'Donoghue 2014). Three wave gauge placed along the wave propagation direction such as (RWP1, RWP3, and RWP5) along the Geotube section and (RWP2, RWP4, and RWP6) along the Geotube with gabion section detailed in Figs. 3 and 4. All the wave gauges are be planned in such manner to measure the combined incident and reflected wave surface elevation at three different locations which positioned at L/10 for the probe 1 & 2, and L/4 probe 1 & 3. The information required in its simplest form is the variation of water surface elevations at the each wave probe position identically of interest.

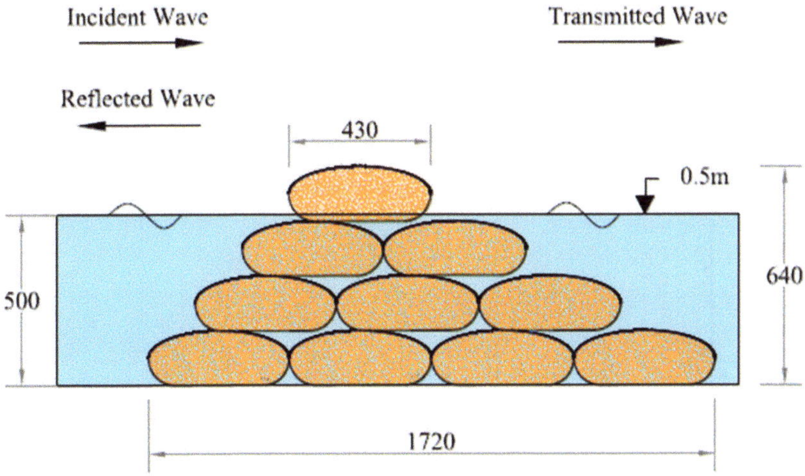

Fig. 6. Typical cross section of geo tube section (GTS)

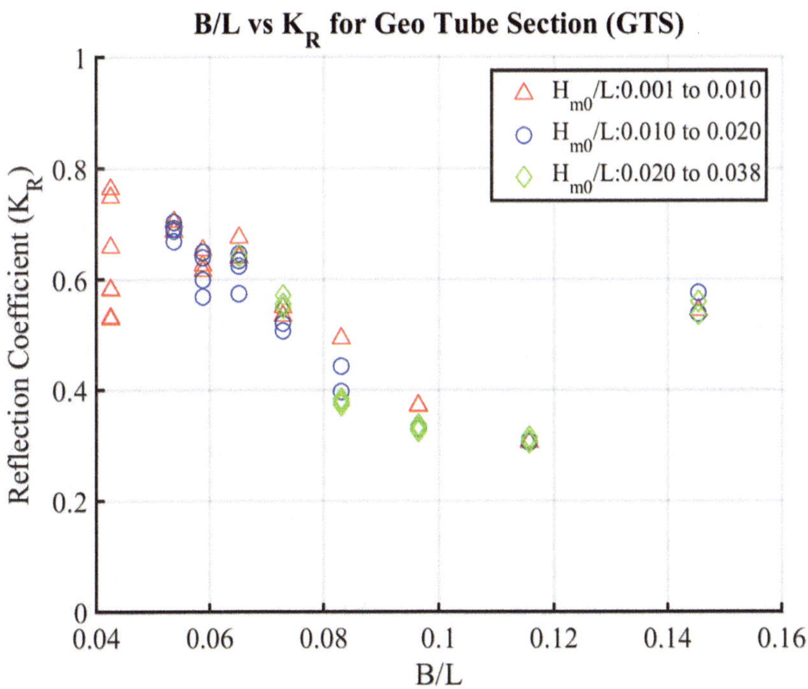

Fig. 7. Typical plots of B/L vs. K_R for geo tube section

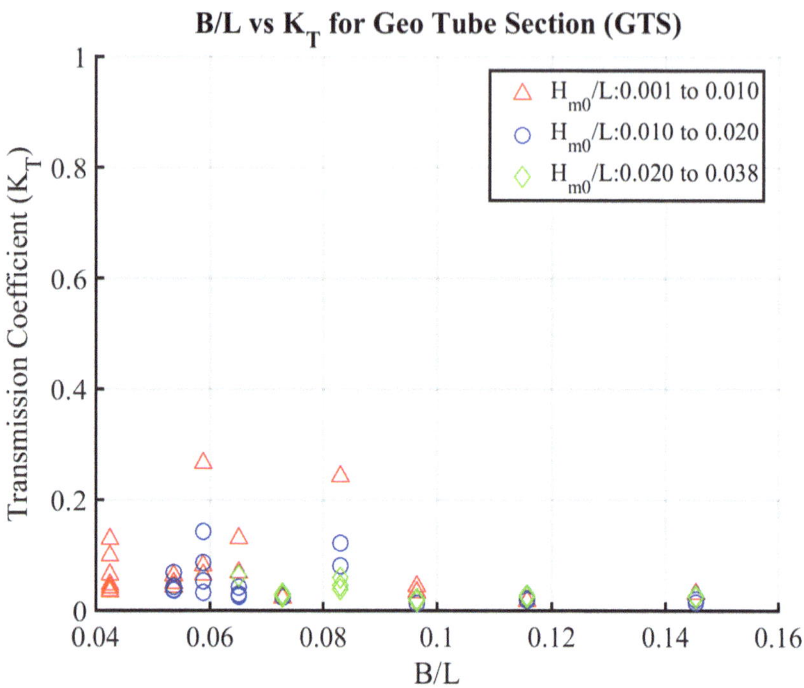

Fig. 8. Typical plots of B/L vs. K_T for geo tube section

For analysis and interpretation of test results, it is often required to separate the measurements in short and long waves. The separation can be either related to a fixed frequency or by detailed analysis of the wave spectrum. For practical reasons, it may be considered to separate at about 20 s wave period; however, in some cases, other values may be more relevant. Reflected waves are unavoidable in wave models. The preferred strategy is to plan the model with a minimum of harmful reflection, those array of wave gauges are positioned to capture the surface elevation on the phase difference in between them.

Further, the measured time series are converted to the frequency domain using Fast Fourier Transform (FFT). Along with their phase spectra, each of these time series has separated into the incident as well as reflected waves. The true incident and reflected waves have then been used to find the reflection coefficient.

Similarly another two probes, viz. TWP7 behind the geo-tube section and TWP8 behind the geo-tube with gabion section, have been used to measure the transmitted wave surface elevation for the purpose of estimating the transmission coefficients.

3.5 Analysis for Hydrodynamic Coefficients

The effectiveness of the design in dissipating the incident wave energy is highly dependent upon the relationship between the wave characteristics, structural characteristics, and

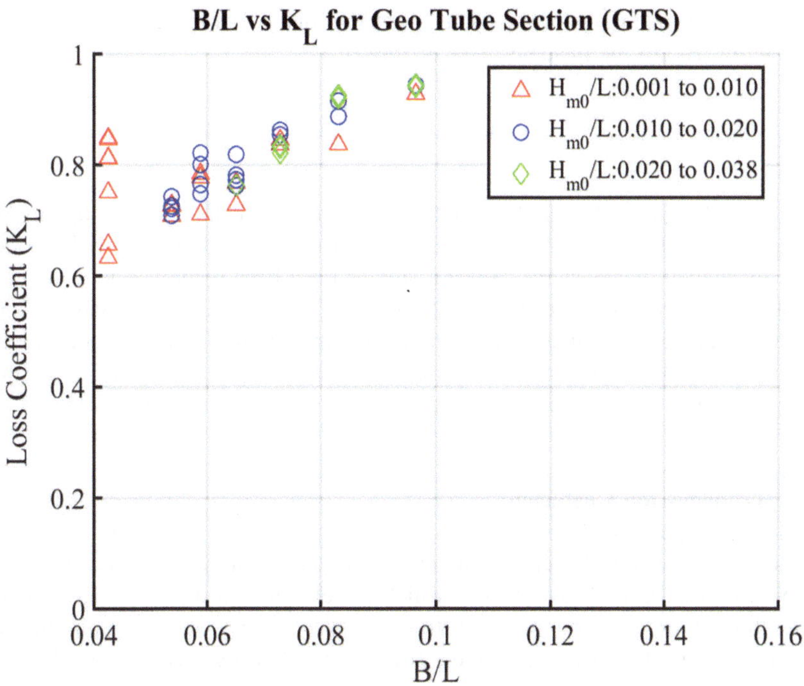

Fig. 9. Typical plots of B/L vs. K_L for geo tube section

water depth. To quantify the wave characteristics, the reflection coefficient (K_R) and transmission coefficient (K_T) are obtained from the wave gauge measurements using three probe method (Mansard and Funke 1980). This approach provides the spectral energy of the incident and transmitted wave. Having obtained the reflection and transmission coefficient, the losses (K_L) are calculated using the conservation principle, i.e.

$$K_R^2 + K_T^2 + K_L^2 = 1$$

For the study the effect of the water depths, an attempt was made in this study to examine the effect of the reduction in the depth of submergence of the structure in attenuating the incident waves. The reduced depth of submergence is expected to reduce the cost of installation of the proposed structure while increasing the water exchange beneath the structure. Further such a measure can provide an insight into the hydrodynamic efficiency of the structure under extreme scenarios.

3.5.1 Reflection Coefficient

Incident waves may be reflected (partially or wholly) from a beach and coastal or harbour structures, depending on the wave characteristics and the structure geometry. The magnitude of the reflection can be represented by a reflection coefficient (K_R), which is defined as the ratio of the reflected wave height (H_R) to the incident wave height (H_I), or else in terms of wave energy, as the square root the ratio of the reflected wave energy (E_R) to the incident wave energy (E_I).

$$K_R = \frac{H_R}{H_I} = \sqrt{\frac{E_R}{E_I}}$$

Impermeable vertical walls fully reflect the majority of the non-overtopping incident waves (i.e., $K_R \approx 1.0$). Beaches and sloped structures, however, reflect only a portion of incident wave energy. Several studies have been employed to estimate the amount of reflected energy regarding reflection coefficient (Harris and Sample 2009). We have employed the three probe method for determining reflection coefficient. It helps in the resolution of the incident and reflected amplitudes using least square technique and two phase difference of the waves at three locations (Mansard and Funke 1980).

3.5.2 Transmission Coefficient

The primary purpose of a breakwater or a coastal structure is to reduce the wave energy on its lee side as well to lessen the attenuation of approaching waves. The term wave transmission used is the wave energy which travels through a breakwater, either by passing through or by overtopping the structure. Wave energy attenuation in the lee of the breakwater is either dissipated by the structure (such as by friction, wave breaking, armour unit movement,) or reflected back as reflected wave energy (Yuliastuti and Hashim 2011). The effectiveness of a breakwater in attenuating wave energy measured by the amount of wave energy is transmitted or pass through the structure. Wave transmission quantified by the use of the wave transmission coefficient.

$$K_R = \frac{H_T}{H_I} = \sqrt{\frac{E_T}{E_I}}$$

Where K_T is the wave transmission coefficient, H_T is the height of the transmitted waves on the leeward side of the structure, and H_I is the height of the incident waves on the seaward of the structure, or else in terms of wave energy, as the square root the ratio of the transmitted wave energy (E_T) to the incident wave energy (E_I).

3.5.3 Loss or Dissipation Coefficient

The effectiveness of a breakwater had to judge by the portion of the energy is dissipates through friction, turbulence and wave breaking. Loss or dissipation coefficient (K_L) can be determined using the following relationship.

$$K_L = \sqrt{(1 - K_R^2 - K_T^2)}$$

4 Results and Discussion

The variation of K_R, K_T, $K_{L,}$ with (B/L) for various wave steepness ratios are filtered and separated on three different wave steepness range, where B is the breath of the crest to which the water plane cuts the structure and L are the wavelengths corresponding to each period. The results for K_R, K_T, and K_L are discussed for three different wave steepness range (Hm0/L) viz. Lowest (0.001 to 0.01), Moderate (0.01 to 0.02), and Highest (0.02 to 0.038). moreover, Comparisons had discussed in the following sections (Figs. 10, 11, and 12).

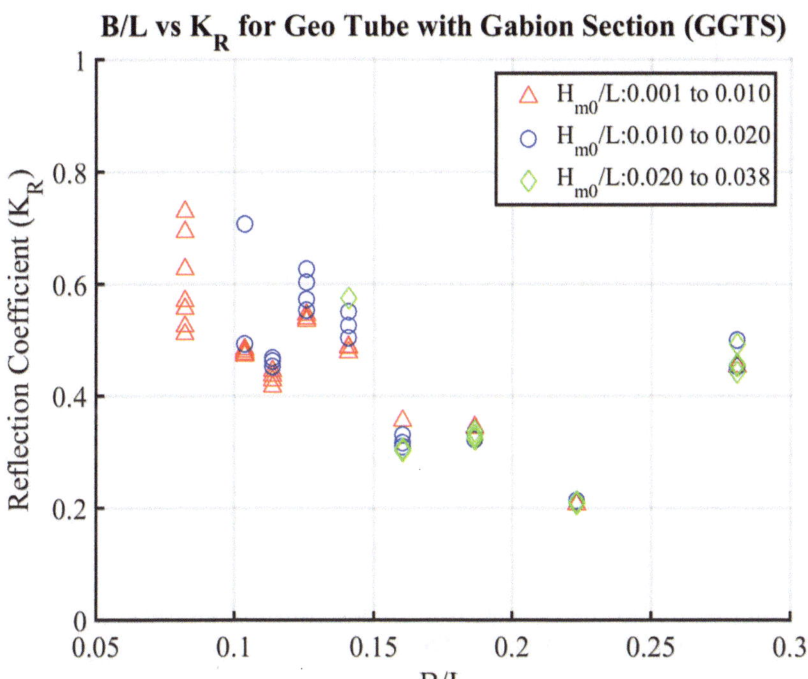

Fig. 10. Typical plots of B/L vs. K_R for geo tube with Gabion section

4.1 Influence of B/L on Hydrodynamic Coefficients for the Geotube Section (GTS)

Let us consider the Geotube section (GTS) the model for the moderately exposed case; wherein the water depth is slightly lower than the structure height. For 0.011 to 0.038 highest wave steepness (Hm0/L) with a B/L range of 0.05 to 0.125, the reflection coefficient (K_R) is decreasing from 70% to 30%; Transmission is also reducing by 15% to 2%. Loss or dissipation coefficient is moderately increasing from 70% to 95% as detailed in plot Figs. 7, 8, and 9 as follows. For 0.001 to 0.01 the lowest wave steepness

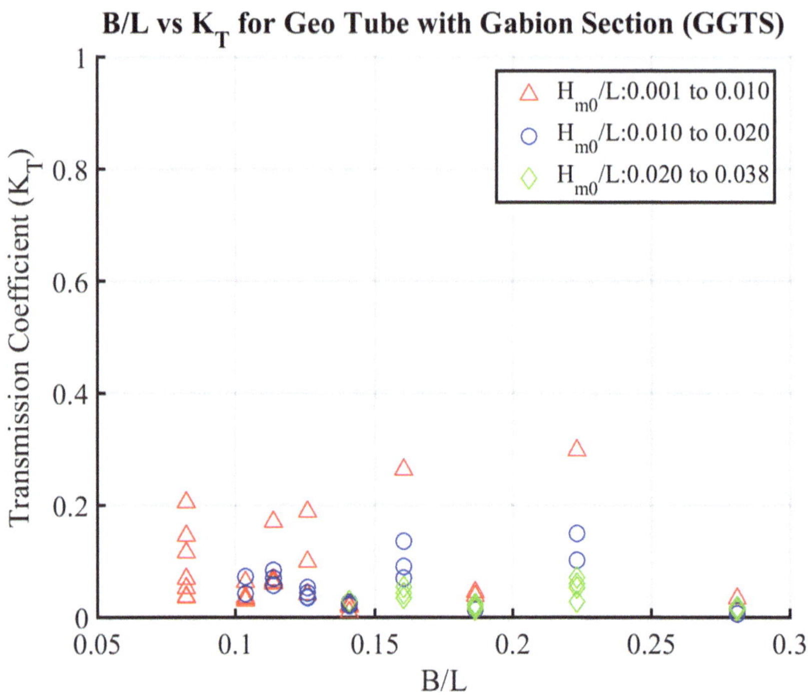

Fig. 11. Typical plots of B/L vs. K_T for geo tube with Gabion section

(Hm0/L) the K_R range is increasing rapidly from 50% to 80% due to long wavelength. Further, the transmission coefficients (KT) for the few cases had noticed to a range of 27% to 2%. For the lower steepness, loss coefficient is increasing 65% to 85%.

4.2 Influence of B/L on Hydrodynamic Coefficients for Geotube with Gabion Box Section (GGTS)

Considering the Geotube with gabion section (GGTS), the model is moderately exposed to the water since gabion boxes protected the Geotube on top. For the lower wave steepness of range 0.001 to 0.01, all the hydrodynamic coefficients increase linearly. E.g., reflection coefficient (KR) by 50% to 75%, transmission coefficient varying from 2.5% to 20% and loss coefficient increasing from 67.5% to 85%. The physical reason for the increase in hydrodynamic coefficients for the lowest wave steepness is due to the effect of longer wavelength. For the B/L range of 0.1 to 0.275, the reflection coefficient is decreasing from 70% to 20%; the transmission coefficient is reducing from 15% to 2%, and loss coefficient is increasing from 70% to 97%. The energy loss coefficient (remains between 0.7 and 0.97), It means that most of the energy lost due to the interactions of the waves with the geo-tube with gabion structure through wave breaking over the structure.

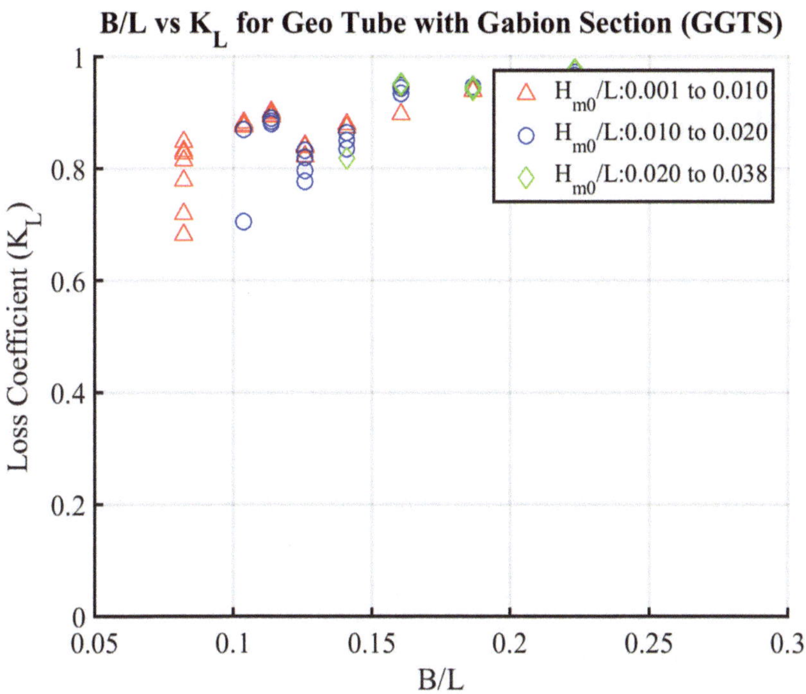

Fig. 12. Typical plots of B/L vs. K_L for geo tube with Gabion section

4.3 Comparison Between Geotube Section and Geotube with Gabion Box Section Cases

Comparing the above two case, i.e. Geotube section and Geotube with Gabion section B/L varies from a minimum of 0.0375 to a maximum of 0.15 for Geotube section; the hydrodynamic coefficients show a peak response for the lowest wave steepness range 0.001 to 0.0. Again for B/L range of 0.04 to 0.225, the reflection and transmission coefficient decreases along with wave steepness while thé loss coefficient increases simultaneously. The B/L range of Geotube with gabion case varies from 0.08 to 0.28. Coefficients, such as reflection coefficient and transmission coefficient (K_T), decrease drastically but at the same time loss coefficient (K_L) increase.

For the both the cases higher reflection and transmission is reported for lowest wave steepness due to the effect of longer wavelengths. From Figs. 13, 14, and 15, one can infer the efficiency on the relative crest. Owing to the wave breaking from reflections, we have chosen small, medium and steep cases based on visual observations. Hence, hydrodynamic characteristics are better for the Geotube with gabion section. As both the reflection and transmission coefficient for the slightly submerged case is about 0.5, this may be the optimum height of the relatively submerged depth to dissipate the incident waves.

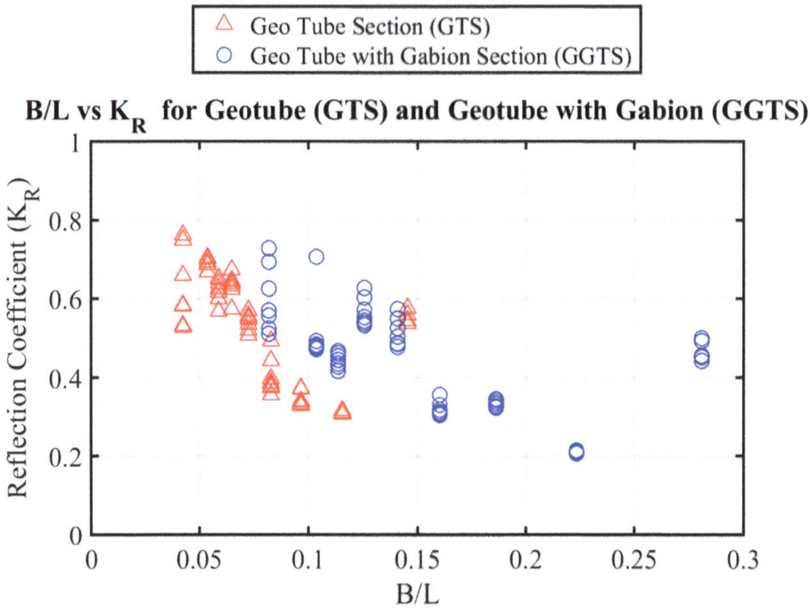

Fig. 13. Typical plots of B/L vs. K_R for GTS and GGTS

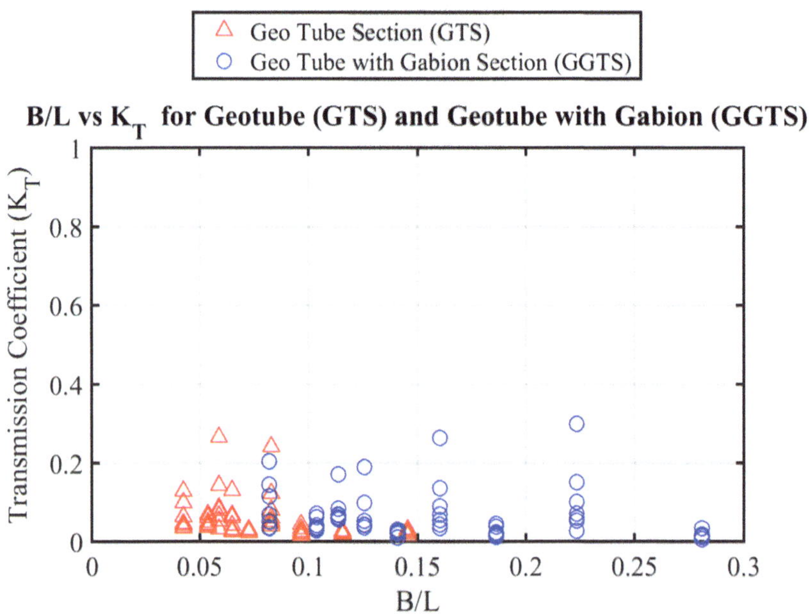

Fig. 14. Typical plots of B/L vs. K_T for GTS and GGTS

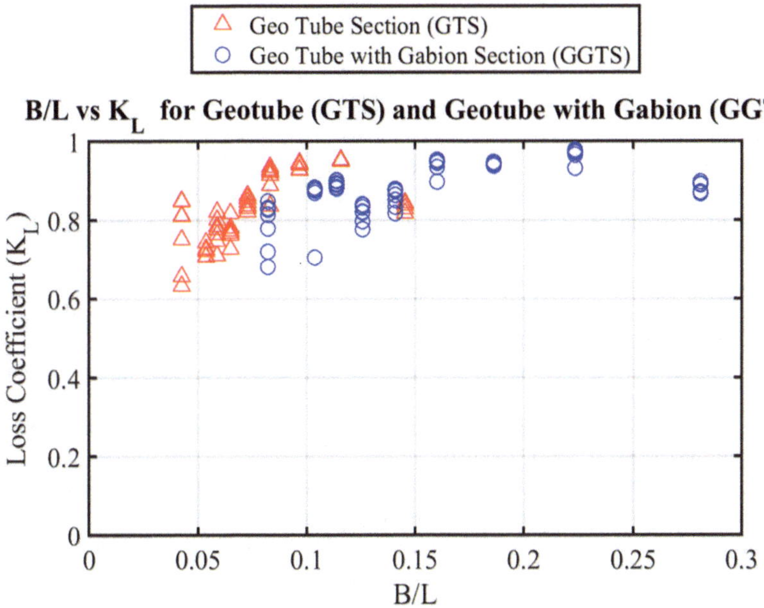

Fig. 15. Typical plots of B/L vs. K_L for GTS and GGTS

5 Conclusion

The hydrodynamic performance of two different structure types has been examined and quantified. The wave reflection, transmission, and energy dissipation characteristics checked for regular waves of different wave heights and wave periods at a constant water depth. Both the models have higher energy dissipation characteristics. Normally, the reflection coefficient will be higher for long period waves. However, Geotube with Gabion (GGTS) model provides a better reduction in reflection coefficients than the Geotube, section (GTS) model.

Considering the limitations of geotextile in coastal protection applications, a major problem with geotextile tube is that they have ultraviolet (UV) stability even though it is eco-friendly to the marine environment. Geotextiles, when exposed to high UV radiation, will fail. Hence it is suggested to provide a model of Geotube with Gabion protection. It will preserve the integrity of the Geotextile tube core. Moreover, Gabion boxes are polypropylene tar coated mesh boxes filled with a smaller range of stones which increases that porosity. Such Gabion boxes are capable of dissipating large kinematic wave forces than the conventional monolithic coastal structures. In addition to these measures, Gabion boxes protect the Geotubes from various hydrodynamic and geotechnical failure.

References

Bezuijen, A., De Groot, M.B., Breteler, M.K., Berendsen, E.: Placing Accuracy and stability of geocontainers. In: Proceedings 3rd European Geosynthetics Conference EuroGeo3, Munich (2004)

Cho, S.-M.: Geotechnical Designs Of The Incheon Bridge. Korea (2009)

Das Neves, L., Lopes, M.L., Veloso-Gomes, F., Taveira-Pinto, F.: Experimental stability analysis of geotextile sand-filled containers for dune erosion control. J. Coast. Res. **56**, 487–490 (2009). Special Issue

Dassanayake, D., Oumeraci, H.: Engineering properties of geotextile sand containers and their effect on hydraulic stability and damage development of low-crested/submerged structures. Int. J. Ocean Clim. Syst. **3**(3), 135–150 (2012). doi:10.1260/1759-3131.3.3.135

Harris, L.E., Sample, J.W.: The evolution of multi-celled sand-filled geosynthetic systems for coastal protection and surfing enhancement. Reef J. **1**(1), 1–15 (2009)

Hudson, R.Y.: Laboratory investigation of rubble mound breakwaters. J. Waterw. Port Coast. Ocean Div. ASCE **85**(3), 93–121 (1959)

Kim, H.-J., Jamin, J., Mission, J.L.: Finite element analysis of ground modification techniques for improved stability of geotubes reinforced reclamation embankments subjected to scouring. In: Proceedings of the 2013 World Congress in Structural Engineering and Mechanics, pp. 2970–2979 (2013)

Lawson, C.R.: Geotextile containment for hydraulic and environmental engineering. Geosynthetics Int. **15**(6), 384–427 (2008). doi:10.1680/gein.2008.15.6.384

Leshchinsky, D., Leshchinsky, O., Ling, H.I., Gilbert, P.: Geosynthetic tubes for confining pressurized slurry: some design aspects. J. Geotech. Eng. **122**(8), 682–690 (1996). doi:10.1061/(ASCE)0733-9410(1996)122:8(682)

Mansard, E.P.D., Funke, E.R.: The measurement of incident and reflected spectra using a least squares method. In: Proceedings of 17th Conference on Coastal Engineering, Sydney, Australia, pp. 154–172 (1980). http://journals.tdl.org/icce/index.php/icce/article/view/3432/3112

Nishold, S.P., Sundaravadivelu, R., Saha, N.: Analysis and design of geotube saline embankment. In: International Conference on Computational and Experimental Marine Hydrodynamics, vol. 2, pp. 292–300 (2014). https://doi.org/10.13140/RG.2.1.4062.9529

O'Donoghue, T.: Book review: users guide to physical modelling and experimentation: experience of the HYDRALAB network. In: Proceedings of the ICE - Engineering and Computational Mechanics, vol. 167 (2014). https://doi.org/10.1680/eacm.14.00013

Pilarczyk, K.: Alternatives for coastal protection. J. Water Resour. Environ. Eng. **23**, 181–188 (2008)

Pilarczyk, K.W.: Design aspects of geotubes and geocontainers (1996)

Restall, S.J., Jackson, L.A., Heerten, G., Hornsey, W.P.: Case studies showing the growth and development of geotextile sand containers: an Australian perspective. Geotext. Geomembr. (2002). doi:10.1016/S0266-1144(02)00030-4

Shin, E.C., Oh, Y.I.: Coastal erosion prevention by geotextile tube technology. Geotext. Geomembr. **25**(4–5), 264–277 (2007). doi:10.1016/j.geotexmem.2007.02.003

Van Steeg, P., Vastenburg, E., Bezuijen, A., Zengerink, E., De Gijt, J.: Large-scale physical Model tests on sand-filled geotextile tubes and containers under wave attack. In: Proceedings of the 6th International Conference, Coastal Structures 2011, pp. 1083–1094 (2013)

Yan, S.W., Chu, J.: Construction of an offshore dike using slurry filled geotextile mats. Geotext. Geomembr. **28**(5), 422–433 (2010). doi:10.1016/j.geotexmem.2009.12.004

Yuliastuti, D.I., Hashim, A.M.: Wave Transmission on submerged rubble mound breakwater using L-Blocks. In: International Conference on Environmental Science and Technology, vol. 6, pp. 243–248 (2011)

Insights into Hydro-mechanical Behavior of Bentonite Based Seals for Deep Geological Repositories

Nadia Mokni[1(✉)] and Agustin Molinero Guerra[1,2]

[1] Institute for Radiation Protection and Nuclear Safety (IRSN), Fontenay-aux-Roses, France
nadia.mokni@irsn.fr
[2] Ecole des Ponts Paris Tech, Laboratoire Navier, Marne la Vallée, France

Abstract. Pre-compacted blocks of bentonite sand mixture are candidate materials for sealing plugs of radioactive waste disposal. Choice of this material is especially based on its swelling capacity allowing all technological gaps existing between the bentonite based core and the host rock to be sealed. Under real repository conditions, the sealing plugs will start to take water from the host rock and swell. Thereby, a swelling pressure will develop in the radial direction against the host rock and in the axial direction against the concrete confining structures. One important characteristic of this type of heterogeneous mixture is the multimodal nature of its porous network. This multimodal porosity and the highly heterogeneous fabric of the mixture result in a quite complex behaviour during hydration. The paper describes numerical analysis of small scale tests performed on compacted samples of bentonite and sand mixture (70/30 in dry mass). The A coupled HM formulation that incorporates the relevant processes involved in the problem under consideration has been adopted. Technological gaps and heterogeneous structural distribution of the mixture were demonstrated to have an effect on swelling pressure evolution.

1 Introduction

Deep geological repository has been considered for High Level and Intermediate Level Long Lived radioactive Wastes (HLW & ILLLW) in several countries such as China, Belgium and France among others. The safety of this disposal concept is based on the multi-barrier system consisting on the geological barrier (host rock) and engineered barriers (waste container, buffer and sealing elements). In the French concept, once the waste containers will be installed, galleries and access shafts will be closed and sealed by swelling clay based cores. In the French concept of deep geological disposal the wastes are emplaced within large diameter boreholes (HLW) and galleries (ILLLW) excavated at great depths in a low-permeability host-rock (Callovo-Oxfordian argillite), Sealing of these underground works to prevent potential pathways for water, gas and radionuclides migration is one of the key points to ensure the long-term safety of the repository. Highly compacted bentonite/sand mixtures are candidate sealing materials for deep underground repositories due to their low permeability, high radionuclide retardation capacity and high swelling ability (Push 1979; Yong et al. 1986).

J. Wasowski et al. (eds.), *Engineering Geology and Geological Engineering for Sustainable Use of the Earth's Resources, Urbanization and Infrastructure Protection from Geohazards*, Sustainable Civil Infrastructures, DOI 10.1007/978-3-319-61648-3_16

Once installed in the repository, these sealing materials will be subject to coupled hydro-mechanical loadings: hydration due to the infiltration of pore water from the natural barrier and mechanical confinement resulting from the engineered barriers. It is therefore essential to understand their behavior under such loadings when assessing the overall repository safety.

In this context, the French Institute of Radiation protection and Nuclear Safety (IRSN) has launched the SEALEX project (SEALing performance Experiments) to investigate the long term performance of swelling clay-based sealing systems. This project relies on a series of in-situ experiments emplaced in IRSN's Underground Research Laboratory (URL) at Tournemire (South France) (Barnichon and Deleruyelle 2009; Mokni and Barnichon 2016a, b). Each experiment consists of a bentonite-based core (diameter 60 cm, length, 120 cm) mechanically confined and hydrated at both ends. One of the objectives of the project is to investigate the impact of technological gaps on the long term performance of bentonite based seals. Thereby, several technological gaps exist within the in situ tests: Gaps between the bentonite blocks and annular gap with variable width between the bentonite-based core (constituted of 8 blocks) and the host rock. Small-scale tests were also conducted in the laboratory (Wang 2012; Saba 2013) to provide first insights into the hydro-mechanical behaviour of the mixture.

One important characteristic of this type of heterogeneous mixture is the multimodal nature of its porous network. This multimodal porosity and the highly heterogeneous fabric of the mixture result in a quite complex behaviour during hydration (e.g. Chen et al. 2014; Saba et al. 2014a; Wang et al. 2012, 2013b; Villar et al. 2012; Alonso et al. 2011; Gens et al. 2011; Imbert and Villar 2006, Lloret et al. 2007). This complexity is even more increased when technological gaps are left between the pre-compacted blocks of sand/bentonite and the host-rock. This paper aims at analysing the effect of the heterogeneous distribution of the structure and of technological gaps on the hydromechanical behaviour of the mixture. The paper starts with the description of microstructural features. Afterwards, a double structure coupled hydro-mechanical formulation required for the analysis of this type of mixture is briefly presented. The formulation is then applied to the modelling of the 1/10 scale mock-up of SEALEX in situ tests.

2 Microstructural Features of Bentonie Sand Mixtures

The tested material is a compacted MX80 bentonite/sand mixture with a proportion of 70/30 in dry mass. The bentonite is from Wyoming, USA, with high montmorillonite content (80%). The cation exchange capacity (CEC) is 76 meq/100 g (Wang et al. 2012). The major exchangeable cation is Na+ (83%). The liquid limit is of 575%, the plastic limit is of 53% and the unit mass is of 2.77 g/cm^3 (Wang et al. 2012, 2013a). The quartz sand used in the mixture comes from Eure and Loire (France) with a unit mass of 2.65 g/cm^3. The quartz sand was sieved at 2 mm prior to being mixed with bentonite. The water used for samples preparation and hydration has the same chemical composition as the pore water of the Callavo-Oxfordian claystone from ANDRA URL in Meuse Haute la Marne (France) (Wang et al. 2012).

The mixture of MX80 and sand (70/30) was characterized in the laboratory in terms of Mercury Intrusion Porosimetry (MIP) tests, and Microfocus X-ray Tomography (μCT) (Wang et al. 2012, 2013a, b, c; Saba et al. 2013, 2014a, b, c). Pore Size Distribution (PSD) of the mixture compacted at $\rho_d = 1.67$ and 1.97 Mg/m^3 are shown in Fig. 1 (Wang et al. 2013c). Two structural levels can be distinguished from this MIP information. The macrostructure refer to the large-scale arrangement of soil particle aggregates and the relatively large pores between them. The microstructure refers to the clay particles and micropores identified as intra-aggregate pores. Here the depression between dominant peaks (0.2 μm) has been used as limiting entrance size separation between intra-aggregate and inter-aggregate porosity (e.g. Wang et al. 2013d).

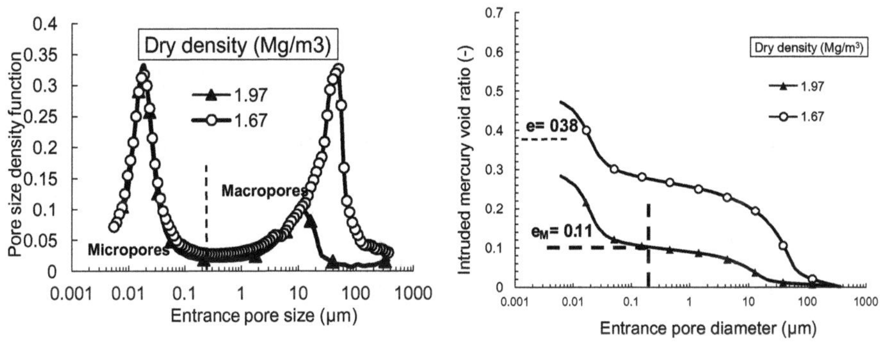

Fig. 1. Mercury intrusion porosimetry tests of MX80 bentonite/sand mixture (70/30) samples compacted at different densities (Wang et al. 2013c).

A complementary insight into the microstructural description is qualitatively provided by microfocus X-ray tomography (μCT). Saba et al. (Saba et al. 2014a, b) used this technique to examine in 3D the internal microstructure of a MX80 bentonite/sand (70/30) sample as compacted at $\rho_d = 1.8\,\mathrm{Mg/m^3}$ and after four months of hydration in constant volume conditions. Figure 2 shows a radial μCT cross section of the as compacted sample (38 mm diameter, 10 mm height) at 4 mm from the top. Heterogeneous distribution of macro porosity is observed on the radial direction. Indeed, more macro voids (in black) are detected at the outer boundary of the sample while at the centre, the bentonite powder grains (in dark grey) appear to be more compacted resulting in a denser macrostructure. Saba et al. (2014b) demonstrated that this heterogeneity has an impact on the swelling pressure measured on the radial and axial directions.

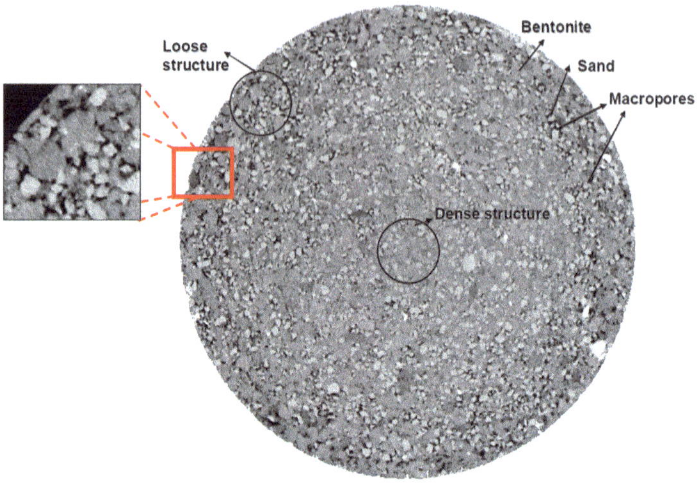

Fig. 2. μCT cross section of a bentonite/sand mixture (70/30) compacted at ρ_d = 1.8 g/cm^3 (38 mm diameter, 10 mm height) taken at 4 mm from the top. Saba et al. (2014a).

3 Analysis of the Behavior of Bentonite Sand Mixtures During Hydration

3.1 Double Structure Formulation

The formulation is based on the distinction within the material of two dominant pore levels (macro and micropores) which allow taking into account the dominant phenomena that affect the behavior of each structural level and the main interactions between them (e.g. Gens and Alonso 1992, Alonso et al. 2011; Gens et al. 2011). The macropores are considered to be partially filled with the liquid phase. The macroscopic transport of liquid and its components takes places in the macropores which are also called flowing pores. The formulation assumes that the microstructural level is saturated even at relatively high suctions, which can be supported by the high affinity of the active clay minerals by water. An important feature of the formulation is that hydraulic equilibrium is not assumed between both structural levels. Water filling the micropores can transfer into free water or conversely depending on water potential differences between micro and macropores. Macroscopic mass balance equations are given for macropores (Eq. 1) as well as micropores species (Eq. 2).

$$\frac{\partial}{\partial t}(\rho_{wM} S_{wM} \phi_M) + \nabla \cdot \left(\mathbf{j}_l^w\right) = -\Gamma^w \tag{1}$$

where ρ_{wM} is water density, S_{wM} the macropores degree of saturation, ϕ_M is the macroporosity and j_l^w the total flux of water with respect to a fixed reference system.

$$\frac{\partial}{\partial t}\left(\rho_{wm}(1-\phi_M)\phi'_m\right)=\Gamma^w \tag{2}$$

Where ρ_{wm} is water density in the microstructure and ϕ'_m is the volume of the micropores with respect to the volume of the aggregates.

The mass exchange between macro and microstructure Γ^w is treated as source/sink for which an additional relation is proposed (Eq. 3) (Mokni 2016).

$$\Gamma^w=\left(\rho_{wm}\frac{k_m}{\mu_m}\right)\frac{1}{\partial V}\frac{s}{l}(p_M-p_m)=\gamma\zeta(s_M-s_m) \tag{3}$$

The ratio $\frac{1}{\partial V}\frac{s}{l}=\zeta$ is a geometrical parameter that may depend on several variables, among them the specific surface of the micropores and the size and geometry of the micro and macropores. This geometrical parameter is kept separated from the transfer coefficient γ, so the physical meaning, range and units of the dependent parameters (ρ_{wm}, μ_m and k_m) are maintained.

Macroporosity ϕ_M and microporosity ϕ_m are defined as the volume of macropores and micropores divided by the total volume of the soil. As a consequence total porosity ϕ equals $\phi_M + \phi_m$.

To represent the mechanical behavior of the material, a double structure elasto-plastic model for expansive clays has been used (e.g. Gens and Alonso 1992; Gens et al. 2011; Sanchez et al. 2005). The model proposes constitutive laws to describe the macro and microstructural levels, and the coupling between them. The model assumes that for the macrostructural level, both elastic and plastic stains can develop as a result of stress and suction changes. The behavior of the macrostructural level is defined by the Barcelona Basic Model (BBM) (Alonso et al. 1990). A fundamental assumption of the framework is that microstructural behavior is independent of the macrostructural state and only responds to changes in suction and stresses at local microstructural level. A detailed description of the mechanical constitutive model could be found in (Mokni 2016).

The numerical analysis has been performed with a modified version of Code_Bright (Olivella et al. 1994) that incorporates the double structure elasto-plastic model for expansive clays.

3.2 Impact of Heterogeneous Structural Distribution of the Mixture

Modelling of Swelling Pressure Tests Under Constant Volume Conditions

The formulation described above has been applied to the modelling of swelling pressure tests at constant volume conditions performed on specimens of sand/bentonite mixture (30/70 in dry mass) compacted $\rho_d = 1.60\,\mathrm{Mg/m^3}$ (Saba 2013). Two Specimens (named CS14C1PV and CS05C2PV) of 10 mm height and 38 mm diameter were placed in a rigid constant volume cells equipped with a total pressure sensor to measure the radial swelling pressure and a force transducer to measure the axial swelling pressure. During the test, the samples were saturated from the bottom. Detailed description of the performed tests is presented by Saba (2013). The estimated initial microstructural and macrostructural porosities using MIP tests are indicated in Table 1. The finite element mesh used in the coupled HM analysis is depicted in Fig. 3 together with the boundary conditions. Advantage has been taken of the basically axisymmetric nature of the test. Hydraulic and mechanical parameters are summarized in Table 1.

Table 1. Mechanical parameters for MX80 bentonite/sand mixture

70/30 MX80 bentonite/sand					Parameter relationships
Parameters	Symbols	Units	ρ_d 1.97 Mg/m^3	ρ_d 1.60 Mg/m^3	
Poisson's ratio	υ (-)	–	0.3	0.3	$d\varepsilon^e_{vM} = \frac{\kappa}{1+e_M}\frac{dp'}{p'} + \frac{k_M}{1+e_m}\frac{ds_M}{s_M+p_{at}}$ Where $d\varepsilon^e_{vM}$ is elastic volumetric strain for macrostructure p' is mean net stress e_M is macro void ratio s_M is macrostructural suction p_{at} is atmospheric pressure
Macrostructure parameter for elastic compressibility against mean stress change	κ	–	0.01	0.005	
Macrostructure parameter for elastic volumetric swelling against suction change	κ_M	–	0.047	0.001	
Microstructure parameter	κ_m	–	0.062	0.044	$d\varepsilon^e_{vm} = \frac{d\hat{p}}{K_m}; \quad \hat{p} = (p+s_m)$ $K_m = \frac{1+e_m}{\kappa_m}(p+s_m)$ Where $d\varepsilon^e_{vm}$ is elastic volumetric strain for microstructure $\hat{p}$ is mean effective stress for microstructure e_m is micro void ratio s_m is microstructural suction
Elasto-plastic volumetric compressibility	λ(0)	–	0.2	0.14	$p_0 = p^c\left(\frac{p_0^*}{p^c}\right)^{\frac{\lambda(0)-\kappa_{i0}}{\lambda(s)-\kappa_{i0}}}$ where $\lambda(s) = \lambda(0)[(1-r)\exp(-\beta s)+r]$
Parameters to define LC yield curve	r		0.7	0.82	
	β	MPa^{-1}	0.1	0.2	
Reference stress	p^c	MPa	0.005	0.005	
Slope of critical state	M	-	1.07	1.07	
Initial pre-consolidation stress for saturated conditions	$po*$	MPa	0.5	0.4	
Interaction functions (microstructural swelling)	f_{s0}		−2	−0.3	$f_s = f_{s0} + f_{si}\left(1-\frac{p}{p_0}\right)^{n_s}$
	f_{si}		1	2.19	
	n_s		2	2.4	

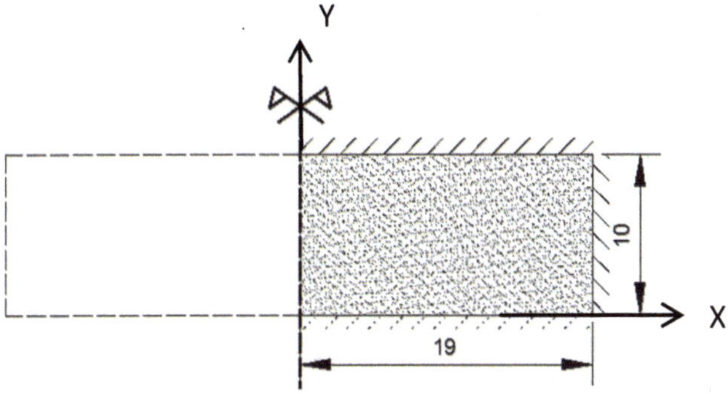

Fig. 3. Boundary conditions (dimensions in mm). Displacements are prescribed in both X and Y directions

Figure 4a shows the computed evolution of axial swelling pressures compared with the measured values for both specimens. Initially upon water injection, a fast increase of swelling pressure is observed until reaching a peak value of 1.3 MPa. Afterwards the swelling pressure decreased again until reaching a stationary swelling pressure of 1.1 MPa. The computed evolution lies within the measured values. It has to be noted that experimental results of both samples tested under similar initial and boundary conditions show some differences. The peak value reached 1.5 MPa for CS14C1PV while it reached a lower value for CS05C2PV (1.37 MPa). The stationary swelling pressure varied between 1 and 1.1 MPa for CS05C2PV and CS14C1PV respectively.

The first increase in selling pressure is related to the swelling of bentonite grains at microscopic scale. The peak occurrence followed by a decrease of swelling pressure corresponds to the reorganisation of the microstructure of the soil characterised by the collapse of macropores. As a result, the vertical stress drops to compensate the tendency of the macrostructure to collapse so the sample volume is kept constant.

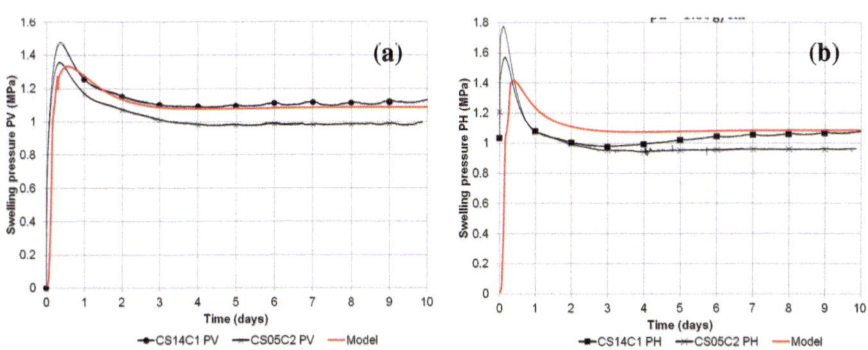

Fig. 4. Evolution of (a) axial and (b) radial swelling pressures. Experimental results and modelling results.

Figure 4b compares the computed and experimental evolutions of radial swelling pressures. The rate of the swelling pressure development at the start of the test, the magnitude of the peak are underestimated by the calculations but the final steady state value is well predicted. Comparison between the evolution of radial and axial swelling pressures shows that peaks are more marked for the horizontal pressures than for the vertical ones. This difference suggests the presence of more macropores along the radial direction than along the axial one as revealed by Microfocus X–ray computed tomography (μCT) images performed by Saba et al. (2014a) (Fig. 2). As a first approach, in order to model this heterogeneity, a zone with lager macropores is considered at the outer boundary of the sample (Fig. 5). Several cases were analysed by varying the thickness and the macroporosity of this outer zone.

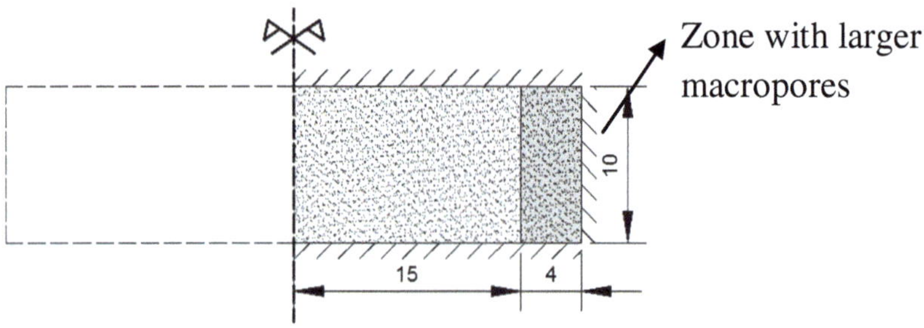

Fig. 5. Heterogeneous modelling case.

In the first case, the thickness of the outer zone was maintained constant while macroporosity were varied. Computed evolutions of vertical and horizontal swelling pressure are shown in Fig. 6. Interestingly, no significant effect of ϕ_M increase is observed for axial swelling pressure, while a large increase of the peak magnitude is observed for the radial swelling pressure. For the case of $\phi_M = 0.5$ the swelling pressure at both radial and axial directions are well predicted.

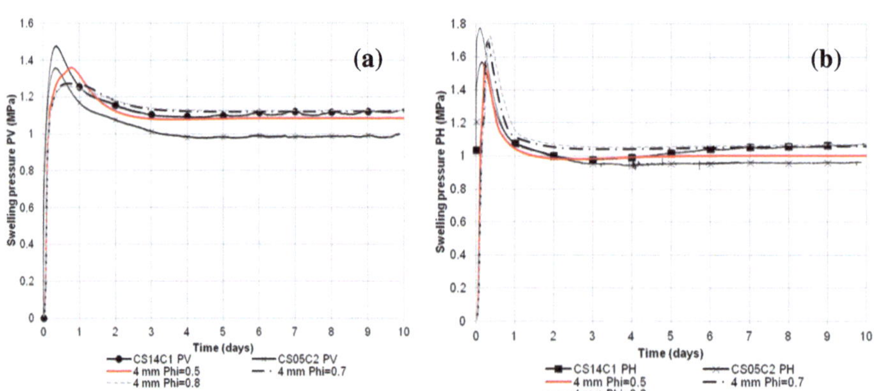

Fig. 6. Evolution of axial and radial swelling pressures considering an outer zone with larger macropores. Effect of the macroporosity of the outer zone on modelling results.

In the second case, the thickness of the outer zone was varied maintaining constant the macroporosity ($\phi_M = 0.5$) (Fig. 7). As the outer zone was larger the magnitudes of the peak for both horizontal and vertical swelling pressure increased, due to the presence of more macro-pores within the sample.

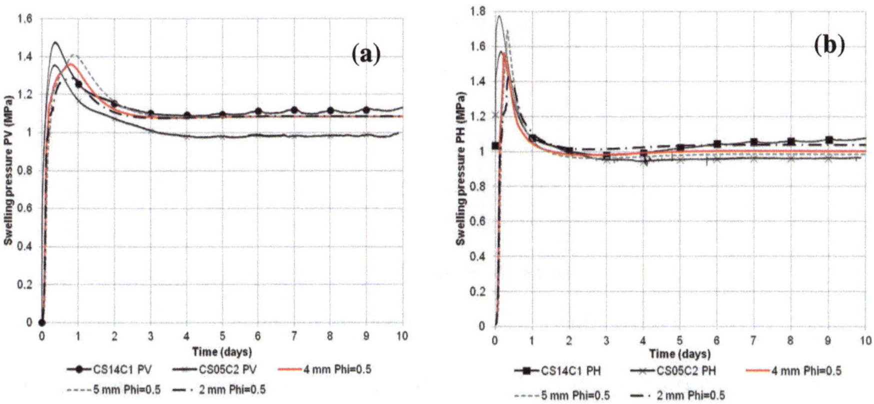

Fig. 7. Evolution of (a) axial and (b) radial swelling pressures considering an outer zone with larger macropores. Effect of the thickness of the outer zone on modelling results.

3.3 Impact of Technological Gaps

Modelling of 1/10 Mock-Up of SEALEX In-Situ Tests

In order to investigate the effect of technological gaps on the hydromechanical behaviour of bentonite sand/mixture Wang et al. (2013b) carried out 1/10th scale mock-up test to reproduce the behaviour of the bentonite-sand mixture of SEALEX experiment. A compacted sample of bentonite-sand mixture (dry density of 1.97 Mg/m^3) of 120 mm height and 55.5 mm diameter was placed in a rigid (stainless steel) hydration cell of 200 mm length and 60 mm inner diameter. The cell diameter was greater than the sample diameter to simulate the technological void which exists in the in situ SEALEX experiment, between the bentonite seal and the surrounding rock-mass. The specimen was hydrated from the bottom and the evolution of axial swelling pressure was recorded. More details of the experiment are reported in Wang et al. (2013b).

The test has been simulated using the theoretical formulation outlined above. The estimated initial microstructural and macrostructural porosities using MIP tests are indicated in Fig. 1. The finite element mesh used in the coupled HM analysis is depicted in Fig. 8 together with the different materials and boundary conditions. Advantage has been taken of the basically axisymmetric nature of the test. A fundamental issue in modeling the swelling pressure test lies in properly representing the initially air-filled gap. As a first approach, the 2.5 cm annular void is modelled as a highly porous, highly permeable material and highly deformable material. Materials properties are summarized in Table 1.

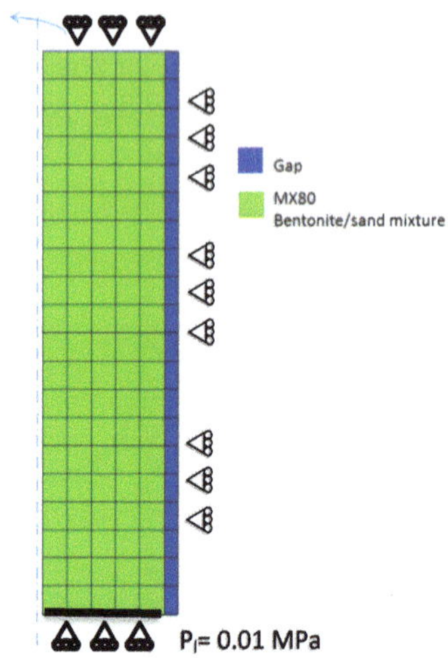

Fig. 8. Model geometry and boundary conditions.

Interesting information not readily available from direct laboratory measurements might be obtained by analyzing in more detail some of the results of the calculations. Figure 9 shows the variations in time of the macrostructural and microstructural suctions at two different positions: near the bottom hydration boundary (Y = 0), and the top of the sample (Y = 12 cm). At the bottom boundary, the model shows a relatively fast decrease of macrostructural suction (s_M) as water floods the macropores. As a result, suctions associated with the macrostructure and microstructure, are no longer in equilibrium. A reduction of microstructural suction is observed but with a delay associated to the local transfer of water from the macrostructure to the microstructure.

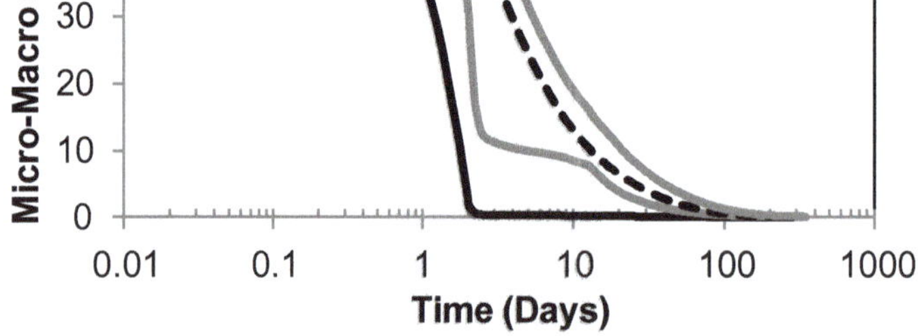

Fig. 9. Evolutions of micro and macro structural suctions near the bottom hydration boundary (Y = 0), and the top of the sample (Y = 12 cm).

At the top of the sample, both suctions remain in equilibrium until reached by the hydration front. The macrostructural suction then decreases rapidly down to a low value (s_M around 12 MPa). Afterwards, s_M decreases with a slower rate. Indeed, at this point, the higher value of microstructural suction needs water which is supplied by the macrostructure. When microstructural suction reaches lower values, both suctions decrease with similar rates.

Figure 10 shows the computed evolution of axial swelling pressure compared with the measured values. Transient drop of swelling pressure is observed at early hydration time suggesting the complex behavior of the mixture. Initially upon water injection, a fast increase of swelling pressure is observed until reaching a peak value. After about 12 days, a significant decrease occurred at a minimum value of 0.7 MPa. Afterwards the swelling pressure increased again until reaching a stationary swelling pressure. The rate of the swelling pressure development at the start of the test, the magnitude of the peak and the time required to reach it are adequately reproduced by the calculations. Subsequent behavior including the observed reduction of swelling pressure and the final rise are overestimated by the analyses but the final steady state value is well estimated by the modelling.

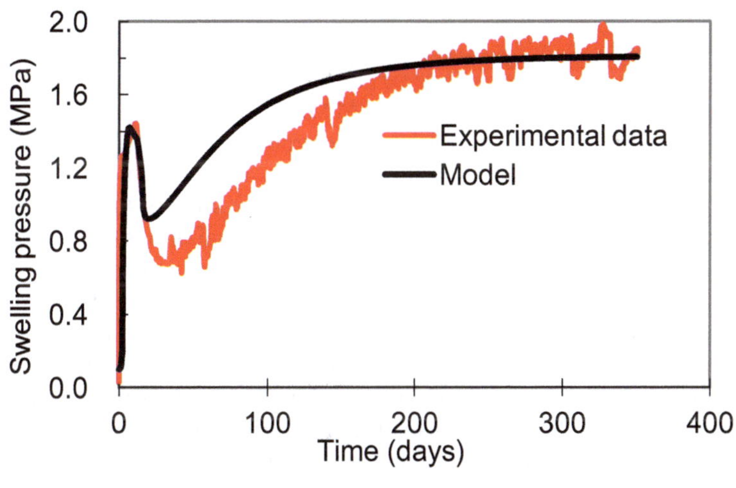

Fig. 10. Evolutions of axial swelling pressure. Experimental (Wang et al. 2013a) and modelling results.

The variations in time of the macrostructural and the microstructural porosities are shown in Fig. 11. To demonstrate the interdependency of the various aspects, suctions variations are displayed on the same Figure. When reached by the hydration front, macrostructural suction decrease induces an increase of macroporosity and swelling of the mixture to fill the technological gap. Further decrease of microstructural suction induces an increase of microporosity. At this stage the macropores are invaded by the micropores. Afterwards, there is a sharp reduction of the macrostructural porosity due to the structural collapse.

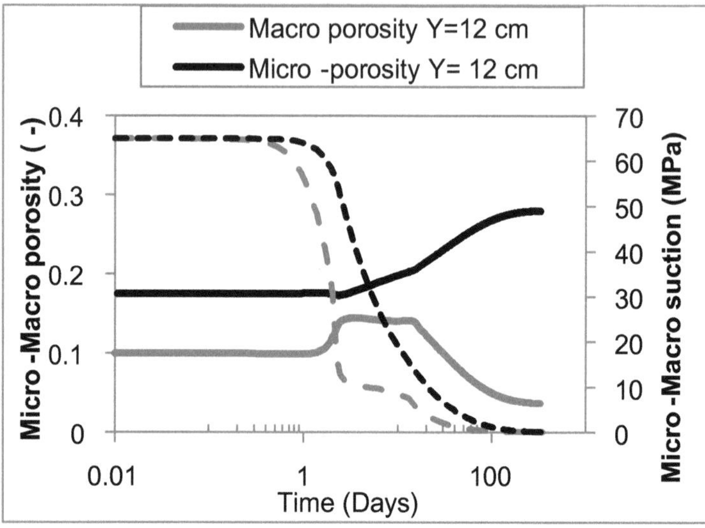

Fig. 11. Evolution of micro and macro porosities.

4 Conclusions

In this paper the HM behaviour of a compacted MX80 bentonite/sand mixture (70/30 in dry mass) intended as sealing material in underground repositories for nuclear waste was investigated using a double structure formulation. Two phenomena were analysed. The first one concerns the anisotropy of the swelling pressure of the compacted mixture. Although, the formulation considers two structural levels and their mutual interaction, the measured differences between the axial and radial swelling pressure were not captured by the model. To reproduce the difference in peaks magnitude between radial and axial swelling pressure kinetics a looser zone with larger macropores were considered. As a result of this simple approach a more pronounced peaks magnitude on the radial direction were predicted by the model. This demonstrates the necessity to consider a heterogeneous structural distribution of the tested samples to well describe the swelling behaviour of the mixture. The second phenomenon concerns the sharp early drop of swelling pressure on wetting considering a technological gap which might exist between the seal and the host rock under real repository conditions. A detailed examination of the computational results provided interesting information on the complex behaviour of the mixture on wetting under these particular conditions. It was observed, that the collapse of the macrostructure induced by the swelling of the microstructure occurs when the technological gap is filled up by the swollen bentonite and constant volume conditions are recovered.

References

Alonso, E.E., Gens, A., Josa, A.: A constitutive model for partially saturated soils. Géotechnique **40**(3), 405–430 (1990)

Alonso, E.E., Romero, E., Hoffmann, C.: Hydromechanical behavior of compacted granular expansive mixtures: experimental and constitutive study. Géotechnique **61**(4), 329–377 (2011)

Barnichon, J.D., Deleruyelle, F.: Sealing experiments at the Tournemire URL. EUROSAFE (2009)

Chen, Y.G., Cui, Y.J., Tang, A.M., Wang, Q., Ye, W.M.: A preliminary study on hydraulic resistance of bentonite/host-rock seal interface. Geotechnique **64**(12), 997–1002 (2014)

Gens, A., Vallejan, B., Sánchez, M., Imbert, C., Villar, M.V., Van Geet, M.: Hydromechanical behaviour of a heterogeneous compacted soil: experimental observations and modelling. Géotechnique **61**(5), 367–386 (2011)

Gens, A., Alonso, E.E.: A framework for the behaviour of unsaturated expansive clays. Can. Geotech. J. **29**, 1013–1032 (1992)

Imbert, C., Villar, M.V.: Hydro-mechanical response of a bentonite pellets-powder mixture upon infiltration. Appl. Clay Sci. **32**(3–4), 197–209 (2006)

Lloret, A., Villar, M.V., Sanchez, M., Gens, A.: Advances on the knowledge of the thermo-hydro-mechanical behaviour of heavily compacted "FEBEX" bentonite. Phys. Chem. Earth **32**, 701–715 (2007)

Mokni, N., Barnichon, J.D.: Hydro-mechanical analysis of SEALEX in-situ tests- impact of technological gaps on long term performance of repository seals. Eng. Geol. **205**, 81–92 (2016)

Mokni, N., Barnichon, J.D., Dick, P., Nguyen, S.: Effect of technological macro voids on he performance of compacted bentonite/sand seals for deep geological repositories. Int J. Rock Mech. Min. Sci. **88**, 87–97 (2016)

Mokni, N.: Analysis if hydro-mechanical behavior of compacted bentonite/sand mixture using double strcuture formulations. Environ. Earth Sci. **75**, 1087 (2016). doi:10.1007/s12665-016-5872-2

Olivella, S., Carrera, J., Gens, A., Alonso, E.E.: Non isothermal multiphase flow of brine and gas through saline media. Transp. Porous Media **15**, 271–293 (1994)

Pusch, R.: Highly compacted sodium bentonite for isolating rock-deposited radio-active waste products. Nucl. Technol. **45**(2), 153–157 (1979). United States

Saba, S.: Hydro-mechanical behaviour of bentonite-sand mixture used as sealing material in radioactive waste disposal galleries. Ph.D. thesis, Paris-Est University, France (2013)

Saba, S., Delage, P., Lenoir, N., Cui, Y.J., Tang, A.M., Barnichon, J.D.: Further insight into the microstructure of compacted bentonite/sand mixture. Eng. Geol. **168**, 141–148 (2014a)

Saba, S., Barnichon, J.D., Cui, Y.J., Tang, A.M., Delage, P.: Microstructure and anisotropic swelling behaviour of compacted bentonite/sand mixture. J. Rock Mech. Geotech. Eng. **6**, 126–132 (2014b)

Saba, S., Cui, Y.J., Tang, A.M., Barnichon, J.D.: Investigation of the swelling behaviour of compacted bentonite-sand mixture by mock-up tests. Can. Geotech. J. **51**(12), 1399–1412 (2014c)

Sanchez, M., Gens, A., Guimaraes, L., Olivella, S.: A double structure generalized plasticity model for expansive materials. Int. J. Numer. Analyt. Methods Geomech. **29**(8), 751–787 (2005)

Villar, M.V., Gómez-Espina, R., Campos, R., Barrios, I., Gutiérrez-Nebot, L.: Porosity changes due to hydration of compacted bentonite. In: Manusco, C., Jommi, C., Donza, F. (eds.) Unsaturated Soils: Research and Applications, vol. 1, pp. 137–144. Springer, Berlin (2012)
Wang, Q.: Hydro-mechanical behaviour of bentonite-based materials used for high-level radioactive waste disposal. Ph.D. thesis, Paris-Est University, France (2012)
Wang, Q., Tang, A.M., Cui, Y.J., Delage, P., Gatmiri, B.: Experimental study on the swelling behaviour of bentonite/claystone mixture. Eng. Geol. **124**, 59–66 (2012). doi:10.1016/j.engeo.2011.10.003
Wang, Q., Tang, A.M., Cui, Y.J., Barnichon, J.D., Ye, W.M.: A comparative study on the hydro-mechanical behaviour of compacted bentonite/sand plug based on laboratory and field infiltration tests. Eng. Geol. **162**, 79–87 (2013a)
Wang, Q., Cui, Y.J., Tang, A.M., Barnichon, J.D., Saba, S., Ye, W.M.: Hydraulic conductivity and microstructure changes of compacted bentonite/sand mixture during hydration. Eng. Geol. (2013b). doi:10.1016/j.enggeo.2013.06.013
Wang, Q., Tang, A.M., Cui, Y.J., Delage, P., Barnichon, J.D., Ye, W.M.: The effects of technological voids on the hydro-mechanical behaviour of compacted bentonite-sand mixture. Soils Found. **53**(2), 232–245 (2013c)
Wang, Q., Tang, A.M., Cui, Y.J., Barnichon, J.D., Ye, W.M.: Investigation of the hydro-mechanical behaviour of compacted bentonite/sand mixture based on the BExM model. Comput. Geotech. **5**, 46–52 (2013d)
Yong, R.N., Boonsinsuk, P., Wong, G.: Formulation of backfill material for a nuclear fuel waste disposal vault. Can. Geotech. J. **23**(2), 216–228 (1986)

Aquifer Distribution and Flow Patterns of Land Jatinangor Educational Area, Sumedang Regency, West Java Province

Andreas Franzona Pangaribuan(✉), Febriwan Mohammad, and Deden Zaenuddin Muttaqin

Department of Geology Engineering, Faculty of Geology, Padjadjaran University, Bandung, Indonesia
andreaspangaribuanl@gmail.com

Abstract. Continuous development in the area of Jatinangor campus is becoming one of the problems threatening the groundwater supply. In order to support the availability of groundwater in the area of Jatinangor campus, geophysical investigation with geo-electric method is conducted to determine the condition of subsurface based on the value of resistivity of rock. This research is aimed to determine potential groundwater zone region to support the availability of groundwater for Jatinangor education region. Based on Bandung's regional geological map, rocks in the Jatinangor area consist of volcanic rock breccia, tuffs and lapilli that makes it possible to contain the groundwater. 32 stations of 1-Dimensional (DC sounding) geo-electric measurement using Schlumberger configuration are performed in Jatinangor area. The results of data processing geo-electric are integrated with geological and hydrogeological observation information with the aim of producing images of subsurface rocks and distribution models. Based on the model, the type of aquifer contained in the study area as well as its potential reserve can be determined. Furthermore, the results of this research are expected to provide insights in implementing conservation strategies for Jatinangor educational area, Sumedang regency.

1 Introduction

Jatinangor Education Area is located in the 6°52′30″–7°00′00″ of south latitude and 107°45′00″–107°52′30″ of east latitude. Administratively, This area lies on the border between The Sumedang Regency and Bandung City, West Java Province, Indonesia. High population growth in this region resulted in a high population density which is around 4,270 people/km^2. We tried determine the potential groundwater zone region in this area to support the availability of groundwater for Jatinangor education region. Furthermore, the results of this research are expected to provide insights in implementing conservation strategies for Jatinangor educational area, Sumedang regency.

The resistivity method is a geophysical method that is very popular and frequently used both in geological survey and exploration. Because, the method is better to be used for knowing the geological condition or structure of the subsurface based on the resistivity variety of the rocks. The value depends on several factors, including the age of rocks, electrolyte contents, rock density, the amount of minerals it contains, porosity,

J. Wasowski et al. (eds.), *Engineering Geology and Geological Engineering for Sustainable Use of the Earth's Resources, Urbanization and Infrastructure Protection from Geohazards*, Sustainable Civil Infrastructures, DOI 10.1007/978-3-319-61648-3_17

permeability and so forth. One of the areas that has a clear contrast resistivity compare to the surrounding area is groundwater exploration. When the direct current flows into the earth through two current electrodes A and B, and the potential difference caused by the flow of the stream in two potential electrode M and N is measured, we can obtain the apparent resistivity.

1.1 Geological Condition

The exposed rocks in study area consists of five main units which all of them are young volcanic products, as can be seen in the Fig. 1. The unit consists of irrecudibly young volcanic deposits which are tuffaceous sand, lapilli, breccia, lava, and agglomerate. These rocks came from The Tangkubanparahu Mountains and also from some parts of The Tampomas Mount, which spread to many places within Sumedang-Bandung Border. These Quarter-age rocks formed a small and flat plains. The rocks also formed low hills covered by yellowish-grey and reddish-grey soils.

Fig. 1. Geological map of study area, modification of Silitonga (Hendarmawan 2014)

1.2 Hydrogeological Setting

The groundwater basin of the study area can be divided into two regions,. The first region occupies the southern part of the study area which is about 60% of the study area. This area has a discharge rate of less than 5l/s and it has a deep widespread moderate productivity aquifer, as can be seen in the Fig. 2. The second region occupies the northern part of the study area with an area of ±40% of the study area. This region has the shallow productive aquifer with a deep water table. There are several springs which was encountered in the area.

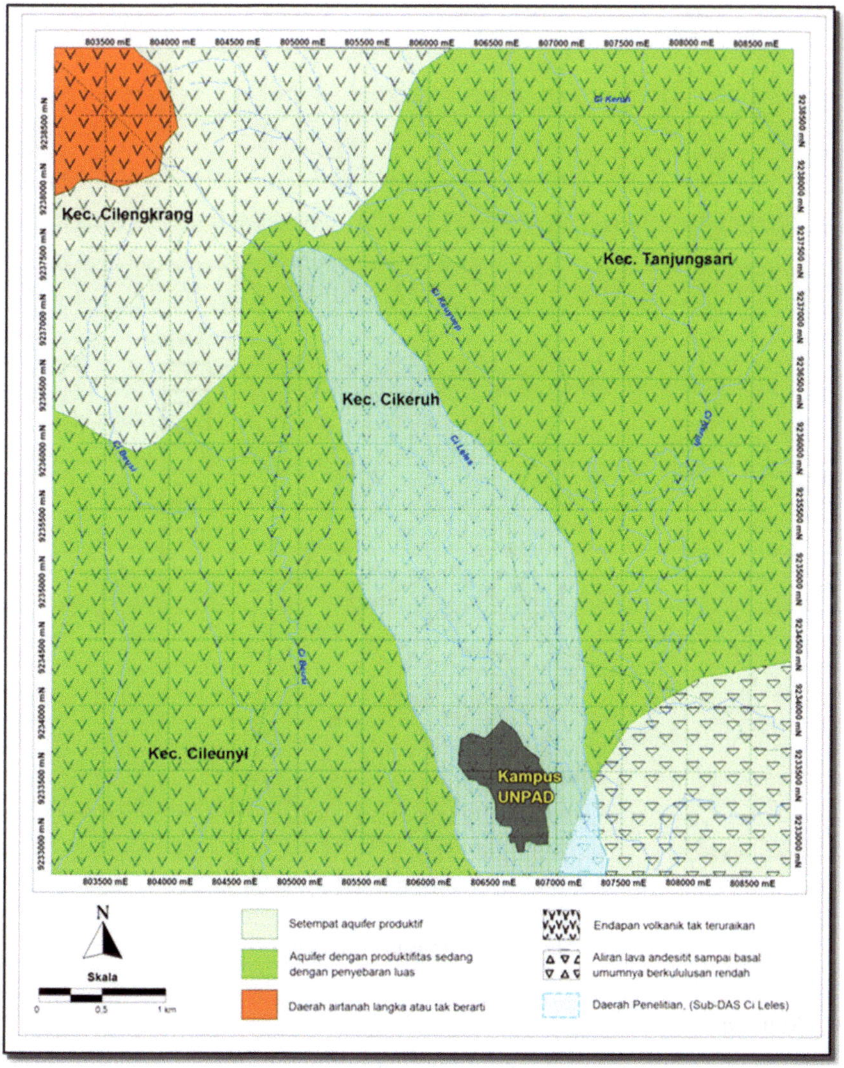

Fig. 2. Hydrogeology regional map, modification of Bandung Quadrangle Hydrogeological map (Soetrisno 1983).

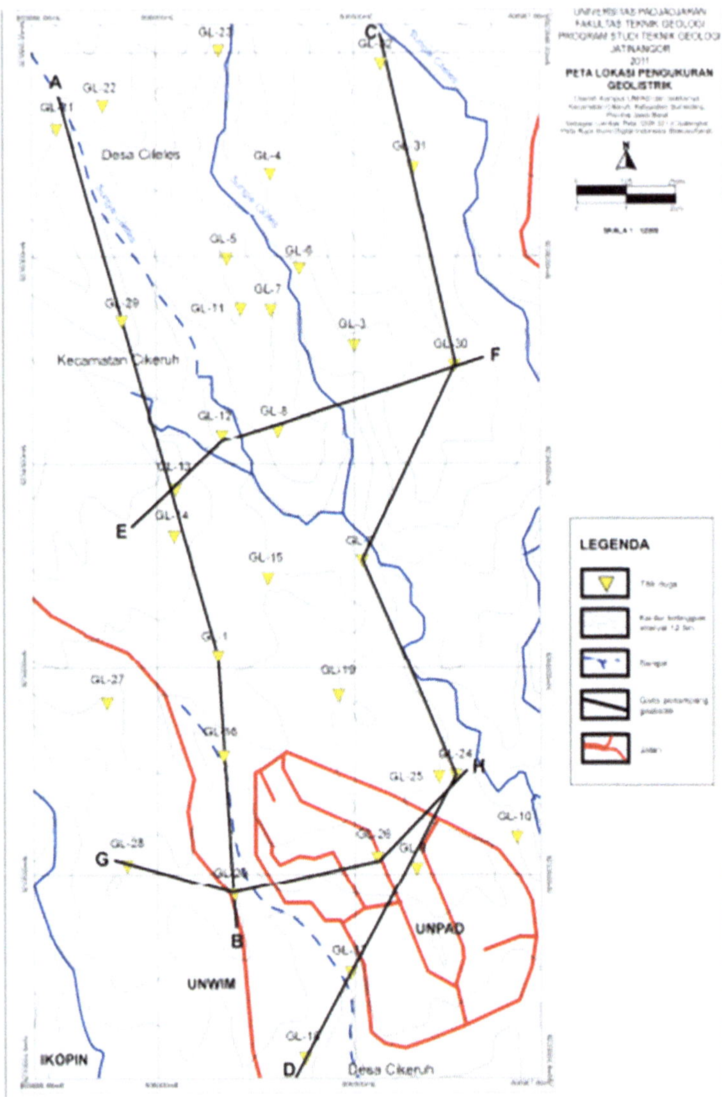

Fig. 3. The measurement points

1.3 Method

The research was conducted on Universitas Padjadjaran Campus by using a single Geoelectric Tool Unit ABEM DC Terrameter - SAZ 2000. The research used The Schlumberger configuration of Resistivity method, where both of potential electrodes M and N are placed between two current electrodes A and B. The Measurements have been carried out at 32 points with a range between electrodes (AB/2) as long as 300 m, thereby the estimation of the depth of each point were expected to reach 175 m.

Every point of measurement provides an estimation model of rock layers in an upright or vertical direction (stratigraphy) based on the resistance value (ρ). Each point has a 12 from each measurement tools were processed and correlated with geological and hydrogeological data from the field observations of previous studies that have been done by Geophysics Laboratory, in order to identify the composition of rock layers in the subsurface.

1.4 Result and Discussion

After the measurement on the field has been done, we obtained the variation of the apparent resistivity values and the current electrode spacing stretch (AB/2). Then, the data were processed by the software Progress ver.3.0 to obtain the true resistivity value and the actual thickness of the layer (Fig. 4). Each resistivity value represent a lithology of rocks at a certain depth because every rock has a certain resistivity value of which varies depending on rock hardness, porosity, permeability, age of rocks, the content of electrolytes, minerals, rock structures and geological phenomena.

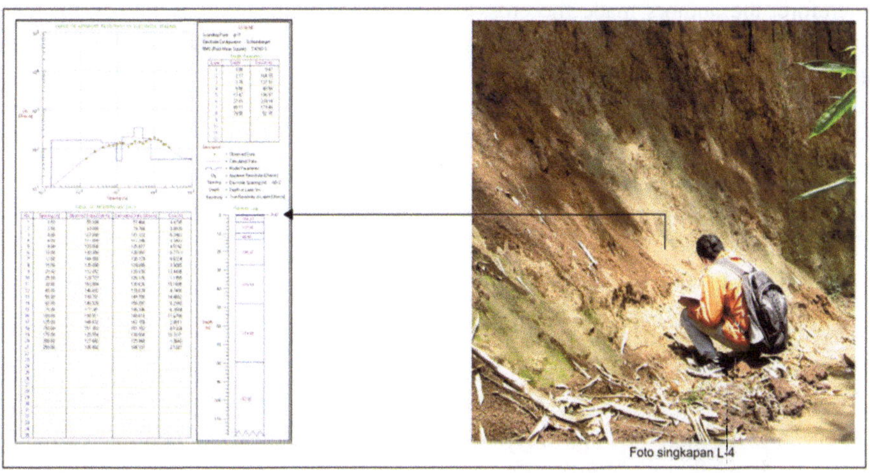

Fig. 4. Resistivity cross section GL-31 and Outcrop Picture L-4

Based on the results of the geoelectrical data processing correlated with the geological conditions around the study site, the study area is mostly composed of volcanic activity deposits. Several variations of the resistivity values can be divided into several groups to indicate the type of rock below the surface. The range of resistivity values for each rock in Table 1.

The Isoresistivity maps were created to determine the distribution of resistivity value at each depth. The resistivity map can be used to delineate the distribution of 1D resistivity laterally (Fig. 5). The resistivity value distributions on the isoresistivity map can provide a picture of the rock layers below the surface by correlating the existing

Table 1. Resistivity variations with their estimated lithology and hydrogeological properties

ρ(Ω.m)	Estimated lithology	Estimated hydrogeology
<30	Fine tuff layer	Impermeable layer
31–60	Sandy tuff	Aquifer
61–100	Tuffaceous sandstones	Aquifer
101–200	Matrix supported breccia (laharic breccia)	Aquifer
201–350	Grain supported breccia (volcanic breccia)	Impermeable layer
>350	Andesitic lava	Impermeable layer

geological data into it. The sameness resistivity values can be expected as the same layer of a rock while the contrast value is expected to be a striking contact between the different rock layers or it may be the emergence of geological structure. In this study, the authors created a map isoresistivity at a depth of 0 m, 1 m, 5 m, 10 m, 15 m, 20 m, 30 m, 40 m, 50 m, 70 m, 100 m, to 120 m.

The Geoelectric cross-section can give us a presence, distribution and thickness of the rock layers below the surface with reference to the geological data around the study site. It can also provide an overview of the aquifer characteristics beneath the surface. Therefore, the authors made the cross-section line passing through the several measurement points (Fig. 6). The authors made at least four cross section lines trough the study area (Fig. 3).

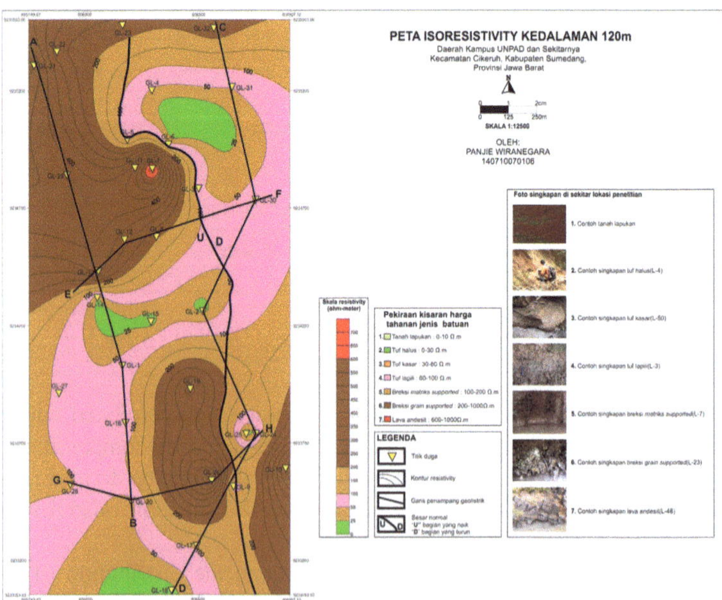

Fig. 5. The Isoresistivity Map at the depth of 120 m

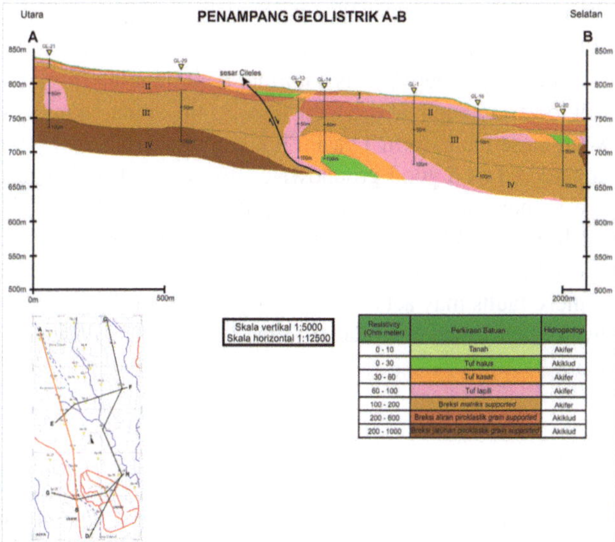

Fig. 6. The Cross section A-B and its lithology and Hydrogeological properties

The Geoelectric block diagram (Fig. 7) also was made to obtain the distribution picture of the rock layers in 3D. The block was created by linking the spread of the rock layers based on the cross-section line reconstruction and, using Rock Work 14 application, producing the distribution of the rock layers in 3D. It is easier to interpret the rock layer distribution in 3D.

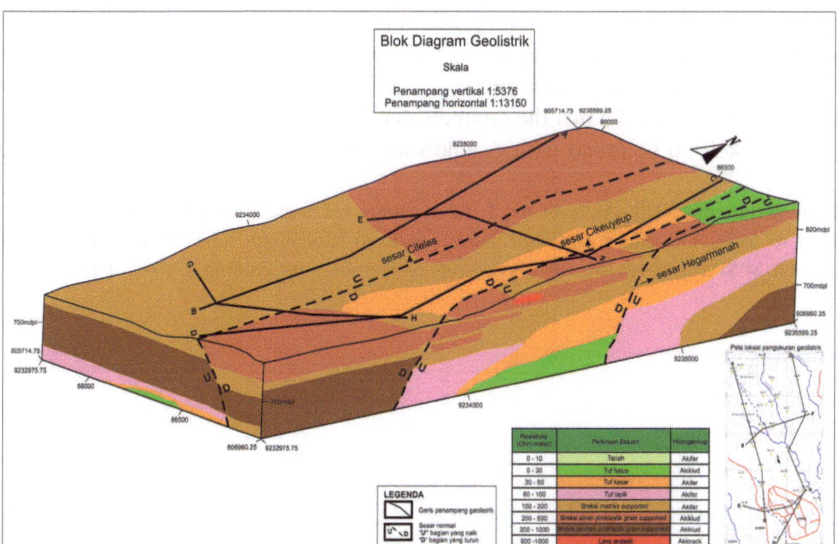

Fig. 7. The geoelectric block diagram

There are Cileles Fault and Cikeuyeup Fault found in the study area. The Cileles fault is a normal fault trending from the northwest to the southeast with lowering depth at the west side. While Cikeuyeup fault is a normal fault trending from the northwest to the southeast with lowering depth at the west side. So the situation under the surface in the middle between these faults form a basin. It is estimated that the groundwater basin in this area is considered as a good groundwater resources. This is because the basin below the surface is more dominated by the aquifer layer (coarse tuff, lapilli tuff and matrix supported breccia). While aquiclude layers (grain supported breccia and fine tuff) are only exposed in some places and their thickness does not dominate those areas. Besides, both of these faults may act as a exit-entry barrier of the groundwater flow. So that the groundwater that is accommodated in the basin can be collected properly and difficult to flow out.

2 Conclusions

There are several locations that are expected to be aquifers. The southwestern and northeastern parts of the study area are expected to have the considerable groundwater resources. This is because the areas are dominated by a rock layer that serves as an aquifer including coarse tuff, lapilli tuff and matrix supported breccia. While the layer of fine tuffs, grain supported pyroclastic flow breccias and grain supported pyroclastic fall breccias that serves as aquiclude only exist in a small part of the entire layers that make up this region.

On the contrary, the northwestern and southeastern parts of the study area are expected to have fewer groundwater resources. This is because the areas are dominated by grain supported pyroclastic flow breccias and grain supported pyroclastic breccias which serve as aquiclude. The Distribution of these rock layers is very dominating and quite thick. So that the space for groundwater flow in these areas are very small. Besides, the distribution of the aquifer layers such as coarse tuff, lapilli tuff and matrix supported breccias and also andesitic lava that serves as bad aquifer spread only in a few places and depths. So that the groundwater flow in the area is very narrow and are much less likely to be found with groundwater resources that have large discharge.

Acknowledgments. The Authors thank to previous researchers and families who gave us support to finish this paper. Also we thank Febriwan Mohamad, S.Si., M.Si and Ir. Undang Mardiana, M.Si as our advisor for suggestions concerning and guiding to finish this paper.

References

Hendarmawan et al.: Pola Aliran Air Tanah Berdasarkan Kandungan Isotop Stabil Deuterium-Oksigen Di Tiga Cekungan Pada Lereng Gunungapi Manglayang, Kawasan Kampus Unpad, Jatinangor. Universitas Padjadjaran, Bandung (2014)

Soetrisno: Hydrogeological Map of The Bandung Quadrangle, Java. Scale 1:250,000 Pusat Penelitian dan Pengembangan Geologi (1983)

Integrating Remote Sensing and GIS for Mapping Garbage Dump Areas in the Kingdom of Bahrain

Khalil I. Al-Joburi(✉)

Department of Civil Engineering, University of Bahrain, Isa Town, Bahrain
kaljuboori@uob.edu.bh

Abstract. Due to land availability and shallow coast line, random garbage dump sites have been used in the past and present in the Kingdom of Bahrain. This has caused economic stress, environmental hazards, and contributed to the failure of many family dwellings throughout the Kingdom. This paper aims to identify, locate, and map possible garbage dump sites. Input data used in this paper includes 300 Borehole data from the Ministry of Work (MoW) LandSat imagery from 1972 to present, agricultural, drainage and soil maps that were produced in 1975, Digital Elevation Models (DEM) from the Surveying and Land Registration Bureau (SLRB) and site investigations. ArcGIS 10.2 and extensions are utilized toward mapping possible sites. The IR band is used to delineate the land borders as the sea depth is very shallow. 1969 Topographic map and aerial photos of 1950s are used to identify agricultural and residential areas and as a benchmark. 3D digital surface is created for the Kingdom based on contour map of 1969 and drainage maps of 1975. Supervised classifications of the blue, green, red, and IR bands is preformed every 5 years to determine the growth of residential areas and the shrinking of agricultural areas. Borehole data provides detailed soil information that contributed significantly in identifying garbage dump sites. Site investigation and citizen interviews indicate that almost all garbage dumps areas are located in lowland areas or in the shallow sea. An algorithm based on least square principles is developed to determine possible sites. Results show that dump sites are spread all over the Kingdom. Empty sites that are located within 100 m from residential areas and are at least 2 m below the average elevation of surrounding areas are highly likely to be used as dump sites. 65 dump sites and 47 possible sites are identified. Dump site areas range between 4,000 to almost more than 10 km^2.

Keywords: Bahrain · Recycling · Municipal solid waste · Landfill · GIS

1 Introduction

The Kingdom of Bahrain is an archipelago of 33 islands; the Northern group of islands consists of five islands, which are the Kingdom's major and inhabited islands: Bahrain, Muharraq, Sitrah, Jiddah, and Umm Nassan (Al-Jeneid et al. 2008). The Southern group consists of a smaller archipelago and is about 20 km off the south east of Bahrain. It consists of six main islands and more than 30 smaller ones, all uninhabited. The Kingdom is located in the middle of the southern coastline of the Arabian Gulf,

J. Wasowski et al. (eds.), *Engineering Geology and Geological Engineering for Sustainable Use of the Earth's Resources, Urbanization and Infrastructure Protection from Geohazards*, Sustainable Civil Infrastructures, DOI 10.1007/978-3-319-61648-3_18

between 25° 22′ and 27° 10′ North and longitude of 51° 07′ and 50° 16′ East (Fig. 1). The Kingdom's land area is continuously changing, due to the land reclamation activities. It was 741 km^2 in 2006 and 774 km^2 in 2014, (CIO, 2016). Territorial waters were 7,428 km^2 in 2006 and presently 7,495 km^2. The Kingdom's population has also increased. It is presently at 1,314,562, (630,744 Bahraini nationals and 683,839 expatriates). Population density 1,700 per km^2 makes the Kingdom the highest in the Middle East and 7th worldwide. The five major islands represent 93% of the total land area. Two-thirds of the population lives within 2.0 km of the coastline (CIO 1970–2001). Bahrain is split into four governorates: Capital, Muharraq, Southern and Northern governorates. The Capital is the most populated and the Southern is the least. Each governorate is managerially and financially autonomous, with responsibility for the environment, public spaces, roads, and beaches. The summer is generally hot with high humidity and the winter is cool. Average temperatures and relative humidity are 35.3 °C and 41.4% in the summer season and 18.4 °C and 67.8% in the winter. Rainfall is rare, with an average between 2004 and 2014 of 74.65 mm/year [minimum of 1.6 mm in 1946 (Bridges and Burnham 1980) and maximum of 211.90 mm in 2006 (CIO 2014)].

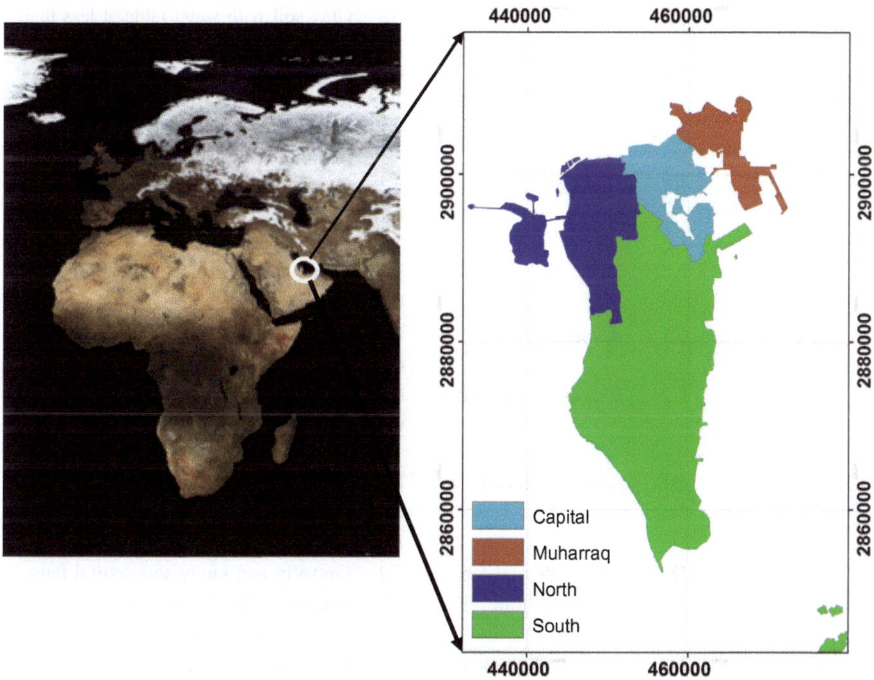

Fig. 1. Bahrain's location

2 Bahrain's Soil

The surface soils of Bahrain were thoroughly investigated by the Bahrain Surface Material Resources Survey (BSMRS) between December 1974 and April 1975. The team was made up of more than 20 scientists, geologists, surveyors, and engineers. Their work was presented in a book titled "GEOLOGY, GEOMORPHOLOGY AND PEDOLOGY OF BAHRAIN" published by Geo Abstracts Ltd. in 1980 (Doornkamp et al. 1980). Later it was published by E.M. Bridges and C.P. Burnham in 1980 in the Journal of Soil Science in an article titled "SOILS OF THE STATE OF BAHRAIN."

Table 1. Bahrain's soil group

Soil Group		Sub-group		Soil description	
A	Cultivated solon chalk	A1	Loamy	A1a	Loamy soil at least 1 m deep
				A1b	Loamy soil with lime stone
				A1c	Loamy soil with water table at less than 1 m
				A1d	Loamy soil with saline
		A2	Sandy	A2	Sandy soil with water table at less than 1 m
		A3	Clay	A3	Clay soil with water table at less than 1 m
B	Natural solon chalk	B1	Gypsiferous solon chalk	B1a	Mostly Loamy soil with water table at less than 1 m
				B1b	Mostly sandy soil with water table at less than 1 m
		B2	Sabkhas	B2a	Quartz- gypsiferous sands
				B2b	Carbonate- rich silt with gypsiferous sand
				B2c	Stable sand
				B2d	Embanked marine mudflats
C	Regosols	C1	Aeolian sands	C1a	Stable sands
				C1b	Dune sands
		C2	Recent beach deposits	C2a	Beach sands with shell
				C2b	Beach sands with shell cemented below 40
D	Raw mineral soils	D1	Soil of interior basin	D1a	Clay and silt clay loams
				D1b	Silts and fine sands
				D1c	Sands overlying playa deposits
		D2	Soil of detrital fans	D2	Gravelly and sandy soil detrital fans
		D3	Soil with stone pavement	D3a	Gravely soils of back slope
		D3		D3b	Gravely soils of interior basin
E	Rock dominated area	E1	Areas of little or no natural soil	E1a	On scarp slopes
				E1b	On other surfaces
		E2	Worked ground	E2	Most top soil removed
F	Miscellaneous	F1	Old Dolman mounds		
		F2	Urban and industrial area		
		F3	Undifferentiated		
		Other			

The team placed Bahrain's soils into 6 groups: Cultivated solon-chalk, natural Solon-chalks, Regosols, Raw mineral, Rock dominated, and unclassified. The soil Sub-groups of these are made-up according to the dominant texture: loamy, sandy or clayey, and the mapping units distinguish soils in which there is a deep profile, or in which solid rock or the water table occurs within 1 m of the surface (Table 1 and Fig. 2)

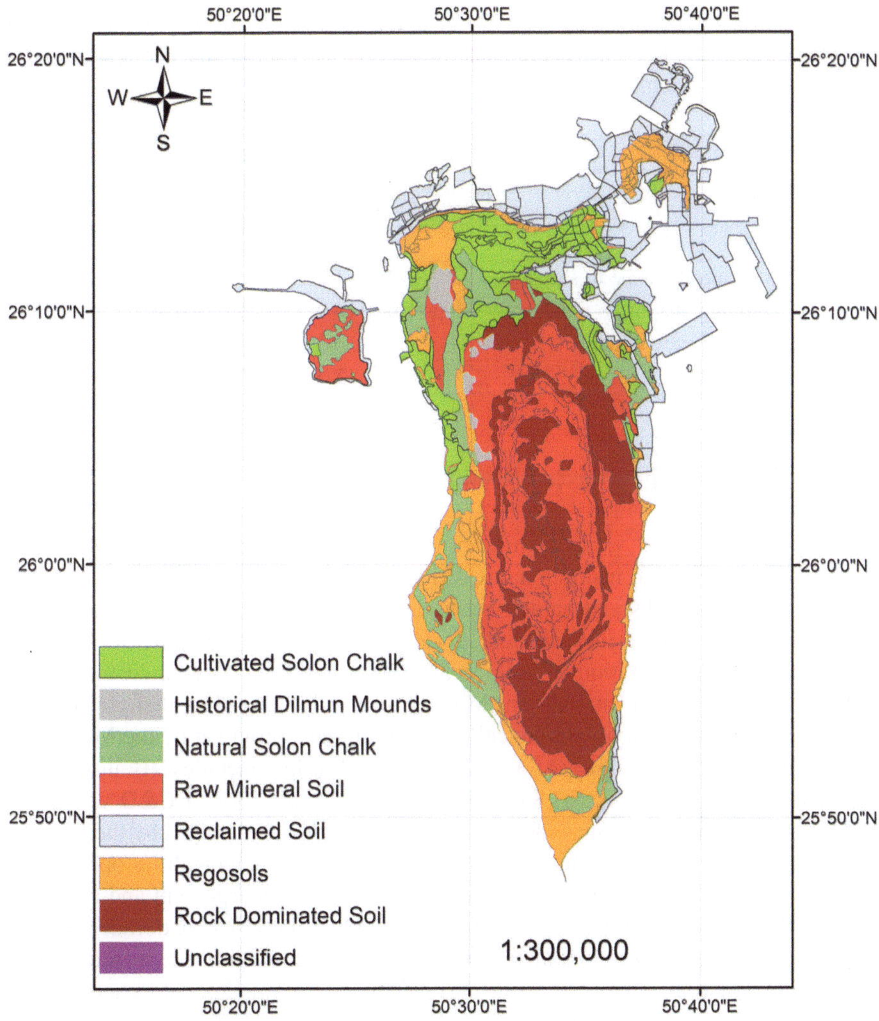

Fig. 2. Bahrain's soil groups

3 Topography

Analysis the 2011 digital elevation data obtained from the SLRB spots elevation shows that islands have a flat sandy coastline More than 13% of the total area of the islands' surface are estimated at 0 to 1.0 m above mean sea level (MSL) (101 km^2) and 391 km^2 (50.5%) are more than 5.0 m above MSL. The United Nations Food and Agricultural Organization (FAO) estimated that 89% is between 0–1% slope and 0.07% above 30% steep to very steep as shown in Table 2 (FAO 1990). The water around Bahrain is shallow ranging between <0.5 and 30 m (Zainal et al. 2012). In some areas, the sea depth is less than 1.0 m stretching up to 2 km offshore. Because of the low depth of the sea water, it is used intensively for fishing and garbage dumping by the local population.

Table 2. Bahrain's terrain slope distribution (FAO 1990)

Type	Slope	Total area km^2	Percent
Flat to nearly level	0–1	559.5	88
Very gently slope to sloping	1–10	150	21
Strongly sloping to moderately steep	10–30	6.2	0.86
Steep to very steep	$\gg$30	0.5	0.7
Total area		716	

4 Municipal Solid Waste (Msw)

The definition of MSW varies significantly. MSW often includes: household waste; waste of commercial and institutional (C&I) and industrial premises; street sweepings; green wastes from parks and garden waste; and dead animals (Sabbagh et al. 2012). Other solid wastes may be included to a greater or lesser extent, including, for example, construction and demolition (C&D) waste and non-hazardous industrial waste (Scheinberg et al. 2010). In Bahrain, MSW is generally defined as that collected by municipalities rather than by private collectors. From data provided by the Ministry of Municipalities and Agriculture Affairs (2009), MSW in fact includes: (i) all domestic and household bulky waste; (ii) 83% of garden waste; (iii) 23% of C&D waste; and (iv) 16% of commercial waste (Table 3). MSW in Bahrain is managed by the Ministry of Municipalities and Agriculture Affairs, under which four municipalities operate, one in every governorate, each of which is responsible for the management and collection of the MSW within the governorate's boundaries. Over the last four decades Bahrain has experienced accelerated globalization, in the form of financial modernization due to crude oil exploitation, which has resulted in increased consumption habits and changes in morphology (Hamouche 2004). A strategic location and government policies to diversify income sources away from oil have led to rapid industrialization and urbanization, making Bahrain one of the main industrial and financial centers in the Middle East, with a per capita gross domestic product (GDP) of US$ 19,736 in 2008

(CIO 2009). Population and economic growth has also contributed to the significant increase in waste generation rate (WGR) per capita^{-1} day^{-1}. As a result, the WGR has increased annually. For example, in 2011, the domestic waste reached 1.2 million tons or a 300% increase compared to 2009. WGR was 2.7 in 2011 and 3.62 kg/capita^{-1} day^{-1} (Municipality report 2011). This put Bahrain's WGR higher than the international estimated WGR of 1.7 kg/capita/day (Municipality report 2011), and higher than USA and Japan (Sabbagh et al. 2012).

Table 3. Municipality waste generated in 2015, source (Ministry of Municipalities and Agricultural 2015)

Description	Weight in kg
Domestic	442,135,130
Building	408,160,522
Commercial	263,141,151
Garden	165,570,580
Sand	164,418,630
Building-Buhair	145,240,130
Industrial	49,735,363
Liquid	34,069,439
Water	26,639,650
Animal	12,751,172
Household Bulky	9,780,915
Tires	5,682,871
Desert Fill	5,398,990
Blood	3,527,900
Expired Food Stuff	1,531,936
Insulation	168,115
Electronic & commercial destruction	125,945
Sponge	5,150
Total	1,738,083,589
Waste generation rate kg/capita/day	3.62

4.1 Municipal Solid Waste Collection and Disposal

For hundreds of years, the people of the Kingdom of Bahrain disposed of their domestic, agricultural, and building debris in the sea. For those who are within walking distance of the shore line, dumping in the sea was relatively easy. People living away from the shore line dumped their wastes in low level uninhabited empty spaces near their dwelling. As things progress with Bahrain's shore changes, the FISHT area decreases and land increases. In fact, Bahrain's shore is dynamic, not due to natural causes but manmade activities. The newly created land is used for agricultural purposes, with the garbage waste changing to natural fertilizer, thereby making the newly created area fertile. This is evident in Central, Muharraq, Manama, and the Northern provinces. The new land is also used for building new houses as the population increases. As a result, this causes both environmental and hazard issues for the newly built houses. Due to the sparse population in previous years, these issues were not causing serious problems. However, with many of these houses will suffer minor to severe structural damages. It is estimated that over 9,000 houses are deteriorated, most of them built on, or next to, the newly created land (Al-Joburi 2015). As these areas were not mapped, people only have a vague idea on the general areas that previously were garbage dump sites.

Recently, the Kingdom started to collect and dispose of all forms of municipal soil waste in the only disposal place available, the Asker landfill located in the Southern governorate. The Ministry of Municipalities and Agriculture Affairs and SLR Consulting prepared an integrated waste management strategy for the Kingdom (Sabbagh et al. 2012). In June 2010, the five municipalities and the Ministry of Municipalities and Agriculture Affairs signed a contract with an international waste management company to construct a waste-to-energy (WTE) facility in the country. The plant, start operating in 2013, will process 390,000 tons of mixed MSW per year and produce 50 MW of electricity. This tonnage is equivalent to 54% of the sum of domestic, C&I and industrial waste disposed of in 2009 (Sabbagh et al. 2012)

5 Data Collection and Preparation

To map these areas the following data was collected:

1. Aerial photographs from the 1950 are in the areas of Muharraq and Manama. LandSat images for every 5 years starting from 1973. Both are Geo-referenced so that it can be used along with other data sources.
2. 350 borehole data from the Ministry of Work for different locations throughout the Kingdom. This data is in PDF or Excel format and their coordinates most of the time are not available. The coordinates are estimated from the location maps that are enclosed within the PDF file, which were mostly drawn to scale 1:100. Location accuracy is within ±5 m, which does not impact site identification. Borehole data was converted into GIS format. it provides the most accurate account on whether or not there was a garbage dump site in their location.
3. Citizen interviews. People of specific localities have a general idea on the area where they believe there was a garbage dump site at one point. This information is used as a guideline on identifying garbage dump sites.
4. A location of deteriorated houses. This data is also used as a guideline to find a possible correlation between house structural issues and garbage dump sites. Unfortunately, such data is unavailable and in order to obtain it, addresses of these house are collected from the local Municipality council and an algorithm is developed to extract (x, y, z) location from the CIO address shape file. The

algorithm is designed to compare deteriorated house addresses that are provided by the local municipality council, with the one listed in the CIO address file. When both addresses match, the coordinates of the CIO is extracted.

5. Digital Elevation Model (DEM). This is spot elevation for more than 400,000 points at 50 m spatial horizontal resolution, obtained from the SLRB for 2007 and 2011. The data was collected using digital photogrammetry. Additional 50,000 points were surveyed with GPS and total station to consolidate the data and to record important feature coordinates. DEM data prior to 2007 is unavailable and, therefore, DEM was created based on contour maps of 1969 that were obtained from the SLRB.

6 Data Analysis

ArcGIS 10.2, and extensions, are used to process and perform analysis. A geospatial database is created based on WGS 1984 UTM zone 39 north. Aerial photographs and satellite images are treated as raster data. IR band is used to identify the shore line and the difference between IR blue bands provides the FISHT areas. This process is repeated every 5 years. The difference between IR and the aerial photographs shows the newly reclaimed land, which may be part of the government's reclamation or land formed by domestic garbage dumping. However, the government's land reclamation started around 1960. As such, the obtained difference is most likely garbage waste dump sites. This is process took place in the Muharraq area, where aerial photos are available. Furthermore, supervised multispectral image classification is performed, using the known garbage sites as training sets.

Citizen testimonies are treated as a point feature, where attributed fields are Eastern and Northern coordinates, and block number. Location of deteriorated houses is treated similarly, since both provide a general idea on possible garbage sites. Borehole data is converted into access database and imported into Geodatabase.

Elevation data for 1969 and 2007 is converted into raster data using topo to raster tool. The objective is to determine empty and low open spaces close to community dwellings. These sites seem to be the most possible areas to be used as garbage dumping sites. To identify garbage dumping sites, an algorithm model is designed. The algorithm is based on least square principles and made-up of two phases. Phase 1 computed the relative weight of each input element. Phase 2 is designed to determine whether or not the site is a garbage dump site.

The analysis revealed the following:

- Dumping sites are spread all over the Kingdom.
- There are 65 confirmed sites that contain domestic or other types of solid waste. 23 sites in the Capital, 17 in Muharraq, 17 in the Northern governorates, and 7 in the Southern governorates (Fig. 3).
- 47 possible MWS dumping sites, 19 in the Central, 3 in Muhrraq, 13 in the

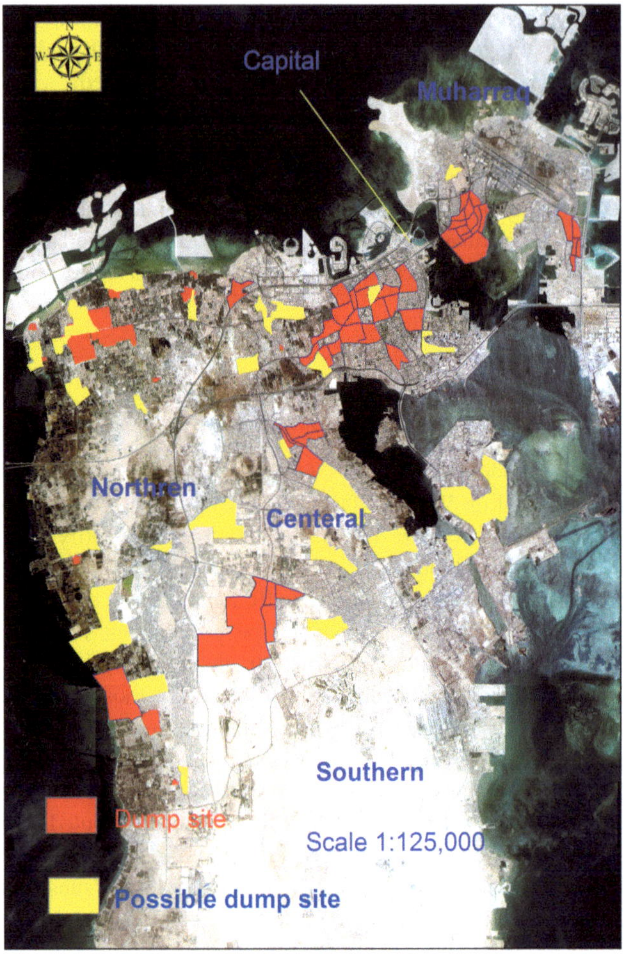

Fig. 3. Location of garbage dumps sites

Northern governorates, and 12 in the Southern governorates (Fig. 3).

- Site size varies significantly ranging from 120,000 m^2 to more than 10 km^2.
- Inland sites are empty spaces that are located within 100 m from residential areas and are at least 2 m below average elevation.
- Depth of dumping ranges from 50 cm to more than 5 m.
- Determining the actual infected size and the depth of dumping area is difficult due to limited the borehole data available.

7 Conclusion

Mapping the location of illegally buried solid waste is a challenge. People all over the globe, especially in poor and underdeveloped countries, are used to dumping domestic solid waste in low land areas. The inhabitants of Bahrain, 70% of whom live within 2 km from one of the shallowest coastlines in the world. Due to limited land, citizens dump domestic and agricultural solid waste in lowland open spaces and in the sea for convenience and as a means to reclaim new land. With time many of these dumps sites are covered and become unrecognizable. They present environmental hazards and have contributed to the failure of many houses. Integrating remote sensing and GIS approaches have been presented, and many sites were identified as dump sites or possible dump sites. Further research is needed to develop and test a more comprehensive approach that may be able to positively identify suspicious sites. Soil investigation is highly recommendable before any construction activities take place.

References

Al-Jeneid, S., et al.: Vulnerability assessment and adaptation to the impacts of sea level rise on the Kingdom of Bahrain. Mitig. Adapt. Strat. Glob. Change **2008**(13), 87–104 (2008)

Al-Joburi, K.: Geospatial approach for analyzing soil type impact on Bahrain deteriorating houses. Int. J. Geosci. Geomatics **3**(1) (2015). ISSN 2052-5591

AL Sabbagh, M., et al.: Resource management performance in Bahrain: asystematic analysis of municipal waste management, secondary material flows and organizational aspects. Waste Manage. Res. **30**(8), 813–824 (2012)

Bridges, E.M., Burnham, C.P.: Soils of the state of Bahrain. J. Soil Sci. **1980**(31), 689–707 (1980)

Central Information Organization-Bahrain (2014). http://CIO.gov.bh

Doornkamp, J.C., Brunsden, D., Jones, D.K.C., Cooke, R.U.: Geology, Geomorphology and Pedology of Bahrain. GeoBooks, Norwich (1980)

FAO: Guidelines for Soils Profile Description, 3rd edn. FAO, Rome (1990). http://www.aims.gov.au/pages/research/project-net/reefs-at-risk/apnet-rar00.html

Scheinberg, et al.: Solid Waste Management in the World's Cities. Water and Sanitation in the World Cities 2010. Earthscan published for UN Habitat, London (2010)

Zainal, K., et al.: The cumulative impacts of reclamation and dredging on the marine ecology and land-use in the Kingdom of Bahrain. Mar. Pollut. Bull. **64**(2012), 1452–1458 (2012)

Geology and Remote Sensing Investigations in Antarctic Environments

Amin Beiranvand Pour[1(✉)], Mazlan Hashim[1], and Yongcheol Park[2]

[1] Geoscience and Digital Earth Centre (Geo-DEC), Research Institute for Sustainability and Environment (RISE), Universiti Teknologi Malaysia (UTM), UTM Skudai, 81310 Johor Bahru, Malaysia
beiranvand.amin80@gmail.com

[2] Korea Polar Research Institute (KOPRI), Songdomirae-ro, Yeonsu-gu, Incheon 21990, Republic of Korea

Abstract. Antarctica remains a remote and logistically difficult region in which to conduct geological fieldwork, making the data collected there of significant value. Actually, many parts of the continent still remain poorly known in regional geological scale and structural architecture. Remote sensing imagery is capable to provide a solution to overcome the difficulties associated with field mapping in the Antarctic. Advanced optical and radar satellite imagery is the most applicable tool for mapping and identification of inaccessible and un-exposed regions in Antarctic. Consequently, an improved scientific research using remote sensing technology would be essential to provide new and more complete lithological and structural data to fill the numerous knowledge gaps on Antarctica's geology. In this study, the Antarctic Peninsula (AP) is selected to conduct a regional remote sensing investigations. The Antarctic Peninsula (AP) is the northernmost part of the continent of Antarctica. Despite more than 50 years of geological mapping of the Antarctic Peninsula there are still significant gaps in the coverage in difficult to access areas and many regions where mapping is based upon sparse or inferred field observations. Recent generation of multi-platform satellite sensors could be investigated to extract geological information for Antarctic environments. Landsat-7 Thematic Mapper (TM), Landsat-8 and the Advanced Spaceborne Thermal Emission and Reflection Radiometer (ASTER) data were used in this study for regional geological mapping. The improvised image processing algorithms and Systematic GIS techniques were implemented to detect structural elements and geological features for producing regional image maps of the Antarctic Peninsula especially for Oscar II coast area, north-eastern Graham Land. The outcomes of the investigation demonstrated the applicability of satellite remote sensing data to produce new revisions of geological maps with high accuracy for the regions with exposed rocks in the Antarctica.

J. Wasowski et al. (eds.), *Engineering Geology and Geological Engineering for Sustainable Use of the Earth's Resources, Urbanization and Infrastructure Protection from Geohazards*, Sustainable Civil Infrastructures, DOI 10.1007/978-3-319-61648-3_19

1 Introduction

Antarctica's geology is extremely various (Elliot et al. 2016). Antarctica remains a remote and logistically difficult region in which to conduct geological fieldwork, making the data collected there of significant value. This is predictable considering the size of the continent and the changing tectonic processes, environments and climates that it has experienced over geological era (Salvatore et al. 2013). Antarctica's geology is not identified in detail because of the extensive ice cover. Actually, many parts of the continent still remain poorly known in regional geological scale and structural architecture. Remote sensing imagery is capable to provide a solution to overcome the difficulties associated with field mapping in the Antarctic. Advanced optical and radar satellite imagery is the most applicable tool for mapping and identification of inaccessible and un-exposed regions in Antarctic.

Remote sensing imagery has been used extensively for lithological mapping, structural analysis and mineral exploration in arid and semi-arid and tropical regions around the world (Mars and Rowan 2006, Pour and Hashim, 2014a, b, c, 2015a, b, c). Consequently, an improved scientific research using remote sensing technology would be essential to provide new and more complete lithological and structural data to fill the numerous knowledge gaps on Antarctica's geology. The Antarctic Peninsula (AP) contains a variety of well-exposed lithologies and areas that have not been mapped directly providing an appropriate test of the use of optical and radar remote sensing data for lithological, structural and alteration mapping. There is good coverage of cloud and seasonal snow free over the Antarctic Peninsula (AP) providing a sound basis for remote sensing investigations.

In this scientific research, the Antarctic Peninsula (AP) is selected to conduct regional remote sensing investigations (Fig. 1). The AP is the northernmost part of the continent of Antarctica, which is located approximately 650 miles of South America. The AP is resulted from ocean-continent collision in subduction zone (Eagles 2004). Despite more than 50 years of geological mapping of the Antarctic Peninsula there are still significant gaps in the coverage in difficult to access areas and many regions where mapping is based upon sparse or inferred field observations. The objective of this study is to map the AP and Oscar coast area in north-eastern Graham Land at regional scale using Landsat-7 Thematic Mapper (TM), Landsat-8 and the Advanced Spaceborne Thermal Emission and Reflection Radiometer (ASTER) satellite remote sensing data.

2 Geology of the Antarctic

The Antarctic continent comprises three primary tectonic regions: (i) East Antarctica; (ii) West Antarctica; and (iii) the Transantarctic Mountains. East Antarctica is thought

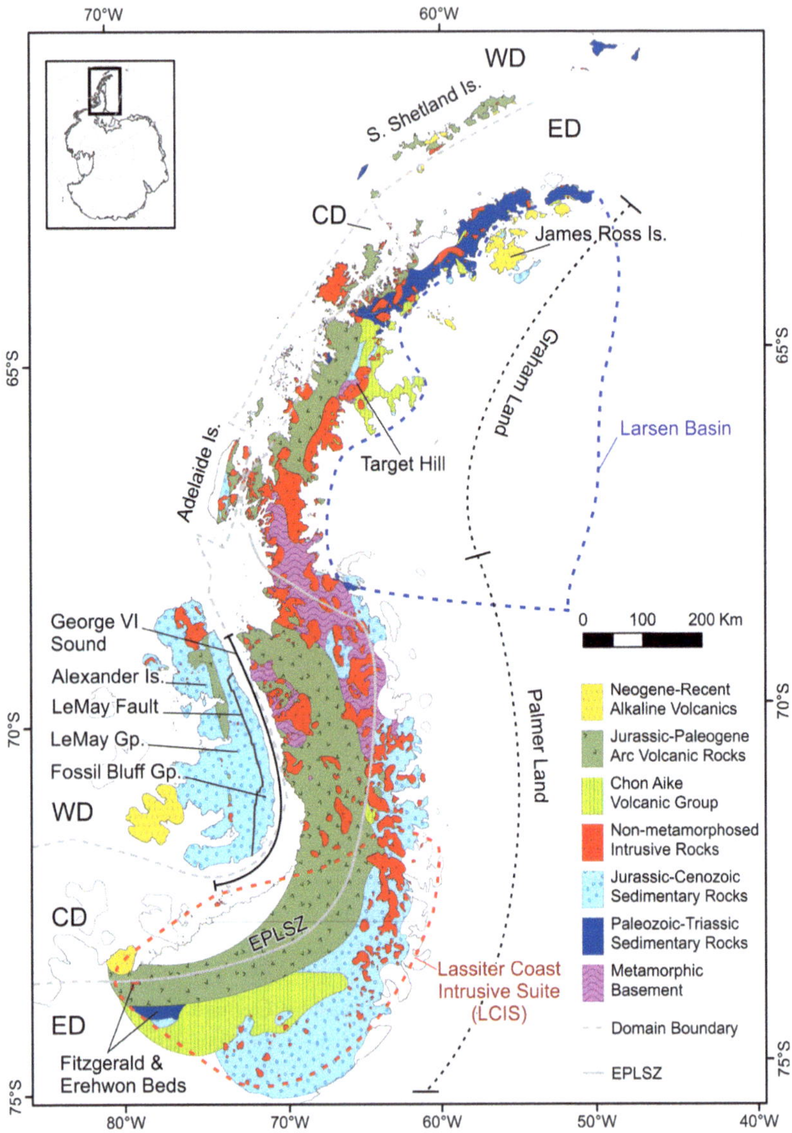

Fig. 1. General geologic map of the Antarctic Peninsula.

to feature Precambrian continental lithosphere 35–45 km thick, stable, coherent and topographically high, that held a central position in the Palaeozoic supercontinent of Gondwana as it did in the Mesoproterozoic supercontinent Rodinia. In contrast, West Antarctica is an amalgamation of low-lying, 20–35 km thick, younger crustal blocks. The Transantarctic Mountains are approximately 2,500 km long and 200 km wide, dividing East Antarctica from West Antarctica with peaks that rise over 4 km above sea level. Crustal thickness estimates under the Transantarctic Mountains vary between 20 and 45 km (Talarico and Kleinschmidt 2009). The AP is the most accessible region of West Antarctica, which occurred along the southeast Pacific continental margin (Fig. 1). The AP consists of a number of large domains and is the largest tectonic block of West Antarctica. The AP has been traditionally regarded as a magmatic arc formed along the palaeo-Pacific margin due to Gondwana breakup (Yegorova et al. 2011).

3 Materials and Methods

In this investigation, Landsat-7 Thematic Mapper (TM), Landsat-8 and the Advanced Spaceborne Thermal Emission and Reflection Radiometer (ASTER) data were obtained from the U.S. Geological Survey's Earth Resources Observation System (EROS) Data Center (EDC). They were acquired on December to February during summer season (2003 to 2007) with low cloud cover and more rock exposures for the AP and the Oscar coast area in north-eastern Graham Land.

Prior to spectral analysis of the satellite data an atmospheric correction model (Fast Line-of-sight Atmospheric Analysis of Spectral Hypercube, FLAASH) was applied to the datasets. Due to areas of snow, ice, sea and cloud in Antarctic environment, only convinced image processing tasks could be restricted to just areas of rock exposure (sunlit rock in the case of the reflective data) (Hall et al. 1995). Contrast-enhanced Red-Green-Blue (RGB) composites were applied to Landsat-8 and ASTER dataset for establishing the spectral separation of the main lithologic groups exposed in the study area prior to band ratios and Relative Band Depth (RBD) image processing techniques. Band ratio is a technique where the digital number value of one band is divided by the digital number value of another band. Band ratios are very useful for highlighting certain features or materials that cannot be seen in the raw bands. High digital number values in the scene indicate spectral signatures similar to those of the particular materials (minerals) they were designed to map (Pour and Hashim 2015a). Relative Absorption Band Depth (RBD) is a useful three-point ratio formulation for detecting diagnostic mineral absorption features. For each absorption feature, the numerator is the sum of the bands representing the shoulders, and the denominator is the band located nearest the absorption feature minimum. Relative Absorption Band Depth (RBD) is a useful three-point ratio formulation for displaying Al-O-H, Fe, Mg-O-H, and CO3 absorption intensities (Pour and Hashim 2015b). In this study, the data sets were processed using the ENVI (Environment for Visualizing Images) version 5.2 and Arc GIS version 10.3 software packages.

4 Results and Discussion

Figure 2 shows AP as merged RGB colour-composite of visible bands (1, 2 and 3) of Landsat-7 TM at regional scale. Antarctic Peninsula has significant areas of exposed rock, varying surface slope and texture that are observable in Fig. 2. The rocks of the

AP in many ways resemble the Andes Mountains, which is dominated by volcanic and plutonic rocks. Different textural patterns are manifested associated with ice shelfs and rocky land background. Larsen Ice Shelf and other ice shelves (George Vi and Wilkins) with smooth to semi-smooth feature are easily recognizable from rough pattern of rocky land background in the image. Palmer Land, Graham Land, Trinity Peninsula and several islands such as South Okrney Islands, South Shetland Islands, Anvers Island, Adelaide Island and Alexander Island have rocky background features, which show different pattern compere to ice shelves (Fig. 2). Initial spectral analysis of rock exposures was implemented using RGB colour composites of near-infrared and shortwave infrared Landsat-8 bands. Figure 3 shows RGB colour composites of band 5 (near-infrared: 0.845–0.885 μm), band 6 (shortwave infrared: 1.560–1.660 μm) and band 7 (shortwave infrared: 2.100–2.30 μm) of the Landsat-8 data for eastern segment of Graham Land. Geological features such as textural characteristics of rocky background, structural features, glacier, ice shelfs and sea water are observable at regional scale. Rock exposure areas appear in greenish blue colour especially in eastern segment of Fig. 3. Ice shelves and glacier show different structural features and manifest as mustard and purple to light whitish blue areas. Sea water is recognizable as dark blue colour in the western part of the image.

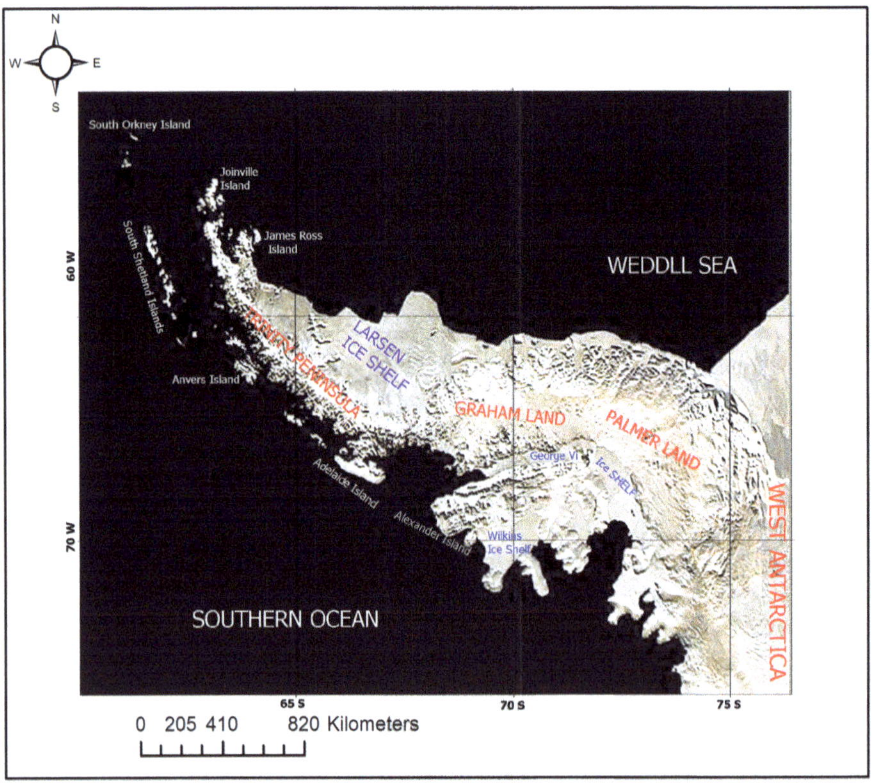

Fig. 2. RGB colour combination of Landsat 7 mosaic image for Antarctic Peninsula (AP).

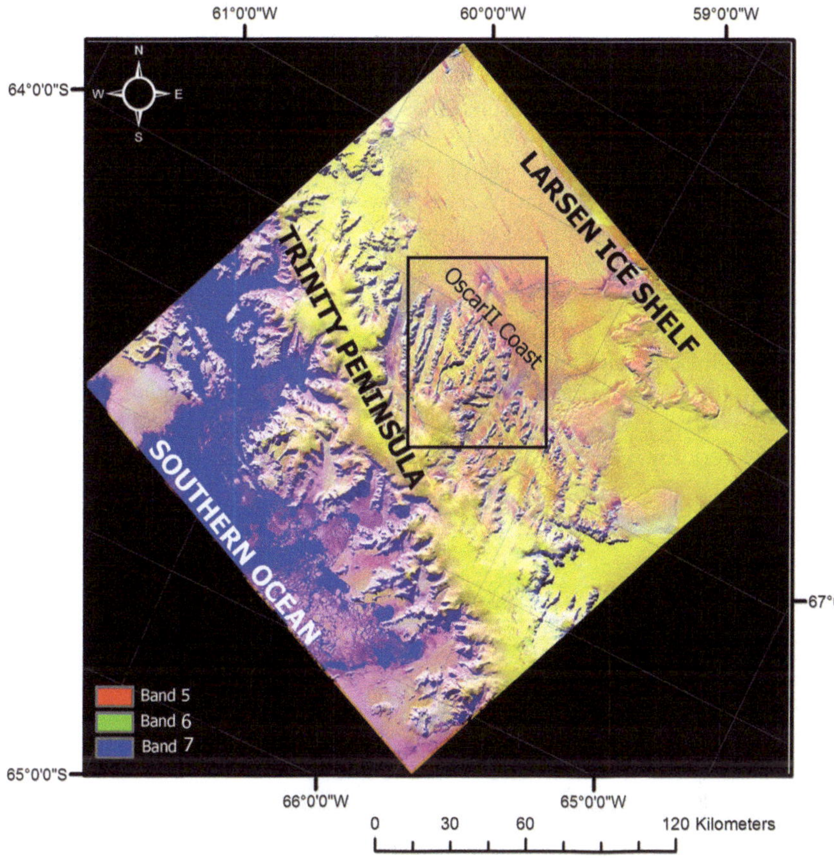

Fig. 3. RGB colour combination of bands 5, 6 and 7 of Lansat-8 for Graham Land, Antarctic Peninsula (AP). Cube shows ASTER coverage for the study area and rock exposure zone.

TIR bands of Landsat-8 have improved the quality and applicability of the Landsat data in a variety of earth-based and atmospheric phenomenon. Two thermal infrared bands (bands 10 and 11) of Landsat-8 have spectral coverage in 10.30–11.30 μm and 11.50–12.50 μm, respectively (Pour and Hashim 2015c). The energy measured by TIR bands from the Earth's surface is a function of temperature as well as the emissivity of the target, which is dependent on its chemistry and texture. High radiometric sensitivity in the Landsat-8 TIRS bands has high potential for mapping exposed lithological units in Polar Regions through variation in temperature as felsic to mafic rocks show a modified response to solar heating due to different mineral compositions (Roy et al. 2014).

In polar context, rock exposure is much warmer than surrounding snow, ice or sea as well as overlying cloud and is readily discriminated due to its high thermal radiance values. Moreover, silicate minerals exhibit spectral features in the TIR. The silica emissivity curve shows significant variation in 8.5 μm to 9.30 μm and 10.30 to 11.70 μm. There are clear minima in 8.5 μm to 9.30 μm while higher emissivity values can be seen in 10.30 to 11.70 μm (Pour and Hashim 2015a, b). RGB color combination image can be produced for band 10 (10.30–11.30 μm), 11 (11.50–12.50 μm) and 7 (2.100–2.300 μm) of Landsat-8 at a regional scale. In this study, band 7 has been selected for RGB color combination image as representative of rocks absorption features in SWIR region due to Al–OH, Fe, Mg–OH, Si–O–H and CO3 in their compositions. Figure 4 shows image map of TIR bands+band of Landsat-8 for eastern segment of Graham Land. This image map shows variety of thermal radiance values and discriminate different geological features such as sea water, glacier, ice shelfs and rock exposures. Highest thermal radiance and emissivity value is associated with rock exposures in the eastern part of the image (Fig. 4) due to warmer constituents and silicate or carbonate in their compositions. On the other hand, sea water also shows high thermal radiance values than surrounding environment caused by summer season, which most of the ice was melted into the sea water with more temperature. Glacier bed and glacier flow have low thermal radiance, while ice shelf and high altitude inland ice regions show very low value emissivity zones in the image map (Fig. 4).

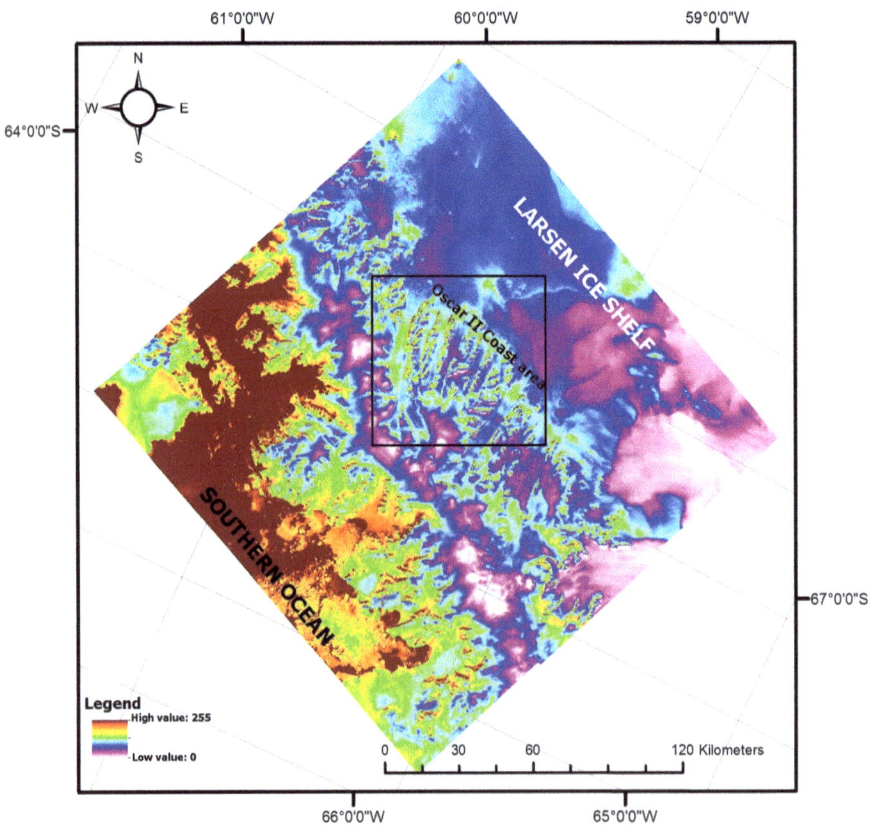

Fig. 4. Image map of TIR bands+SWIR band of Landsat-8 for eastern segment of Graham land.

Band ratios derived from image spectra were used to map rock exposure in the eastern part of the image (Fig. 4). Because ASTER has 14 spectral bands, many permutations of ratio images, and thus more lithologic and mineralogic indices can be derived from ASTER data. An image map was produced (Fig. 5), using VNIR+bands of ASTER to map iron oxides and hydroxides mineral (band1/band2), clay minerals (band4 +band7/band6) and FeOH/MgOH minerals (band7+band9/band8). Rock exposures are manifested as red, purple and light green colures in a light blue ice-background. The ratio for ferrous iron (band1/band2) records absorption in the VNIR region associated with the Fe ion that is present in mafic minerals such as chlorite and epidote. The band ratio for the AlOH absorption feature (band4+band7/band6) centred on ASTER band 6 was used to map clay minerals and micas such as smectite, muscovite, and illite. The FeOH/MgOH band ratio (band7+band9/band8) that records the absorption feature centred on ASTER band 8 was used to discriminate minerals such as chlorite, hornblende, and epidote. Accordingly, it seems that intermediate/basic rocks volcanic and volcanoclastic rocks and chlorite-bearing sediments are dominated in the study area. Geological map of the Oscar II coast area was used as a reference data information for comparison and

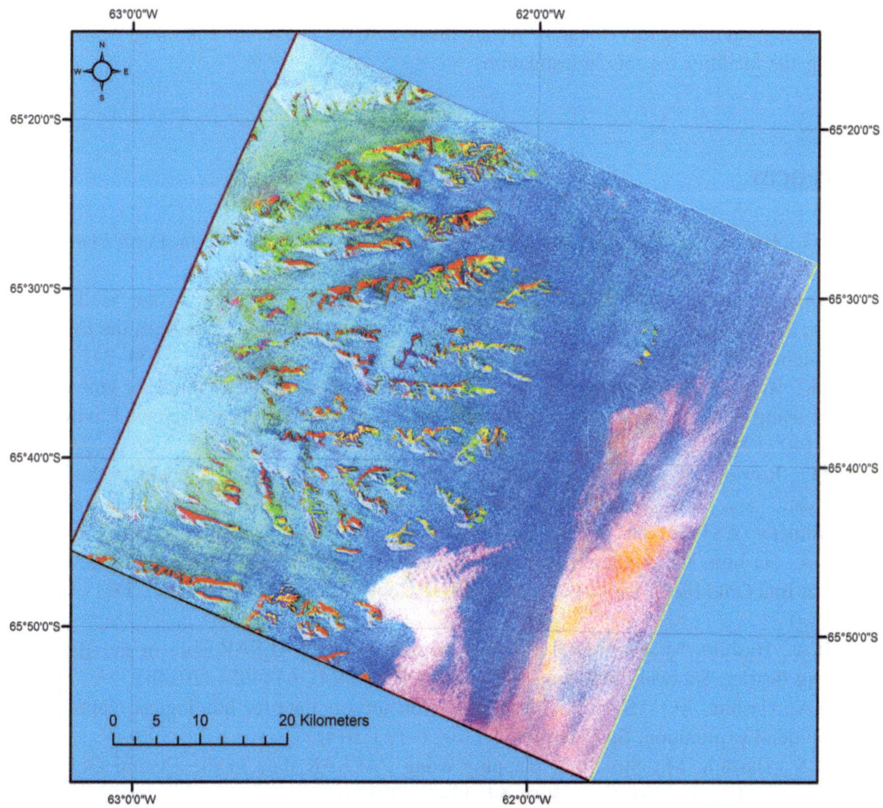

Fig. 5. Image map of (band1/band2 = R), (band4+band7/band6 = G) (band7+band9/band8 = B) of ASTER VNIR+SWIR bands.

verification of the resultant image maps derived from image processing algorithms for the study area at regional scale.

5 Conclusions

Landsat-7 Thematic Mapper (TM), Landsat-8 and ASTER satellite sensors have been investigated for geological information extraction in the Antarctic environments. Oscar coast area in Graham Land, AP has been mapped at regional and district scale using VNIR, SWIR and TIR bands of selected data sets. Results indicate that important geological information could be achieved such discrimination of major lithological group within the study area. The outcomes of this investigation shows the applicability of satellite remote sensing data for revisions of existing geological map of AP and mapping of non-investigated regions further east.

Acknowledgements. This study was conducted as a part of Yayasan Penyelidikan Antartika Sultan Mizan (YPASM) research grant (Vote no: R.J130000.7309.4B221), Sultan Mizan Antarctic Research Foundation, Malaysia. We also would like to express our great appreciation to KOPRI (Korea Polar Research Institute) Asian Polar Science Fellowship Program 2016 for their great support during this research. We are thankful to the Universiti Teknologi Malaysia for providing the facilities for this investigation.

References

Eagles, G.: Tectonic evolution of the Antarctic-Phoenix system since 15 Ma. Earth Planet. Sci. Lett. **217**, 97–113 (2004)

Elliot, D.H., Fanning, C.M., Laudon, T.S.: The Gondwana plate margin in the Weddell sea sector: Zircon geochronology of upper Paleozoic (mainly Permian) strata from the Ellsworth mountains and Eastern Ellsworth Land, Antarctica. Gondwana Res. **29**(1), 234–247 (2016)

Hall, D.K., Riggs, G.A., Salomonson, V.V.: Development of methods for mapping global snow cover using moderate resolution imaging spectroradiometer data. Remote Sens. Environ. **54**, 127–140 (1995)

Mars, J.C., Rowan, L.C.: Regional mapping of phyllic- and argillic-altered rocks in the Zagros magmatic arc, Iran, using Advanced Spaceborne Thermal Emission and Reflection Radiometer (ASTER) data and logical operator algorithms. Geosphere **2**(3), 161–186 (2006)

Pour, B.A., Hashim, M.: Alteration mineral mapping using ETM+ and hyperion remote sensing data at Bau Gold Field, Sarawak, Malaysia. IOP Conf. Ser. Earth Environ. Sci. **18**(1), 012149 (2014a)

Pour, B.A., Hashim, M.: Structural geology mapping using PALSAR data in the Bau gold mining district, Sarawak, Malaysia. Adv. Space Res. **54**(4), 644–654 (2014b)

Pour, B.A., Hashim, M.: ASTER, ALI and hyperion sensors data for lithological mapping and ore mineral exploration. Springerplus **3**(130), 1–19 (2014c)

Pour, B.A., Hashim, M.: Structural mapping using PALSAR data in the Central Gold Belt, Peninsular Malaysia. Ore Geol. Rev. **64**, 13–22 (2015a)

Pour, B.A., Hashim, M.: Integrating PALSAR and ASTER data for mineral deposits exploration in tropical environments: a case study from Central Belt, Peninsular Malaysia. Int. J. Image Data Fusion (2015b). http://dx.doi.org/10.1080/19479832.2014.985619

Pour, B.A., Hashim, M.: Hydrothermal alteration mapping from Landsat-8 data, Sar Cheshmeh copper mining district, South-Eastern Islamic Republic of Iran. J. Taibah Univ. Sci. **9**, 155–166 (2015b)

Roy, D.P., Wulder, M.A., Loveland, T.A., Woodcock, C.E., Allen, R.G., Anderson, M.C., et al.: Landsat-8: science and product vision for terrestrial global change research. Remote Sens. Environ. **145**, 154–172 (2014)

Salvatore, M.R., Mustard, J.F., Head, J.W., Cooper, R.F., Marchant, D.R., Wyatt, M.B.: Development of alteration rinds by oxidative weathering processes in Beacon valley, Antarctica, and implications for Mars. Geochim. Cosmochim. Acta **115**, 137–161 (2013)

Talarico, F. M., Kleinschmidt, G.: The Antarctic continent in Gondwanaland: a tectonic review and potential research targets for future investigations. In: Florindo, F., Siegert, M. (eds.) Developments in Earth & Environmental Sciences, vol. 8 (2009). Elsevier B.V. doi:10.1016/S1571-9197(08)00007-4. All rights reserved 2009

Yegorova, T., Bakhmutov, V., Janik, T., Grad, M.: Joint geophysical and petrological models for the lithosphere structure of the Antarctic Peninsula continental margin. Geophys. J. Int. **184**, 90–110 (2011)

Author Index

J. Wasowski et al. (eds.), *Engineering Geology and Geological Engineering for Sustainable Use of the Earth's Resources, Urbanization and Infrastructure Protection from Geohazards*, Sustainable Civil Infrastructures, DOI 10.1007/978-3-319-61648-3

MIX
Papier aus verantwortungsvollen Quellen
Paper from responsible sources
FSC® C105338

If you have any concerns about our products,
you can contact us on
ProductSafety@springernature.com

In case Publisher is established outside the EU,
the EU authorized representative is:
Springer Nature Customer Service Center GmbH
Europaplatz 3, 69115 Heidelberg, Germany

Printed by Libri Plureos GmbH
in Hamburg, Germany